Heavy Ground

Heavy Ground

William Mulholland and the St. Francis Dam Disaster

Norris Hundley jr. and Donald C. Jackson

UNIVERSITY OF NEVADA PRESS | *Reno & Las Vegas*

Heavy Ground: William Mulholland and the St. Francis Dam Disaster was first published by the Huntington Library and the University of California Press. Copyright © 2015 by Huntington Library, Art Collections, and Botanical Gardens and University of California Press. In 2017 the copyright reverted to the Estate of Norris Hundley jr. and Donald C. Jackson. The University of Nevada Press edition is published by agreement with the Estate of Norris Hundley jr. and Donald C. Jackson.

University of Nevada Press, Reno, Nevada 89557 USA

All rights reserved
Interior design and composition by Doug Davis

LIBRARY OF CONGRESS CATALOGING-IN-PUBLICATION DATA
Names: Hundley, Norris, Jr., author. | Jackson, Donald C. (Donald Conrad), 1953- author.
Title: Heavy ground : William Mulholland and the St. Francis Dam disaster / Norris Hundley jr. and Donald C. Jackson.
Description: University of Nevada Press paperback edition. | Reno ; Las Vegas : University of Nevada Press, 2020. | Includes bibliographical references and index. | Summary: "Heavy Ground explores the social, political, and technological history of the St. Francis Dam Disaster, the worst civil engineering disaster in 20th century-American history. Some 400 people died in March 1928, when the concrete gravity dam built by Los Angeles engineer William Mulholland suddenly and tragically collapsed, releasing over 12 billion gallons of water into the Santa Clara River Valley"—Provided by publisher.
Identifiers: LCCN 2020028474 (print) | LCCN 2020028475 (ebook) | ISBN 978-1-948908-88-7 (paperback) | ISBN 978-1-948908-89-4 (ebook)
Subjects: LCSH: Floods—California—Santa Clara River. | Water-supply engineers—California—Los Angeles—Biography. | Water-supply—California—Los Angeles—History. | Saint Francis Dam (Calif.) | Mulholland, William, 1855–1935.
Classification: LCC TC557.C3 S245 2020 (print) | LCC TC557.C3 (ebook) | DDC 363.34/930979494—dc23
LC record available at https://lccn.loc.gov/2020028474
LC ebook record available at https://lccn.loc.gov/2020028475

The paper used in this book is a recycled stock made from 30 percent post-consumer waste materials, certified by FSC, and meets the requirements of American National Standard for Information Sciences—Permanence of Paper for Printed Library Materials, ANSI/NISO Z39.48-1992 (R2002). Binding materials were selected for strength and durability.

University of Nevada Press Paperback Edition, 2020
FIRST PRINTING

Manufactured in the United States of America

24 23 22 21 20 5 4 3 2 1

To Carol and Carol

The foundations on which the [St. Francis] dam was built were not good. I was intimately connected with the driving of a series of tunnels for our aqueduct through the range of mountains on which the left or east abutment of the dam rested...The rock that we encountered was a broken schist and a good deal of it expanded when it came in contact with the air and was what the tunnel men called "heavy ground." We had great difficulty in holding this ground [for the aqueduct tunnel] before it was lined with concrete.

Joseph B. Lippincott, March 26, 1928

Contents

St. Francis Dam, 1927. [DC Jackson]

Preface

In the early hours of March 13, 1928, more than 12 billion gallons of water surged through the Santa Clara Valley of Southern California, killing some 400 people before spilling into the Pacific Ocean at daybreak.[1] Precipitated by the collapse of the St. Francis Dam, this catastrophic flood is considered the greatest civil engineering disaster in California history, and one of the worst to occur anywhere in the United States. In terms of storage dam failures in the United States, only Pennsylvania's Johnstown Flood, which claimed over 2,000 lives in May 1889, exceeded its toll in death and devastation.[2]

The person facing the greatest scrutiny in the wake of the St. Francis tragedy was William Mulholland, famed chief engineer of Los Angeles' municipal water system and the individual responsible for designing and constructing the ill-fated dam. In 2004 the authors of this book published an article on the St. Francis disaster that focused on the design of the dam, the way it related to dam engineering practice in the 1920s, and the question of Mulholland's responsibility for the collapse.[3] We now expand upon our article and, in addition to describing the dam's construction and disintegration in greater detail, tell of the horrific effect of its failure on the residents of the Santa Clara River Valley. In broadening the context of our research and analysis, we also focus on key issues such as the political dimensions of water-control technology and legislation in twentieth-century California and the influence of the impending Boulder Canyon Project Act on investigations of the dam's collapse. We also direct special attention to William Mulholland's background and training and to the professional and political context of his work as Los Angeles' most prominent hydraulic engineer of the twentieth century.

A seminal figure in the history of Southern California, Mulholland has achieved almost mythic status among a citizenry that intuitively appreciates the role of water supply in creating the culture and political economy of greater Los Angeles. While the character of his accomplishments may have been distorted in the public mind by the 1974 film noir classic *Chinatown*, Mulholland nonetheless occupies a central place in the collective memory of Los Angeles. Thus, it is important that this account of his association with the St. Francis Dam disaster be free of hagiography, and that it objectively assess the skills and knowledge of other gravity dam designers of his era. The issue of responsibility—and specifically Mulholland's responsibility for the disaster—looms large.

Among those who linked Mulholland to the St. Francis Dam collapse was Charles F. Outland, a longtime resident of Santa Paula who, as a teenager, directly experienced the flood's devastation. He later drew upon his recollections, supplemented by extensive research, in writing *Man-Made Disaster: The Story of St. Francis Dam*, a book first published in 1963 and expanded in a revised edition fourteen years later. Although he was careful not to extrapolate beyond the available evidence, and refrained from demonizing the one-time chief engineer, Outland was nonetheless unequivocal in ascribing to Mulholland responsibility for the disaster. As he tersely phrased it, "In the final analysis...the responsibility was his alone."[4]

Not everyone has shared Outland's perspective on Mulholland's culpability. In the 1990s J. David Rogers, a geologist who has researched the dam's collapse, put forward the thesis that Mulholland was guilty of little more than an excusable ignorance he shared with a civil engineering community that, among other things, "did not completely appreciate or understand the concepts of effective stress and uplift" at the time of the dam's construction.[5] Rogers presents a useful analysis of the geological character of the St. Francis dam site and a thoughtful forensic description of the mechanics of the failure.[6] But he also downplays the nature of Mulholland's responsibility for the tragedy. In recent years his work has been relied upon by others claiming that Mulholland can reasonably be absolved of responsibility for the disaster.[7]

A widely held perception that Mulholland should be exonerated for the St. Francis disaster prompted us to write our article and, in no small part, spawned *Heavy Ground*. But our intention here is not to simplistically blame Mulholland for the tragedy that brought his career to an ignoble end. We hope to tell a larger story that intertwines a range of important issues, including the relationship of the St. Francis Dam to the Los Angeles Aqueduct (that is, how and why the dam came to be built); how Mulholland was able to design and build the structure free from state regulation; how the design failed to adhere to contemporary gravity dam design standards (especially in regard to "uplift" pressure); the

William Mulholland, not long after the completion of the Los Angeles Aqueduct in 1913. [*Complete Report on Construction of the Los Angeles Aqueduct*, 1916]

effects of the flood upon the Santa Clara Valley; the way that Los Angeles assumed responsibility for reparations to victims and downstream landowners; the politically driven—and often less than enlightening—post-disaster engineering investigations and their relationship to the Boulder Canyon Project; and the manner in which dam safety legislation evolved following the flood.[8]

The historical record is far from complete, and much has been lost or obscured over the past century. Some of this loss was probably purposeful, as professional colleagues of Mulholland, proponents of concrete gravity dam technology, advocates of the Boulder Canyon Project Act, and city officials and civic leaders in Southern California sought to quickly close the door on the disaster, cleaning up paper trails that someday might prove embarrassing or remind people of a tragedy that city officials would prefer be forgotten.[9] Some of the loss is doubtless less deliberate, the result of vagaries of document survival common in the records of engineers and engineering projects. But regardless of where and why documents survived, finding new data and information on the disaster has proved to be an engaging exercise in historical detective work. At times it has been frustrating and unfulfilling; at other times, new insights have come to light from both familiar and previously unexamined sources.

We have spent considerable time reviewing records retained by the Los Angeles Department of Water and Power (a successor agency to the Mulholland-led Bureau of Water Works and Supply). There we uncovered data on the planning and construction of the dam and information on legal matters associated with damage claims and on the way city officials worked with Ventura County leaders to manage the city's financial liability. Numerous newspaper stories as well as documents in the collections of the Huntington Library in San Marino and the Museum of Ventura County (formerly the Ventura County Museum of History and Art) helped us define the human dimensions of the tragedy. We tracked down sources dealing with the state's regulation of dam safety and with the way impending congressional consideration of the Boulder Canyon Project prompted rapid completion of engineering investigations of the dam failure. In particular, we uncovered revealing letters, telegrams, and memos in the St. Francis Dam File retained by the Division of Safety of Dams in Sacramento. Perhaps most important, we have made extensive use of the only known copy of the transcript of the Los Angeles County Coroner's Inquest into the disaster.[10] This transcript, running more than 800 pages, provides an invaluable record of testimony by Mulholland and many others given in the tumultuous days following the flood. But it also leads to frustrating questions about key issues—such as the raising of the dam's height during construction—that were never addressed by the inquest.

The more than 150 images that illustrate and complement the text of *Heavy Ground* were drawn from a wide range of libraries and archives, including the Huntington Library, the Los Angeles Department of Water and Power, the *Los Angeles Times* photograph collection at UCLA, the Los Angeles Public Library, the Bancroft Library, the Automobile Club of Southern California, the United Water Conservation District in Santa Paula, and the Santa Clarita–based SCVhistory.com. Author Jackson also collected scores of images over the past fifteen years through eBay and other online commerce sites, some of which (such as the 1926 California gubernatorial campaign brochure, illustrated in chapter 6, that heralded soon-to-be Governor C. C. Young as the "Champion of Boulder Dam") offer unique insight into the historical context of the St. Francis Dam. On their own terms, the images serve as an essential source of data documenting the disaster.

We can never know everything that we might wish to about William Mulholland or the tragedy brought by the collapse of the St. Francis Dam. But through extensive illustration and documentation, this book presents a far-ranging history of the ill-fated dam and offers insight into the political aspects of dam construction and the ways in which large-scale engineering projects are embedded in—and expressive of—the culture that sustains them.

PROLOGUE

"A Misty Haze over Everything"

It was a few minutes before midnight when Lillian Curtis awoke in her bed. Gazing out the window she witnessed a scene of calm beauty: "There was a large full moon with big white clouds rolling over it." The serenity of the moment, in contrast to the tumult to come, lingered long in her memory.[1] That evening Lillian, her husband, Lyman, and their three children had retired for the night in their modest frame bungalow. It was one of a group of houses built for married employees of the Bureau of Power and Light, tucked in a small ravine off the main San Francisquito Canyon. They lay close by the city's recently built St. Francis Dam. Other city employees and transient workers lived in bungalows adjoining the city's San Francisquito Power House No. 1, about five miles above the dam, or clustered near San Francisquito Power House No. 2, located a mile and a half downstream from the reservoir. The married employees' compound where the Curtises lived lay but a short walk west from Power House No. 2.[2]

Also residing in the married employees' compound were Ray Rising, his wife, Julia, and their three young daughters, Delores, Eleanor, and Adaline. Rising had worked for the city for nine months; Lillian's husband had been on the job a few months more. Neither family had been living in San Francisquito Canyon when the St. Francis Dam was completed in May 1926, but the Curtises were in residence when the reservoir, with a capacity of more than 12 billion gallons, came close to filling in May 1927. And—not quite a year later, in early March 1928—both families called the canyon home when water first filled the reservoir and rose within three inches of the dam's spillway crest. "Everything seemed perfectly safe," recalled Lillian Curtis.[3]

Seconds after midnight on March 13, Lillian was startled by a strange apparition in the night sky. "I sat up in bed and looked out the windows toward the dam, which was northeast of us, and I seemed to see a misty

Lyman and Lillian Curtis, soon after their wedding in Bakersfield in 1921. Lyman worked for the municipally owned Bureau of Power and Light, and he and his family resided near San Francisquito Power House No. 2. Lillian survived the flood, but Lyman perished along with their two young daughters. [SCVHistory.com]

haze over everything." Nearby, Ray Rising was jolted awake by a roaring sound that reminded him of tornadoes he had experienced as a youth in Minnesota. He grew fearful about what might be occurring at the dam. Lillian Curtis was now also "hearing a strange noise," and she grabbed her husband and screamed, "The dam has broken!" A 125-foot wall of water, moving at about eighteen miles per hour, soon engulfed Power House No. 2, demolishing the plant and killing the on-duty crew.[4]

Floodwaters surged up the ravine where the Curtis and Rising families lived. Lillian, several months pregnant, grabbed her three-and-a-half-year-old son, Daniel. In turn, they were picked up by her husband who, after pushing them through a window with orders to "run up the hill," went back for Daniel's two sisters, Mazie and Marjorie. Lillian never saw Lyman or her infant daughters alive again. For the moment, she concentrated on getting to higher ground, a punishing task amid the brush, silt, and debris filling the rising water. "Mommy," her son called out, "don't let the water get us." With the boy in her arms, she persevered, telling herself, "I must get him out." Breaking free from the tempest near the crest of the hill, she heard a voice. Convinced it was her husband, she crawled over the knoll to safety.[5]

The voice she heard was not her husband's, but Ray Rising's. Rising later recalled that his wife had shouted, "What's that—a wind?" Soon "the sound grew louder and louder, then we heard trees snapping. We went to the door and looked out. Water was coming. We hurried back

Mazie, Daniel, and Marjorie Curtis (left to right), in happier times, before the St. Francis Dam disaster. The two Curtis daughters drowned in the flood but Daniel was miraculously saved by his mother. Both mother and son attended ceremonies in 1978 marking the fiftieth anniversary of the disaster. [SCVHistory.com]

to get the children. When we got back to the door and tried to open it we could do nothing, as the force of the water held it shut." Then he felt himself "thrown into the air with a force like from an explosion," the water knocking him aside and crushing the timber bungalow. Fighting for air in the blackness and entangled in electrical wires and an uprooted tree, Rising somehow found refuge atop a floating roof. Holding tight until the impromptu raft backed into a canyon wall, he jumped to safety, all the while shouting for his wife and children.[6] But they were gone, joined in death by twenty-three of his fellow city workers and forty-two of their family members who, for at least a few months, had created and shared the community at Power House No. 2. The settlement's only survivors were Lillian Curtis, her son, Daniel, and Ray Rising.[7]

The torrent sowed death and destruction throughout the night. Not until daybreak did it finally wash into the Pacific Ocean south of Ventura, leaving in its wake millions of dollars in property damage and some 400 lost lives. By then, William Mulholland—the man responsible for building the dam—had been roused from his bed in Los Angeles and driven forty-five miles to the site of the now empty reservoir. Acclaimed as a master engineer for his success in creating the modern city's expansive water supply system, the seventy-two-year-old Mulholland had long reigned supreme in the world of Southern California water. But now, as Lillian Curtis and Ray Rising—along with hundreds of law enforcement and civic officials, citizen volunteers, and family members with ties to the Santa Clara Valley—set out in search of survivors, it would be left to Mulholland to explain how his once great dam had turned to rubble.

What follows is an account of how the St. Francis Dam came to be built, the causes of the collapse, and the magnitude of destruction visited upon the people of the Santa Clara River Valley. Also recounted are efforts to restore the valley after the floodwaters had passed, political factors influencing engineering investigations of the collapse, and the effect of the disaster upon subsequent dam safety regulation. Underlying all is a consideration of how the dam—and the disaster—were inextricably intertwined with the life and career of William Mulholland.

CHAPTER ONE

Mulholland: A Man and an Aqueduct

He spoke of criticism from various sources as nothing more than an annoyance, as of the barking of a dog in the night.

The *Los Angeles Times* reporting on remarks by William Mulholland at banquet celebrating the Los Angeles Aqueduct, 1913[1]

Since time immemorial every profession, every line of human pursuit, has had its outstanding character, its shining light, its great leader. In the profession of water works engineering there is an outstanding figure, a leader who...has proved to be a builder of an empire—an empire of unsurpassed progress in municipal development—William Mulholland.

William Hurlbut, *Western Construction News*, 1926[2]

One of many sad truths about the St. Francis Dam disaster is that residents of the Santa Clara Valley—the people who bore the brunt of suffering and death—played essentially no role in the conception or operation of the dam. The vast bulk of the floodwater was artificially fed into the Santa Clara River watershed with no intention that it would ever benefit citizens of the valley. True, some water in the reservoir originated within the thirty-seven square miles of foothills lying above the dam, but far more fell to earth as rain and snow 200 miles to the north, along the lofty eastern slope of the Sierra Nevada. This imported water entered San Francisquito Creek through an astounding feat of human ingenuity and technological bravado, underwritten by individuals with little interest in the people or prosperity of the Santa Clara Valley. To them, the valley was a mere corridor and way station for an aqueduct carrying water to Los Angeles.

The use of San Francisquito Canyon for a major reservoir was largely an accident of history, a by-product of decisions and initiatives undertaken by Los Angeles officials over a period of twenty years. St. Francis Dam was but one component of a much larger project—the Los Angeles Aqueduct—designed to slake the thirst and light the homes of the rapidly growing city. Stretching over 230 miles from the Owens Valley to the San Fernando Valley, the Los Angeles Aqueduct still holds great prominence in the public mind because of its role in defining the political economy of Southern California's premier city. To appreciate how the St. Francis Dam came to be built requires an understanding of both the Los Angeles Aqueduct and the man most responsible for its construction.

As the aqueduct's chief engineer and its public face, William Mulholland stood as a colossus over early twentieth-century Los Angeles. Diverting the flow of the Owens River across desert wasteland and through mountain escarpment, the aqueduct created the hydraulic foundation of a world-class city. By bringing copious quantities of fresh water to the arid coastal plain, Mulholland led the way to long-term survival and prosperity for the citizenry of a rapidly growing Los Angeles. Of course, for people living outside the city—especially the farmers in the Owens and Santa Clara river valleys—his attitude could appear imperious and even abusive. But to those in his adopted home city, and to many who held no particular stake in Southern California water quarrels, his efforts stood as a paragon of technological achievement and progressive efficiency.

The Ladder of Authority

William Mulholland's prominence and renown sprang from inauspicious beginnings. Born in Ireland in 1855, he joined the British merchant marine in the late 1860s but, after four years as a sailor before the mast, he recognized that a mariner's life "would get him nowhere in a material way." Coming ashore in New York City in 1874 he headed west to work on a Great Lakes steamer and in a Michigan logging camp. Joined by his brother Hugh, he moved to Pittsburgh to work for an uncle at a dry goods emporium. When two of his uncle's children died of tuberculosis, the extended family left for California in late 1876, seeking a more salutary climate. Passage across the Isthmus of Panama and a horseback trip from San Francisco brought Mulholland to Los Angeles in January 1877.[3]

Apparently unimpressed by Southern California, Mulholland headed for the port of San Pedro to ship out. Along the way he met Manuel

Dominguez, nephew of the original owner of Rancho San Pedro, who offered him a job drilling artesian wells near the town of Compton, a few miles south of downtown Los Angeles. Mulholland accepted and, while working his first well, discovered fossil remains that, he later claimed, "changed the whole course of my life...These things fired my curiosity. I wanted to know how they got there and so I got hold of Joseph Le Conte's book on the geology of this country. Right there I decided to become an engineer."[4] For whatever reason, he did not immediately act upon this apparent epiphany. Instead, he trekked to Arizona Territory in search of gold. But riches proved elusive, and Army troops warned of hostile Apaches. Concluding that "presence of mind was best secured by absence of body," he returned to Los Angeles by the spring of 1877. Once back, he joined a ditch gang as a laborer for the Los Angeles City Water Company. Following from his aspiration to become an engineer, his unlikely employment as a lowly ditch digger nonetheless proved to be a major turning point both in his life and in the city's history.[5]

As in many towns and cities in nineteenth-century America, Los Angeles leaders looked to private capital to finance a municipal water-supply system. After some faltering short-lived arrangements, in 1868 the city entered into what appeared a more promising agreement. For the privilege of building and operating a complex of pumps, pipes, valves, ditches, and small distribution reservoirs, the Los Angeles City Water Company paid the city $1,500 a year (soon reduced to $400 in exchange for building a plaza fountain). In return the company received a thirty-year lease protecting it from competition and allowing it to charge residents a usage fee set by the city, the key restriction being that the fee could never fall below what was authorized in 1868. To safeguard the city's claims to water in the Los Angeles River (the so-called pueblo right), the lease prohibited the company from taking more than ten miner's inches directly from the river. (As calculated in Southern California, fifty miner's inches equals a flow of one cubic foot per second, so ten miner's inches is an exceedingly small amount.) The assumption was that additional water needed by the company would be taken from Crystal Springs, a swampy tract near the river with a high water table.[6]

Mulholland's future employers determined that ten miner's inches, even when combined with water from Crystal Springs, fell far short of what they needed to prosper. To protect itself, the company secretly drove a tunnel into the riverbed capable of drawing upwards of 1,500 miner's inches—150 times more than authorized by the lease. Several years passed before the ploy became public and, as Vincent Ostrom has

FIGURE 1-1. Los Angeles in the mid-1870s. This view of First Street documents the scale of the city when William Mulholland arrived in 1877. The streambed of the Los Angeles River is visible in the background. This channel comprised the primary source of the municipal water supply before construction of the Los Angeles Aqueduct. [Huntington Library]

noted, "the city was at a loss as to what to do about it." To shut down the company would be to shut off the city's domestic water supply. City leaders and residents seethed, all the more when they discovered another unsavory gambit: the company had created a subsidiary corporation that claimed a right to the Los Angeles River on grounds that, as a separate entity, it was not bound by the lease. With growing anger, the city fought these and similar subterfuges, winning some battles and losing others. For its part, the company resisted public scrutiny of its corporate affairs and valued employees who protected this insularity.[7]

FIGURE 1-2. A work crew laying track for the Los Angeles street railway, about 1875. This scene reflects the character of Mulholland's early work as a ditch digger for the Los Angeles City Water Company. [Huntington Library]

In a story later recounted by Mulholland's colleagues, his early tenure with the water company found its defining moment in an unanticipated confrontation with its president, William Perry. Riding by a *zanja* (ditch) where the new employee was clearing debris, Perry noticed the single-minded attention Mulholland brought to his job. Asked what he was doing, the future chief engineer kept his head down and growled: "It's none of your damned business!" Perry backed off, but he liked what he saw in Mulholland's attitude and soon promoted him to foreman. While perhaps apocryphal, the anecdote underscores the distinctive relationship forged between Mulholland and the water company. It also highlights two fundamental attributes at the core of Mulholland's professional character: a willingness to work hard in the service of his employers and a disposition to resist (perhaps even resent) outside review of his work. The latter would prove particularly significant when it came time to build the St. Francis Dam.[8]

Over the next several years, Mulholland's field promotion to foreman was followed by steady professional advancement. The first significant step up the ladder came in 1880 when he took charge of a major pipe-laying project in the Buena Vista district near downtown. This move also brought him close to the city's public library, where he embarked on a journey of self-education. He voraciously read texts on civil engineering, hydraulics, geology, and mathematics, as well as Shakespeare, Pope, Carlyle, and other classics, and later famously exclaimed, "Damn a man who doesn't read books...The test of a man's mind is his knowledge of humanity, of the politics of human life, his comprehension of the things that move men." This hunger for self-education was motivated by a lack of schooling. Unlike many civil and hydraulic engineers who later worked with him and attained prominence in the early twentieth century—including his colleagues Arthur P. Davis, John R. Freeman, and J. B. Lippincott—Mulholland was essentially self-taught. His hydraulic engineering knowledge derived from a reading of technical books and articles combined with extensive on-the-job experience. To bring it all together, he relied upon a quick mind and a remarkable memory.[9]

At Buena Vista, Mulholland honed his skills by overseeing expansion of the pipeline and reservoir and, two years later, by supervising a large crew in building a major flume and ditch. After a brief sojourn in 1884 when he and his brother visited the state of Washington "to study rivers" and an equally brief stint as an independent contractor working on city construction projects, he returned to work for Perry and the private water company. In 1885, Perry sent him to Ventura County to help build an irrigation system for the town of Fillmore. In itself this assignment was not particularly noteworthy, but the job brought him into the Santa Clara Valley. If in his off hours he took time to investigate the upper reaches of the watershed, he may have made his first visit to San Francisquito Canyon and the future site of the St. Francis Dam.[10]

When the Fillmore job ended, he returned to Los Angeles to supervise flume- and tunnel-building at Crystal Springs. By this time, Mulholland's technical knowledge and his acumen in managing workers were much appreciated by company officials. Thus, when the superintendent of the Los Angeles City Water Company died suddenly in November 1886, the way opened for advancement. A younger employee serving as assistant superintendent allegedly declined to step up, averring that Mulholland was better qualified for the post. Perry agreed, and the one-time ditch digger was promoted to superintendent of the city's water works. A mere decade had passed since he first set foot in Los Angeles.[11]

When Mulholland started his new job, Southern California was exulting in a major land boom. Ignited by the Santa Fe Railway's arrival in late 1885, which offered competition to the Southern Pacific, the real estate frenzy was stoked by cheap transcontinental fares. Easterners swarmed into the region, and within three years the population of Los Angeles more than quadrupled, from 11,000 to 50,000. The newcomers, as well as local boosters and real estate developers, expected a reliable water supply in their newfound Eden. Mulholland met the challenge by replacing antiquated water mains and extending water pipes to new neighborhoods and commercial districts. When torrential rains hit hard during the winter of 1889–90, the Los Angeles River threatened to jump its banks and knock out the water system. Working long shifts, Mulholland and his devoted crew kept the system from collapse. As a reward, the water company bestowed upon him a gold watch—material evidence of his value to the shareholders.[12]

Through the 1890s, Mulholland's stature continued to grow, but changes loomed as the water company's thirty-year franchise neared expiration. As in many other cities in the United States during the Progressive Era, a powerful political reform movement had emerged in Los Angeles. A key reform sought by civic leaders focused on restoring municipal control over the water system. Perry and the water company resisted divestiture by eminent domain and, for several years, delayed municipal takeover. Nonetheless the tide finally turned and, in 1902, the water system came under direct city control. At this juncture, Mulholland could have remained with the company's owners and pursued other water projects financed by their capital, but he was loath to relinquish control over the system he had worked so hard to create. Steadfastly loyal to the water company while in its camp, he now transferred his allegiance to the city of Los Angeles.[13]

For the city's political leaders there were practical reasons to welcome Mulholland into municipal employment. Paramount among them was the knowledge he brought to his job. When the city bought the Los Angeles City Water Company, detailed design and construction records were not always available. Mulholland compensated for the gaps by having memorized many features of the complex distribution system. When challenged during the purchase negotiations, Mulholland reportedly called for a map and identified details about pipes throughout the city. This only prompted new challenges. Mulholland responded with a call for excavations that, when carried out, corroborated his testimony. This show of bravado—perhaps embellished in later retellings—helped ensure his continuation as superintendent after the municipal takeover. Knowledge often translates into power and, as he later wryly commented, "the city bought the works and me with it."[14]

From the beginning of his tenure as municipal superintendent in 1902, Mulholland held the respect—if not the absolute deference—of his nominal superiors. Over the years he reported to a series of supervisory groups whose names and responsibilities may have changed, but whose managerial authority derived from the Los Angeles City Charter. Following acquisition of the water system, the Board of Water Commissioners became his first boss; then in 1911 the Board of Public Service Commissioners succeeded to that role; fourteen years later—while construction of the St. Francis Dam was underway—authority passed yet again, this time to the newly formed Board of Water and Power Commissioners.[15] No matter who was nominally in charge, or how they were designated, from 1902 through 1928 Mulholland exercised almost unfettered control over the city's network of water supply and distribution.

Once in charge of the public water system, Mulholland moved rapidly to make improvements and take on new initiatives. In concert with reduced water rates, he installed meters as a way to precisely record use and encourage conservation. Within two years water consumption had fallen by a third and, despite nominally lower rates, the proliferation of meters actually raised revenue. This income allowed him to rebuild much of the water system in his first three years on the city payroll. With the water system turning a profit (some $1.5 million in four years), he expanded the pumping capacity at Buena Vista, built Solano Reservoir, and purchased the West Los Angeles Water Company (a remnant of the former Los Angeles City Water Company). Public acclaim for these accomplishments set the stage for a sympathetic response when he announced plans for a major new source of municipal water supply.[16]

Secrecy and Deal-Making

Mulholland's first dramatic exercise of authority came in 1904 when he addressed fears of an imminent water shortage. In 1895 many residents thought that such concerns had been rendered moot when the California Supreme Court ruled that Los Angeles held a pueblo right to the full flow of the Los Angeles River. Rooted in the settlement's Spanish origins, this right allowed the city to take "all the waters" of its namesake river. But as important as this legal victory was, it did not solve long-term supply problems.[17] City leaders began annexing nearby towns and communities that, in the wake of the pueblo-right ruling, were willing to trade political sovereignty for access to the region's most dependable water supply. By 1900 the city had nearly doubled in physical size and, thanks to annexations and throngs of new residents, the population reached 100,000. The boom continued and, by 1905, the city's population had again doubled, to over 200,000. Such rapid growth aroused concern that the Los Angeles River could not long sustain the city's needs, an anxiety Mulholland shared. "The time has come," he publicly announced in 1904, "when we shall have to supplement its flow from some other source."[18] That declaration set in motion events that would transform the city into a major metropolis.

By the time of this general announcement in 1904, Mulholland had already selected the Owens River, a sizable stream some 200 miles to the north, as the best source for a new municipal water supply. But for almost a year he and city leaders kept the plans secret, fearing a rush of speculators to the eastern slope of the Sierra Nevada that would inflate land prices and threaten the project's feasibility. Significantly, the idea of transporting water from the Owens Valley to Los Angeles did not originate with Mulholland, coming instead from Fred Eaton, an engineer who had served as Los Angeles mayor in the 1890s. "I was born here and have seen dry years," Eaton warned Mulholland, "years that you know nothing about. Wait and see." Only after Eaton took him on a surreptitious trip to the Owens Valley in the late summer of 1904 did the once skeptical Mulholland embrace the Owens River as a viable water source.[19]

In terms of building a municipally owned aqueduct, Fred Eaton proved a difficult partner, not least because he sought sizable personal profits from realizing his vision. His scheming took two forms, the first involving a plan to split the aqueduct water with Los Angeles so that he could privately sell his half to companies and customers beyond the city limits. This proposal may have proved palatable to city leaders, but it provoked the ire of federal officials who saw direct private exploitation of the proposed aqueduct as anathema to the greater public good. The U.S. Reclamation Service was developing plans for a federally sponsored irrigation project in the Owens Valley, but the federal bureaucracy also appreciated that Los Angeles had water needs that could be met by tapping into the Owens River. Federal officials hinted at abandoning the Owens Valley reclamation project so that the city's aqueduct plan could prevail—but only if the city completely controlled the aqueduct as a true public works project. In the face of such objections, Eaton abandoned his ploy to sell water to private developers. But he remained involved in helping the city secure land and water rights options in the Owens Valley necessary for the aqueduct.[20]

FIGURE 1-3. A ranch in the Owens Valley, about 1905, with the snowcapped Sierra Nevada in the background. There was bitter and enduring conflict between the city of Los Angeles and Owens Valley ranchers and farmers over the use of snowmelt feeding the Owens River. [Huntington Library]

Eaton's second initiative proved more troublesome, and it plagued the city for years to come. While he was purchasing Owens Valley options for the city, he also kept on the lookout to buy especially valuable parcels for himself. As a result, he came to own land encompassing the largest reservoir site in the upper Owens River watershed. Located in Long Valley north of Bishop, the expansive Rickey Ranch offered the possibility of storing more than 200,000 acre-feet of water at 6,500 feet above sea level. After buying the land, Eaton offered the city a "perpetual right and easement" to build a small dam (100 feet high) below the reservoir site and inundate half the ranch. The price: $450,000—close to what he paid for the entire Rickey Ranch—and Eaton would maintain control over the remaining 22,850 acres, including 5,000 head of cattle, 100 horses and mules, and all of Rickey's farm equipment. In addition, any plans—and compensation— for a larger dam and reservoir would require further negotiation.[21]

FIGURE 1-4. William Mulholland and Fred Eaton, about 1910. Eaton, who served as mayor of Los Angeles in the 1890s, recognized the limits of the Los Angeles River for municipal water supply. He was the first to look to the Owens Valley as a source for water, later convincing Mulholland that a major aqueduct across the Mojave Desert was possible. [*Pictorial History of the Aqueduct*, 1913]

Mulholland bristled at Eaton's audacity but realized the offer could not be ignored. Los Angeles may have had no immediate plans to impound a reservoir at Long Valley, but Mulholland understood that large-scale water storage at the site would ultimately prove vital to the city's interest. As owner of the reservoir site, Eaton held the advantage and, after two days wrangling "at swords' points and arms' lengths," Mulholland acquiesced. As he later explained to the Los Angeles Aqueduct Investigation Board, a deal was struck because Mulholland believed the city risked losing even more by standing pat. Claiming that a group of "very strong men of this town" were prepared to offer Eaton "a great deal more money" than he would get from the city, Mulholland contended that he had little choice but to accept Eaton's terms. When pressed for specific names he demurred, "I don't want to testify who they were."[22]

While Mulholland resented the terms of the deal made with Eaton, he and other city leaders recognized that the easement to build even a small dam virtually eliminated the possibility that the Reclamation Service would ever sponsor an irrigation project in the Owens Valley. "There was a fear that if other parties obtained the rights of the Long Valley reservoir," observed John M. Elliott, a member of the Board of Water

Commissioners at the time, "they might, in the future, interfere with the City's water supply, and for that reason alone, if for no other, it was wise for us to take every precaution to protect the City in the future."[23] Mulholland shared this view, but never forgot that Eaton's financial ambitions were the root of a lingering political embarrassment for the city.

Throughout the months of secret planning and deal-making, apprehension smoldered among the residents of the Owens Valley. From 1904 through the summer of 1905, they had been purposely kept in the dark, but complete secrecy was impossible to maintain as Eaton negotiated with farmers and ranch owners to secure land and water rights options. Hints of collusion between Eaton and the Reclamation Service emerged because Eaton was frequently seen in the company of Joseph B. Lippincott, supervising engineer for the Reclamation Service. Although Lippincott's arrival in the valley had initially been greeted as the harbinger of a federal reclamation project, he proved to be a man of divided loyalties. Along with being a government employee, he also operated as a private consulting engineer whose biggest client was the city of Los Angeles. In the end, Lippincott resolved the conflicting interests posed by his two jobs by simply deciding that the valley's water should go to Los Angeles rather than to a local irrigation project.

With Eaton serving as the city's agent in the Owens Valley, Lippincott quietly provided him access to Public Land Office records, facilitating identification of key land parcels and associated water rights sought by the city. Lippincott did not reveal all his actions to his superiors in the Reclamation Service, but he told them enough to set off alarms. They warned him, as Abraham Hoffman has noted, that he "courted conflict-of-interest charges and rendered the Reclamation Service vulnerable to scandal."[24] Despite some disciplinary saber-rattling, the agency declined to take any action against him, presaging the federal government's ultimate support for the city's plans.

Continued sightings of Eaton and Lippincott fueled uneasiness, and by the summer of 1905 Owens Valley residents suspected something ominous. Mulholland got word of their disquiet and, fearful that the city's effort to maintain secrecy verged on collapse, raced to the valley. In concert with Eaton, he secured options on a last batch of land and water rights deemed necessary to ensure the aqueduct's success.[25] On July 29, 1905, the charade ended when the *Los Angeles Times* finally broke the news in a headline, "Titanic Project to Give City a River." In the same issue, Mulholland could proudly proclaim that "the deal by which Los Angeles city becomes the owner...of the purest snow water has been nailed."[26] Los Angeles leaders were delighted, but Owens Valley residents who had not made handsome profits selling their land and water rights were outraged. "Los Angeles Plots Destruction," asserted the Bishop *Inyo Register*. "Owens Valley is to be made the victim of the greatest water steal on record." Bitterness toward Los Angeles swept across the Owens Valley, breeding resentment so deep that it eventually erupted in violence.[27]

Aqueduct Approval and Construction

Shrugging off the valley's anger over the proposed aqueduct, the city and Mulholland focused on three immediate goals: raising $1.5 million to pay off land options and secure water rights, raising an estimated $23 million to construct an aqueduct connecting the city to the Owens Valley, and winning congressional approval to build the aqueduct across federal land lying between Los Angeles and the valley.

The money came in the form of bonds endorsed by the city electorate, first in 1905 for the land and water rights (voters were 14 to 1 in favor) and then in 1907 to finance construction (approved by a margin of 10 to 1). The financial landscape also brightened when the city realized that significant hydropower could be harnessed where the aqueduct descended from the high desert to the coastal plain. Private power companies opposed the city's plans to enter the electric power business, but the public's appetite for electrical energy (first introduced in Los Angeles in 1882) was voracious, resulting in 1910 in a vote almost 8 to 1 in favor of authorizing bonds for a municipally owned generating system.[28]

Once the first bond issue won approval in September 1905, the city entreated Congress for a right-of-way where the aqueduct would cross federal lands. Owens Valley residents fought to kill the measure or amend it to include a local reclamation project, but their efforts foundered, as the city cultivated a powerful ally in President Theodore Roosevelt. "It is a hundred or thousand fold more important to the State and more valuable to the people as a whole," proclaimed Roosevelt, "if [this water is] used by the city than if used by the people of the Owens Valley." Congress concurred, approving the right-of-way legislation in June 1906.[29]

Los Angeles' victory in Congress incensed Owens Valley residents, especially after it became known that Mulholland had hired a certain engineering specialist to help build the aqueduct—Joseph B. Lippincott. After plans for the aqueduct became public in 1905, valley residents assailed Lippincott, the Reclamation Service official who had encouraged their hopes for a federal irrigation project and then dashed them through collusion with Eaton and Mulholland. The U.S. secretary of the interior had ordered a special investigation; it affirmed that Lippincott's behavior was "indefensible" and recommended that "for the good of the Reclamation Service, he should be separated from it." Though tempted to fire him, the secretary backed off, fearing that dismissal would only further embarrass the service. Shortly after Roosevelt signed the right-of-way legislation, Lippincott resigned from the Reclamation Service and, in the summer of 1906, became the aqueduct's assistant chief engineer working directly under Mulholland. With this, a final symbolic stake was driven through the heart of any remaining hope that the federal government might someday sponsor an irrigation project in the Owens Valley.[30]

Seemingly secure funding, federal approval of the right-of-way, and the engineering leadership of Mulholland and Lippincott augured well for the aqueduct's success. But hurdles remained. During the campaign for the first bond vote in 1905, Los Angeles leaders acknowledged that, if the aqueduct venture failed, a projected financial burden of $24.5 million in bonds for land, water rights, and construction would likely bankrupt the city. Apprehension intensified when Reclamation Service officials accused Mulholland of underestimating construction costs by $13 million. And then consultants for the *Los Angeles Examiner* projected even higher costs. All this led to questions about Mulholland's lack of formal training as an engineer. To be sure, he had read books on the subject and had gained experience supervising crews of workmen for the Los Angeles City Water Company and now the city of Los Angeles. However, none of the projects built to date remotely approached the scale or complexity of the proposed aqueduct.[31]

To estimate the cost reliably and assuage public fears, city leaders appointed an independent panel of experts to appraise the aqueduct's design and financing.[32] After a national search, Los Angeles authorities engaged three prominent engineers to serve on the Board of Consulting Engineers: John R. Freeman, a past vice president of the American Society of Civil Engineers with experience as a consultant for both the Boston and New York City water supply systems; Frederic Stearns, a past president of the American Society for Civil Engineers and chief engineer of the Metropolitan Water Board of Boston; and James D. Schuyler, former California assistant state engineer and design engineer for two major dams in Southern California (Hemet and Sweetwater) as well as multiple water projects elsewhere in the American West.[33]

In November 1906, the consulting board spent eleven days in the field collecting water samples, examining reservoir sites, and carefully

FIGURE 1-5. In November 1906, Mulholland (far right), his Assistant Chief Engineer J. B. Lippincott (center), and the Board of Consulting Engineers set out to inspect the aqueduct's proposed right-of-way. John Freeman (far left), James Schuyler (second from left), and Frederic Stearns (second from right). [*Pictorial History of the Aqueduct*, 1913]

inspecting the route of the aqueduct "from the proposed point of diversion on Owens River, all the way to the San Fernando Valley." Submitting its findings a month later, the board gave a generally positive assessment (chemical analyses found the water "good...for domestic purposes" and "much softer than the water now supplied to the city"). But for financial reasons the board called for major engineering modifications. These included a reduction in the number of expensive steel siphons and elimination of three reservoirs, one in Long Valley and two in the San Fernando Valley. They also rerouted the lower end of the aqueduct, bringing it into the San Fernando Valley via the upper Santa Clara River watershed. In Mulholland's original plans it was to have entered through Big Tujunga Canyon near the eastern end of the San Fernando Valley; the modified route would enter the valley's western end near Sylmar. This change shortened the aqueduct by twenty miles and eliminated more than a mile of hard-rock tunneling. It also allowed the city to take advantage of the hydropower potential "of the power plants to be built on the drop in San Francisquito Canyon."[34]

It is important to note that the consulting board did not propose building any storage dam across San Francisquito Creek; this possibility was not contemplated by Mulholland until the 1920s. Nonetheless, without the board's recommendation to realign the aqueduct away from Big Tujunga Canyon, the city would never have had any reason to impound a large reservoir within the upper Santa Clara Valley.

Construction on the aqueduct officially began in September 1907. Six years later, on November 5, 1913, Owens River water first entered the San Fernando Valley at the Cascades near Sylmar. Mulholland had built the aqueduct on time and—thanks both to changes made by the Board of Consulting Engineers and to modifications of his own—within budget. He brought together a diverse workforce of some 4,000 men to build the 233-mile marvel, the longest aqueduct in the world at the time and a wholly gravity-flow system requiring no pumps. He laid out the general plan and, with Lippincott as his assistant chief engineer, oversaw its execution. Together they completed the engineering surveys, devised bonus incentives for arduous hard-rock tunneling, dealt with problems of cement supply and quality, combated thievery, and tried to rein in rowdy saloons that flourished along the right-of-way. A hospital department oversaw nine fully staffed field hospitals; sleeping accommodations, as well as a commissary, were provided at nominal cost. Mulholland, William Kahrl wrote, "was never cold or distant in dealing with his men. He visited frequently in the work camps...His constant presence, accessibility, and confidence in his fellow workers generated a remarkably cohesive spirit within the aqueduct's work force."[35] People might dispute the efficacy and the justice of the city's effort to bring water from the Owens River into the coastal south. But Mulholland completed the aqueduct on schedule and within budget. This triumphant delivery of the water of the Owens River to the city transformed him into an iconic figure among his staff, the commissioners who supervised him, and—minus a few knockers—the people of Los Angeles.

That the aqueduct was built to meet more than the city's immediate needs is clear. When the project was announced publicly in 1905, the population of Los Angeles stood at 209,000; upon its completion in 1913, the tally was about 500,000. But the aqueduct was designed to deliver 260 million gallons daily, enough water, in Mulholland's estimate, "for a population of two and a half million people." The future beckoned and a great metropolis was in the making. At the aqueduct's public dedication on November 5, Mulholland brought the gala to a close with his famous invitation, "There it is. Take it!"[36]

FIGURE 1-6. The Los Angeles Aqueduct intake about fifteen miles north of Independence, at 3,814 feet above sea level, where the Owens River is diverted out of its natural streambed. [Los Angeles Department of Water and Power]

FIGURE 1-7. The aqueduct wending down the western slope of the lower Owens Valley. This seemingly modest-sized open concrete channel could carry a flow of over 400 cubic feet per second—equal to an annual maximum volume of some 300,000 acre-feet (or about 100 billion gallons). [Los Angeles Department of Water and Power]

FIGURE 1-8. South of Owens Lake, the aqueduct cuts across rough terrain, requiring several large siphons. The most visually impressive is Jawbone Siphon, located on the western edge of the Mojave Desert about ninety miles below the aqueduct's intake. [Los Angeles Department of Water and Power]

FIGURE 1-9. The Consulting Board of Engineers recommended that the aqueduct descend from the western Mojave Desert into San Francisquito Canyon, thus allowing for large-scale hydroelectric power development. This view shows San Francisquito Power House No. 1, completed in 1917, where the flow drops 937 feet and can generate 60,000 kilowatts of power. [Huntington Library]

FIGURE 1-10. San Francisquito Power House No. 2, completed in 1921. After turning the power house turbines, water from the aqueduct continues its journey to the San Fernando Valley, about twenty miles to the south. The St. Francis Dam was later built about a mile and a half upstream from Power House No. 2. [Los Angeles Public Library Photo Collection]

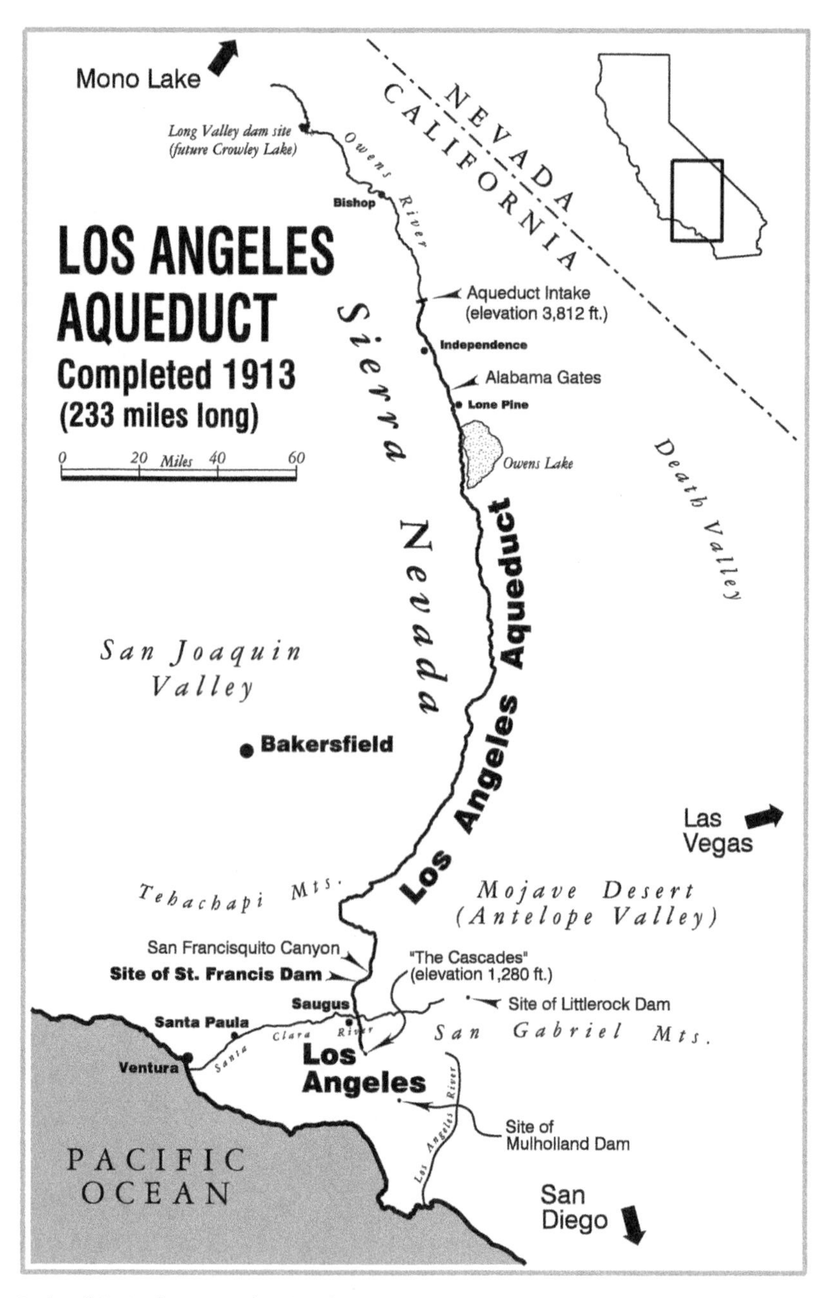

Richard K. Anderson and DC Jackson

FIGURE 1-11. A view showing the civic celebration on November 5, 1913, when Owens Valley water arrived at the Cascades, near Sylmar in the San Fernando Valley. [Los Angeles Department of Water and Power]

A Hero Cheered and Criticized

In 1913 the aqueduct officially terminated at the Cascades, but Mulholland and the city had earlier recognized a need to construct two reservoirs in the San Fernando Valley. In 1906 the Board of Consulting Engineers had excised these reservoirs on the grounds of cost, but they were nonetheless needed to meet fluctuating seasonal demands. In order to maintain the financial integrity of the original aqueduct project, the new dams and reservoirs were to be built by the water department rather than the Bureau of the Los Angeles Aqueduct. Thus, while Mulholland served as chief engineer responsible for both the aqueduct project and the city water department, legally and financially these two enterprises remained distinct. As early as 1908 Mulholland had developed plans for the Lower San Fernando Valley Reservoir (completed in 1915) and the Upper San Fernando Valley Reservoir (built a short time later). Moving forward with an earth-fill design to impound the lower reservoir, in 1912 Mulholland engaged an outside consultant to evaluate the work. Arthur P. Davis, chief engineer for the U.S. Reclamation Service, was hired to examine the dam site and the proposed design. After a field inspection and review of the plans, Davis endorsed the scheme and Mulholland and the Board of Public Service Commissioners proceeded with construction. What is most noteworthy about this episode, as further discussed in chapter 2, is that Mulholland sought the advice of an independent expert to corroborate his plan—a practice he dispensed with in later years.[37]

Why Mulholland backed away from the practice of consulting outside experts and increasingly relied on his own judgment is one of the enduring mysteries underlying the St. Francis Dam disaster, but his growing fame and the hubris it fostered seem likely explanations. So, too, does the criticism that he received in the wake of his aqueduct triumph—sniping he did not enjoy and considered baseless. He was hailed as a hero by most, and the attacks of a few only encouraged him to embrace the fawning sentiments of the majority. A reporter captured the public's infatuation and faith in Mulholland's judgment with a fanciful boast: "If Bill Mulholland should say that he is lining the aqueduct with green cheese because green cheese is better than concrete, this town would not only believe the guff but take the oath that it was so." And there seems little doubt that most people in early twentieth-century Los Angeles agreed with W. W. Hurlbut (an engineer who devoted much of his career to working for Mulholland) when he declared that "the public at large realizes…his untiring efforts in providing the city with the most essential element of its growth—nay, its very life blood." Mulholland's stature among his staff prior to the St. Francis debacle is reflected in Hurlbut's encomium in *Western Construction News*: "Since time immemorial every profession, every line of human pursuit, has had its outstanding character, its shining light, its great leader. In the profession of water works engineering there is an outstanding figure, a leader who... has proved to be a builder of an empire—an empire of unsurpassed progress in municipal development—William Mulholland."[38]

Sharing in the adulation were the Public Service commissioners to whom Mulholland reported. None of the commissioners were engineers with training or practical expertise in the building of water supply systems. Instead, they were lawyers, businessmen, doctors, and investors in real estate—"citizens [with]...part-time responsibilities," as Vincent Ostrom noted, and unable to "undertake...[or] even assume the initiative in the formulation of policies."[39] None possessed the credentials or knowledge to challenge Mulholland even if they had sought to do so—which none ever did. Their attitudes were epitomized by those of R. F. del Valle, an attorney who served on the Board of Public Service Commissioners and later chaired the Board of Water and Power Commissioners. After the St. Francis disaster, Del Valle affirmed:

> Mr. Mulholland has had charge of the department ever since its inception...During that time he conceived the construction of the aqueduct, built it, has built nineteen dams for the department, and during that whole time, the board has found that he has used the proper judgment, has been competent, efficient in every manner, and therefore the matter...as to whom he should consult or what he should do in detail has been left entirely to his judgment, because the board has had the utmost confidence, and has now, in his ability as an engineer.[40]

Widespread respect for Mulholland's judgment—particularly among his nominal superiors—allowed him to weather rumors of involvement with a San Fernando Valley land syndicate that profited handsomely from the water brought by the Los Angeles Aqueduct. These criticisms first emerged during the 1905 and 1907 bond campaigns and resurfaced during the Los Angeles Aqueduct Investigation Board hearings in 1912. In part, this board examined a business syndicate that had acquired large tracts of land before public announcement of the aqueduct. Finding no evidence of a relationship between Mulholland and the syndicate, the board nonetheless concluded that "the Aqueduct affords opportunities for graft, and if this Board had had the necessary time to develop all facts along lines suggested by individuals, a knowledge of human nature indicates that men would have been found who had succumbed to temptation." Historians have confirmed that a San Fernando Mission Land Company bought valley land with the help of insider information (a syndicate confidant was a member of the Board of Water Commissioners). Land owned by the syndicate—along with most of the San Fernando Valley—was annexed to the city in 1915, and it skyrocketed in value because of access to the water flowing in from Owens Valley.[41] In the public mind, memory of secret land deals and collusion among city elites has long lingered (witness the popularity of the film *Chinatown*), but a direct connection between Mulholland and such opportunism remains mere supposition.

While syndicate skullduggery prompted no legal action, it did embolden critics of "The Chief," whose attacks often coincided with the bond campaigns and fed off the hostile atmosphere fomented by the Los Angeles Aqueduct Investigation Board. Among the most vocal faultfinders was Andrae Nordskog, publisher of the Los Angeles *Gridiron*, a newspaper known for championing the rights of Owens Valley landowners and for excoriating Mulholland, Eaton, and Lippincott. Another harsh critic was Samuel Clover, editor of the *Los Angeles News* and a vigorous opponent of the 1907 bond campaign, who feared that the aqueduct was a financial disaster in the making and protested that aqueduct water would be both impure and unneeded. Mulholland dismissed Clover's criticism as a "screed...from a source absolutely ignorant." But doing more to undermine Clover were findings issued by the Board of Consulting Engineers (which endorsed the hygienic suitability of Owens Valley water), and public understanding that his newspaper was controlled by power companies opposed to a municipally owned electric power system.[42]

Perhaps Mulholland's most persistent critic was Frederick Finkle, a consulting engineer who also had ties to private power interests. Like Clover, he opposed the first aqueduct construction bonds in 1907, but he was employed at the time by Edison Electric (later renamed Southern California Edison Company), so his motives could be seen as partisan, or at least less than disinterested. Finkle proved tenacious in his criticism of Mulholland and in 1912, with the support of the Association of American Portland Cement Manufacturers, he publicly lambasted the quality of the cement used in building the aqueduct. Seeking to cut expenses, Mulholland and Lippincott had built a plant to manufacture cement that, if purchased at market prices from private firms, would have brought the aqueduct to the brink of insolvency. The aqueduct

FIGURE 1-12. In this informal snapshot, a young man proudly poses atop a section of the Los Angeles Aqueduct. Although some criticized it, among the citizenry of Los Angeles Mulholland's 233-mile-long conduit was widely celebrated as a great and wondrous achievement. [DC Jackson]

cement worked well in broad canals where water moved slowly, but in places subject to high velocity the cement tended to spall and required replacement. Finkle kept up his criticism as he discerned new problems to expose, in addition to the poor quality of the cement. Among them were expensive caterpillar-type tractors, originally designed for use on the soils of the Sacramento–San Joaquin Delta. But they were poorly suited to a high desert environment and, in the hands of inexperienced operators, quickly broke down.[43]

The criticisms leveled by the Aqueduct Investigation Board and the carping of Nordskog, Clover, and Finkle left Mulholland uneasy and distrustful. Despite his denials of any problems, the difficulties that did exist—such as faulty cement, broken-down tractors, and the controversial agreement with Eaton over Long Valley—could not be kept from the public arena. Circling the wagons, Mulholland projected a dismissive air, doing little to change his modus operandi and expressing no interest in opening up his work to anyone outside his department or beyond his control. Disparagement of the aqueduct simply reinforced Mulholland's proclivity to keep his own counsel and, if possible, avoid independent review of work undertaken by his Bureau of Water Works and Supply. This point was artfully expressed at the banquet held on the evening the aqueduct opened in November 1913. Here, Mulholland was caustically indifferent to the sniping of naysayers, with the *Los Angeles Times* reporting that "he spoke of criticism from various sources as nothing more than an annoyance, as of the barking of a dog in the night."[44]

Fred Eaton and Long Valley

In 1920 Mulholland turned his attention to damming Long Valley, hoping to keep high-volume, regulated flow in the aqueduct while simultaneously satisfying water rights held by Owens Valley irrigators and hydroelectric power developers. His goal: a high dam capable of storing 260,000 acre-feet of water. At the time, this would have been the largest reservoir in California. But a formidable obstacle blocked his way—Fred Eaton. As Mulholland had agreed fifteen years earlier, Los Angeles could not build a dam at Long Valley higher than 100 feet or create a reservoir holding more than 68,000 acre-feet without Eaton's permission.[45]

The Aqueduct Investigation Board of 1912 had castigated Mulholland over the terms of the Long Valley agreement and also urged indictment of Eaton for his duplicity in brokering Owens Valley land purchases. Nonetheless, the board acknowledged that "no direct evidence of graft has been developed" tying Eaton to Mulholland.[46] In the years following the report, acrimony festered between the two men, setting the stage for less than friendly negotiations when Mulholland set out to dam Long Valley at the 140-foot-high contour. When Eaton learned of Mulholland's plan, he proposed $900,000 for use of lands above the 100-foot contour. Mulholland valued the concession at no more than $255,000 and rejected the price as excessive. The result: Eaton increased his demand to $1.5 million, and soon after to $3 million. Incensed by Eaton's audacious dictates, Mulholland cut off talks, determined to go ahead with the 100-foot-high dam. But pursuit of this option provoked the ire of irrigators and hydropower interests downstream from Long Valley.

Along with the city, they held rights attached to the upper Owens River. A large dam at Long Valley would hold enough water to let Los Angeles make diversions to meet the city's growing needs without transgressing on both downstream irrigation rights (held by Bishop-area farmers) and power rights in the Owens River Gorge (held by the Southern Sierras Power Company). But the smaller dam would impound only a modest-sized reservoir, one incapable of simultaneously meeting the desires of the three parties. Because the smaller dam would interfere with the rights of irrigators and private power interests while not providing any sizable benefits, they threatened court action to block the project. Beset by legal headaches and annoyed that Eaton might win their game of high-stakes reservoir poker, Mulholland stepped back and decided that he would not build any dam—small or large—at the north end of the Owens Valley.[47]

In the short term, Mulholland defended his decision by proposing that increased groundwater pumping in the Owens Valley could easily replace supplies lost by his abandonment of plans for Long Valley Dam. Groundwater did hold one advantage over surface reservoir supplies because it meant less evaporation. But pumping requires energy, adding operational costs year after year. And pumping also made Owens Valley residents fear that Los Angeles was bent on draining all water out of the valley for use in the south.

Further complicating matters, neither groundwater pumping nor reservoir storage in the headwaters of the aqueduct provided a way to accommodate seasonal variation in water demands at the lower end of the system. If demand in Los Angeles had been relatively constant over the course of a year, then the lack of large-scale storage would have been a nuisance more than a systemic problem. But the city's water needs fluctuated dramatically. In summer, newly irrigated tracts in the San Fernando Valley required huge quantities of water for citrus production. During winter months the San Francisquito power plants required enough flow to meet the city's power needs regardless of whether water was for irrigation or municipal consumption. In trying to accommodate operation of the aqueduct to these very different demand protocols, Mulholland appreciated the importance of increasing storage capacity at the southern end of the aqueduct.[48] Reservoirs there, he explained, could collect the "great volumes" of water released at the San Francisquito plants that would otherwise "go to waste," or at least not meet any compelling municipal demand.[49]

Reservoirs near the city would have been needed even if Long Valley storage had come online in the 1920s. But in the absence of a Long Valley reservoir, Mulholland's search for increased storage capacity took on added urgency. This was driven by rapidly growing anger in the Owens Valley, where the failure to build a large dam at Long Valley had reignited discontent over Los Angeles' invasion of the rural enclave. Mulholland's concern about the city's "extraordinary growth" (between 1905 and the early 1920s its population had tripled) prompted the municipality to buy additional land and water rights in the Owens Valley. Sellers pleased with the prices made no complaint, but anger grew among those who did not get top dollar, as it did among businessmen, merchants, and residents who saw in-town property values and incomes plummet as farmers and ranchers sold out and left the valley.[50] This discontent eventually spawned vigilantism and in May 1924—almost contemporaneously with the start of work on St. Francis—a section of the aqueduct near Haiwee was blown up with dynamite. Repairs were quickly made and the aqueduct suffered no permanent damage, but relations between city and valley were never the same again. In November 1924 valley residents famously seized control of the aqueduct at the Alabama Gates and for a short time diverted (or wasted) the city's water supply into Owens Lake. Further dynamite attacks came in 1926 and 1927, none with lasting impact. But these assaults were of great concern to Mulholland and his civic brethren, who feared that they might someday become even more devastating.[51]

Mulholland's decision to build a large-scale storage dam in San Francisquito Canyon took shape as agitation in the Owens Valley grew in the early 1920s. Despite problems with the San Francisquito Creek dam site, a great and undeniable attribute of the proposed reservoir location—at least for city officials—was that it lay far removed from the Owens Valley and the perfidious meddling of Fred Eaton. And any water making it into San Francisquito Canyon could never be used for the benefit of Owens Valley farmers. As far as Mulholland was concerned, critics be damned. They had carped about his success in completing the Los Angeles Aqueduct, but to what end? It was his aqueduct, it would be his dam, and he would build it on his terms. Who was to say otherwise?

CHAPTER TWO

The Dam: Site Selection and Design

> It is a crime to design a dam without considering upward pressure.
>
> Edward Godfrey, *Engineering News*, 1913[1]

> There are older slides which may be recognized in the faces, spurs, and ravines of the mountains [on the east side of San Francisquito Canyon]. They are large and old. They are so large that they are easily mistaken for firm spurs of the mountain; but once a slide, always a slide.
>
> Bailey Willis, *Western Construction News*, 1928[2]

Before the 1920s, neither Mulholland nor any city official publicly mentioned San Francisquito Canyon as a reservoir site.[3] It arose quickly as an ad hoc possibility, driven more by difficulties with other sites than by a carefully conceived plan. Originally, Mulholland and Fred Eaton had targeted Long Valley to provide long-term, large-scale storage for the Los Angeles Aqueduct. By 1922 their relationship had soured, pushing Mulholland to abandon those plans. But pressure to provide Los Angeles with more storage capacity did not subside simply because he and Eaton could not come to terms. Increased demand from irrigation and electric power generation was particularly pressing, and finding a way to boost storage capacity near the aqueduct's lower end weighed heavily on Mulholland.[4]

His search to increase storage proved difficult because, while several small reservoir sites existed near the city, they offered only minimal capacity. For example, earth-fill dams built by Mulholland's Bureau of Water Works and Supply in the early 1920s at Stone Canyon, Encino, and Lower Franklin provided a combined capacity of less than 15,000 acre-feet.

And the most prominent of the dams built within the city limits, the 200-foot-high Weid Canyon Dam in the Hollywood Hills, could store a maximum of 8,000 acre-feet, while a large dam at Long Valley could impound more than 250,000.[5]

Before settling on San Francisquito Canyon as a storage site, Mulholland looked to Big Tujunga Canyon at the northeastern end of the San Fernando Valley, about ten miles east of where the aqueduct enters the valley at Sylmar. In the earliest plans for the aqueduct—before the 1906 Board of Consulting Engineers recommended a right-of-way passing through the Santa Clara River watershed—Big Tujunga was projected as the aqueduct's entry point into the valley. It is possible that these earlier investigations spurred Mulholland's interest in the canyon as a reservoir site. But using Big Tujunga would require condemnation of large tracts of privately owned land and likely foster "excessive demands on the part of property owners."[6] In a July 1922 report to the Board of Public Service Commissioners, Mulholland described the need to develop additional storage capacity "below the power plants." Without specifically naming Big Tujunga, he cryptically referred to "reservoir sites already surveyed," adding that "this storage will prove expensive."[7] In view of the headaches of condemnation and onerous costs of land acquisition, Mulholland's interest in a Big Tujunga storage reservoir soon evaporated.

In the late summer of 1922 Mulholland turned his attention to an expansive area in the upper San Francisquito Canyon lying between the city's two hydroelectric power plants.[8] A decade earlier, this stretch of canyon bottomland had been used for a camp and commissary during construction of the aqueduct. The dam and reservoir site (at this point anglicized to St. Francis) encompassed land either already owned by the city or controlled by the U.S. Forest Service.[9] It was therefore available for public use, and could be developed without transgressing private property rights. The city applied to use Forest Service land for the dam and reservoir on December 29, 1922; the permit was issued less than a month later, on January 18, 1923.[10] The one remaining bureaucratic hurdle involved relocating the Southern California Edison Company's 60,000-kilovolt Borel-Lancaster transmission line, which connected to the company's substation in Saugus. Originating at the Borel hydroelectric power plant on the Kern River, this line ran the length of San Francisquito Canyon, but only the segment upstream from the dam needed to be moved to prevent inundation from the new reservoir. Although this was not a complicated engineering problem, negotiations between the city and the power company over the realignment dragged

FIGURE 2-1. The future site of the St. Francis reservoir, used for a camp during construction of the Los Angeles Aqueduct, about 1912. The Pacific Light and Power Co. (later Southern California Edison) power transmission line running down San Francisquito Canyon is visible at center left. This line was relocated along the east ridge of the canyon when the dam was constructed in the 1920s. [*Pictorial History of the Aqueduct*, 1913]

on through the summer of 1925.[11] The placement of the transmission line above the dam and reservoir along the canyon's east ridge ultimately proved significant: close to midnight on March 12, 1928, the cutoff of the current in the Borel-Lancaster transmission line marked the exact time of the dam's cataclysmic collapse.

"A Detriment to Power"

In the city's view, there were two important advantages of the St. Francis site: first, proximity to the aqueduct right-of-way, whereby an emergency water supply could be maintained in the event of drought or failure of the aqueduct; and second, a drainage area of 37½ square miles in the upper San Francisquito Creek watershed that could provide a supplemental water supply for the city estimated at 5,000 acre-feet per year.[12]

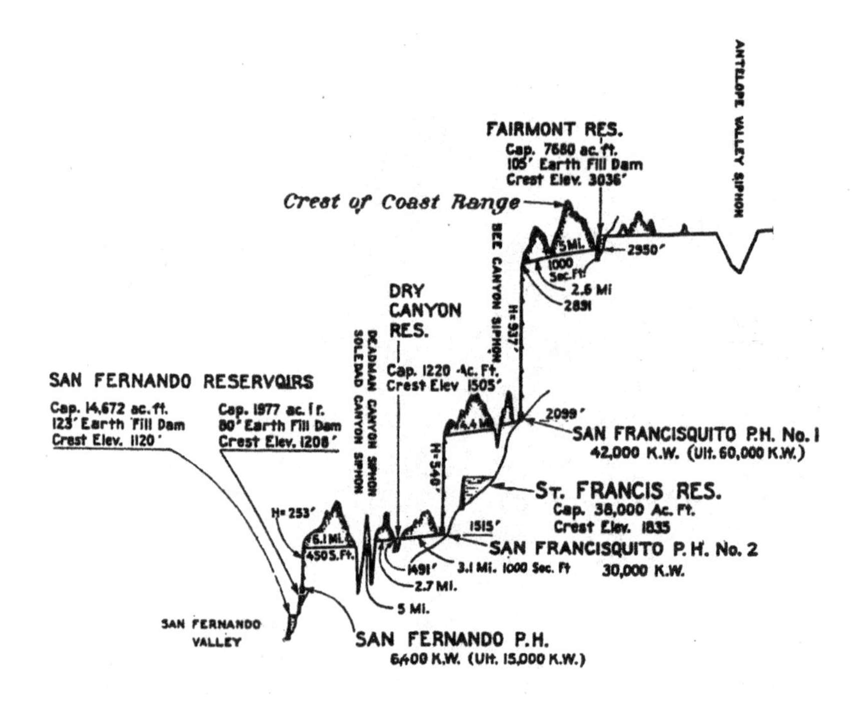

FIGURE 2-2. Schematic drawing of the lower end of the Los Angeles Aqueduct showing the placement of the St. Francis Dam and reservoir between San Francisquito Power House No. 1 and Power House No. 2. This drawing clearly illustrates that water stored behind the dam could not be used to generate electricity at Power House No. 2. [Governor's Commission *Report*, 1928]

In terms of topography, the St. Francis site was certainly suitable for a reservoir. Encompassing a wide, flat valley lying upstream from a relatively narrow gorge, it did not require a lengthy or excessively high dam to provide storage capacity exceeding 30,000 acre-feet. In other ways the location was less than ideal, including the geology of the site, to be discussed later in this chapter. But there was another reason the site should have given Mulholland pause, especially because during the 1920s the city was seeking to increase its municipal power production.

As recommended by the Board of Consulting Engineers, the city had built two hydroelectric power plants in San Francisquito Canyon below the Fairmont Reservoir and the Elizabeth Tunnel. Coming online in 1917, Power House No. 1 captured a hydraulic drop of over 900 feet and lay about five miles upstream from the St. Francis dam site. Power House No. 2 came online in 1921 about a mile and a half below the St. Francis site. Water discharged from Power House No. 1 was directed into a

FIGURE 2-3. A 1920s photograph showing Mulholland (at the table, second from left) and Ezra Scattergood (first at right) meeting with the Board of Public Service Commissioners. Despite the detrimental effect that a dam constructed at the St. Francis site would have on the city's hydroelectric power system, Mulholland was authorized by the commissioners to locate the reservoir in San Francisquito Canyon. [Los Angeles Department of Water and Power]

seven-mile-long tunnel extending through the canyon's eastern ridge, from there dropping more than 500 feet to turn the turbines in Power House No. 2.

The problem presented by Mulholland's proposed dam was simple: water exiting Power House No. 1 and diverted into the reservoir (by allowing it to flow down the natural streambed of San Francisquito Creek) would not pass into the tunnel serving Power House No. 2. Consequently, *no water stored behind the St. Francis Dam could ever be used to generate electricity at Power House No. 2*. This was of no minor consequence. Ezra F. Scattergood, the city's chief electrical engineer, understood the financial impact of the St. Francis reservoir, estimating that it would cost at least $100,000 to replace the power lost by a single filling with electricity purchased from Southern California Edison Company. And this would not be a one-time expense. The dam represented a long-term burden on the city's power system because, over time, use of the reservoir would inexorably result in the loss of more revenue.

While not publicly airing his concerns, internally Scattergood criticized the St. Francis dam site because of the deleterious effect on power generation.[13] As an alternative, he urged either that a reservoir be built in upper Bouquet Canyon (lying to the east of the San Francisquito watershed and requiring a lengthy steel siphon to transport water under pressure); or that a dam be built to increase the capacity of Elizabeth Lake (near the existing Fairmount Reservoir). As he later explained, both of these sites lay above Power House No. 1, and each could provide emergency storage while also adding "to the security and reliability of San Francisquito Power Houses No. 1 and 2 instead of being a detriment to power as was the St. Francis Reservoir."[14] From a hydropower perspective, the logic of Scattergood's objection is undeniable, but in the early 1920s his argument did not hold sway. Without public discussion or debate, the city's political leadership gave Mulholland carte blanche to build the St. Francis Dam where and how he saw fit, regardless of its impact on the municipally owned electric power system.[15]

Mulholland's Concrete Dams

Built in the early 1920s, the Weid Canyon Dam (renamed Mulholland Dam in March 1925 and often called Hollywood Dam) bears a special relation to St. Francis, representing the first time that Mulholland constructed a concrete gravity design. The designs of the St. Francis and Weid Canyon dams were intertwined to a remarkable extent, as became evident in testimony given after the disaster. At the Coroner's Inquest, engineers working under Mulholland made frequent reference to Weid Canyon (usually calling it Hollywood Dam) when discussing St. Francis. Even now, it is not always easy to keep the two projects separate when reading the inquest transcript. Some aspects of the Weid Canyon and St. Francis designs differed—for example, in the architectural character of the dam crests and the degree of upstream curvature (Weid Canyon was built with a radius of 550 feet while St. Francis featured a radius of 500 feet). But overall they are quite similar: almost vertically faced, curved concrete gravity designs developed with little consideration for uplift, an issue to be discussed later in this chapter. In addition, both Weid Canyon and St. Francis feature a stepped (not smooth) downstream face, and both were built using a tower-and-chute system to distribute the wet concrete.

Neither in his inquest testimony nor anywhere else did Mulholland directly state why he chose a concrete gravity design for St. Francis—it

FIGURE 2-4. Weid Canyon Dam nearing completion in late 1924. In March 1925 the structure was formally dedicated and named Mulholland Dam in honor of the chief engineer. It was Mulholland's first concrete gravity dam, and it served as a template for the design of St. Francis Dam. [DC Jackson]

apparently amounted to a simple extension of the Weid Canyon project. So a question arises: Why did Mulholland opt to build a concrete design at the Weid Canyon site in the Hollywood Hills? Or, conversely, why did he break from his long-standing practice of building large-scale earth-fill dams such as the Fairmount Dam, and both the upper and lower San Fernando dams, constructed the prior decade? On a practical level, material for an earth-fill dam did not appear to be readily available in large quantities near Weid Canyon. But beyond this, the rationale for his shift from earth-fill to concrete is subject to speculation.[16]

One possible influence was the partial failure of the Calaveras Dam in Northern California. In 1912, Mulholland—while working for

FIGURE 2-5. Mulholland Dam, located in the Hollywood Hills, is often referred to as Hollywood Dam. The imposing concrete structure was a civic monument that attracted visitors, as reflected in this snapshot from the 1920s. [DC Jackson]

the city of Los Angeles simultaneously—began paid service as a consulting engineer for the Spring Valley Water Company (San Francisco's private water supply franchisee). He soon took charge of building the company's large storage dam across Calaveras Creek south of Oakland. As the 200-foot-high hydraulic-fill Calaveras Dam neared completion in March 1918, it suffered a major "slip" along the upstream face. About 800,000 cubic yards of earth fill slid upstream into the partially filled reservoir and toppled the large reinforced-concrete intake tower. Mulholland and the Spring Valley Water Company escaped widespread public censure, as there were no fatalities and no floodwater or debris surged downstream. Nonetheless, the collapse proved professionally embarrassing and perhaps induced Mulholland to adopt concrete gravity designs for the Weid Canyon and St. Francis projects.[17]

Another factor contributing to the design choice for Weid Canyon was the visual potency of masonry and concrete gravity dams compared to the seeming ordinariness of earth embankment structures. At

FIGURE 2-6. The Lower San Fernando Dam near Sylmar, not long after its completion in the early 1920s. This earth-fill embankment structure represents the type of dam that Mulholland specialized in prior to construction of the Weid Canyon/Mulholland and St. Francis dams. [DC Jackson]

FIGURE 2-7. Remains of the Calaveras Dam, south of Oakland, in March 1918, following a slide of 800,000 cubic yards of earth fill. Mulholland served as consulting engineer in charge of the dam's construction, and he used a hydraulic-fill earthen design similar to that of his San Fernando dams. This partial failure may have influenced his choice of concrete gravity designs for the Weid Canyon and St. Francis dams. [DC Jackson]

FIGURE 2-8. The U.S. Reclamation Service (later the Bureau of Reclamation) prided itself on building massive dams that served as symbols of the agency's technical prowess. A good example is the 354-foot-high concrete curved gravity Arrowrock Dam in Idaho, completed in 1916. [DC Jackson]

ceremonies marking the completion of Weid Canyon Dam in March 1925, Mulholland endorsed this view in his public remarks. He denigrated his earlier work, claiming that earth-fill dams resembled "an old woman's apron—an object of utility but not of beauty." Weid Canyon Dam (now officially named for Mulholland by the Board of Public Service) represented something quite different. "For this job" Mulholland proclaimed, "I think I may take a little pardonable pride."[18] Some of this pride was likely driven by a realization that many of the world's major municipalities already featured large masonry and concrete gravity dams as centerpieces of their water supply systems. These included Liverpool's Vyrnwy Dam and Birmingham's Craig Goch Dam in Britain; Boston's Wachusett Dam; New York City's New Croton, Ashokan/Olive Bridge, and Kensico dams; Denver's Cheeseman Dam; and San Francisco's Hetch Hetchy/O'Shaughnessy Dam, all constructed by the early 1920s. With his design for Weid Canyon—and later for St. Francis— Mulholland could create massive concrete monuments attesting to the stature of Los Angeles as a world-class city. And if these structures visually affirmed his own prowess as an engineer, so much the better.

FIGURE 2-9. Completed in 1915, the straight-crested gravity Olive Bridge (or Ashokan) Dam was an essential component of New York City's Catskill Aqueduct. John R. Freeman, who served as a consulting engineer for New York City, championed the idea that major municipal dams should project an aura of monumentality to inspire public confidence. Mulholland's comments on the "beauty" of the Weid Canyon/Mulholland Dam are consistent with this concept. [DC Jackson]

Parsing Mulholland's choice of a concrete gravity design for Weid Canyon requires conjecture, but the rationale for selecting the site is clear. As early as 1912 the city had perceived a need for a reservoir in the canyon and carried out preliminary surveys. In early 1922 detailed survey and dam design work was underway, prompted by a growing population in districts west of downtown. "The location and elevation of this reservoir," Mulholland declared, "is so situated as to be able to furnish a gravity supply to the fast growing hill section of Hollywood" as well as to other "lower gravity systems."[19] Despite such blandishments, Hollywood residents were less than thrilled when they learned of plans for a dam in the midst of their community. "A break in the projected dam from an earthquake or other cause," declared the chamber of commerce, "would release 7,500 acre-feet of water down upon a densely populated portion of the city." In his rejoinder, Mulholland dismissed such fears as "without the slightest foundation...One does not need to be an engineer to realize that a gravity type, arched concrete dam... is, of course, many times stronger than an earth-filled dam." Later, after the collapse of St. Francis, the safety of Weid Canyon/Mulholland Dam would resurface as a political issue. In the meantime, local opposition had little effect on Mulholland's authority.[20] While construction of the dam perched above Hollywood was still underway—and with similar lack of concern for local interests—Mulholland and his Bureau of Water Works and Supply staff started work at St. Francis.

Geology of the St. Francis Site

The St. Francis Dam site is bounded by two geological formations. The east side canyon wall (including the central gorge of the canyon) consists of a gray-colored mica schist—sometimes referred to as shale and frequently termed Pelona schist—formed some 60 million to 70 million years ago. The west abutment and ridge consist of a red sandstone, often referred to as a conglomerate, deposited atop the schist formation about 20 million years ago.[21] The interface between the two formations creates a fault line generally parallel with the streambed. To the trained eye, this fault is readily visible because the two formations exhibit very different colors (gray to the east, red to the west). However, it is not in any way an active fault (such as the famous San Andreas Fault lying about ten miles to the northeast), where sliding movement is, in geologic time, ongoing. The so-called fault running through San Francisquito Canyon simply represents the contact plane or boundary between the two rock formations. Shortly after the disaster in the spring of 1928, geologist Bailey Willis of Stanford University published an insightful description laying out the geological character of the dam site:

> The gray formation is that mass of mica schist and quartzite which underlies the red formation and constitutes the greater part of the [dam's] foundation. It is much older than the red formation and its upper surface was the land surface on which the latter was deposited. The contact between the red and the gray is one of simple deposit; it is not a major fault plane. It is clearly recognizable because of the striking difference in color.[22]

FIGURE 2-10. Looking upstream at the St. Francis Dam site before construction began. The geologic "contact plane" extending through the foundation is visible in the center-left foreground. [Los Angeles Board of Public Service Commissioners, "Twenty-Second Annual Report," 1923]

Willis further described the red formation as "sandstone and conglomerate [whereby] the sand grains are coated with red iron oxide, and this gives the formation the notable dark red color throughout." Although presenting a hard surface when dry, the red conglomerate was susceptible to separation, softening, and swelling when wet. Small fractures on the surface would also allow subsurface seepage and, in Willis' view, "water could intrude by capillary action and might percolate in small quantity. The leakage would be small, but the confined sheets of water would exert hydraulic pressure [uplift] proportioned to the head."[23]

The red formation did not present a particularly desirable foundation for a dam, and the same was true for the gray formation—but for different reasons. The mica schist was hard and durable, as Willis noted: "the rock is firm. Moreover, the minerals are dense and insoluble. They are therefore not affected by water." But the mica schist was "thinly laminated" and contained "numerous transverse fractures" susceptible to

FIGURE 2-11. Post-disaster view of the dam site highlighting the geologic "contact plane" extending through San Francisquito Canyon. The red conglomerate (which appears dark) of the west abutment is to the left and the gray mica schist formation (lighter) is at center. [DC Jackson]

seepage and percolation. Just as significant, the mass of schist making up the canyon's east wall had been formed by an ancient landslide that long predated the red conglomerate formation. Sometimes referred to as a paleomegaslide, this landslide formation had remained in place for eons but was always vulnerable to surface slides. Discussing this distinctive terrain, Willis observed:

> There are older slides which may be recognized in the faces, spurs, and ravines of the mountains [on the east side of San Francisquito Canyon]. They are large and old. They are so large that they are easily mistaken for firm spurs of the mountain; but once a slide, always a slide. They are so old that the breaks have been

FIGURE 2-12. Post-disaster view showing the east abutment formed by the gray mica schist. This view was annotated by geologist Bailey Willis, who first published it in *Western Construction News* in June 1928 to illustrate the ancient landslide that was reactivated after the dam was built. The aqueduct tunnel between Power House No. 1 and Power House No. 2 is buried a few hundred feet back in from the canyon wall. [St. Francis Dam Disaster Papers, Huntington Library]

> washed down and vegetation has covered them. The characteristic features can be seen, however, in the profiles of the landscape.[24]

Ancient landslides are not uncommon across the American landscape, and geologist J. David Rogers has pointed out that more than one hundred U. S. dams have been constructed abutting such geological features. The presence of an ancient landslide does not automatically mean that a dam will fail if built at such a site, but it does require engineers to take precautions—for example, reducing the possibility of percolation, and draining water that seeps under the structure—to ameliorate the dangers. In essence, these dangers arise from the way in which the slide affects the rock structure. As Rogers explains it, "When the ground or the rock has slid in a landslide, it dilates or increases in volume [and] that increase in volume sets up a whole bunch of cracks, and water can go through those cracks quite easily."[25] Eventually, saturation of the fractured and dilated east side abutment, and the absence of measures to counter this subsurface water flow, would prove crucial to the St. Francis disaster.

Prior to the dam's failure, Mulholland never wrote about or publicly described the geology of the St. Francis site. But during construction of the Los Angeles Aqueduct in 1910–12, when a tunnel was excavated through the mica schist formation lying between Power Houses No. 1 and No. 2, city work crews experienced the character of the rock first-hand. With respect to the excavation, the city's 1911 report on aqueduct construction described the rock as "exceedingly rough, and the dip and the strike of the slate [mica schist] such as to threaten slips."[26] J. B. Lippincott, Mulholland's assistant chief engineer during aqueduct construction, long remembered the challenging geological character of the east canyon ridge and described it to fellow engineer John R. Freeman shortly after the dam collapsed. Acknowledging that he had been "intimately connected with the driving of a series of tunnels for our aqueduct through the range of mountains on which the left or east abutment of the dam rested," Lippincott recalled: "The rock that we encountered was a broken schist and a good deal of it expanded when it came in contact with the air and was what the tunnel men called 'heavy ground.' We had great difficulty in holding this ground before [the aqueduct tunnel] was lined with concrete."[27]

In selecting the St. Francis site for a large dam and reservoir, Mulholland either had forgotten what was encountered in the aqueduct construction a decade earlier or dismissed it as irrelevant to the task at hand. In addition, Mulholland was apparently so intent on pushing the St. Francis project forward that he did not consult, either formally or informally, with his former assistant chief engineer. Even a knowledgeable and skilled insider like Lippincott, who had long since demonstrated his loyalty to "The Chief"—and who had direct experience with San Francisquito Canyon and the troublesome character of its "heavy ground"—was excluded from the design and construction process.[28]

Concrete Gravity Design

St. Francis Dam followed quickly on the heels of Weid Canyon, and for the new project Mulholland apparently never considered building anything other than a concrete gravity dam.[29] In choosing such a design for these projects, the sixty-seven-year-old Mulholland did not prove particularly innovative or technologically adventuresome: he operated within familiar engineering practice, and furthermore exhibited little appreciation of recent developments in the theory and practice of concrete gravity dam design. As an architectural historian might phrase it, St. Francis was a *retardataire* design. It lacked features common in other

FIGURE 2-13. Aqueduct work crew about 1912, during construction of the tunnel that carried water from San Francisquito Power House No. 1 to the penstocks above Power House No. 2. This tunnel extends through the mica schist that forms the east wall of San Francisquito Canyon. Following the dam's collapse, J. B. Lippincott referred to this schist as "heavy ground"—difficult to excavate safely. [*Pictorial History of the Aqueduct,* 1913]

major gravity designs of the prior decade and, in current parlance, fell well short of "best practice."

The modern form of masonry and concrete gravity dams originated with French engineers in the 1850s and 1860s, and they embodied a simple guiding principle: place enough material (either stone masonry or concrete) in the dam so that the horizontal water pressure exerted by the reservoir will not tip the structure over or push it downstream. In a gravity dam, the amount of material necessary to provide structural stability is directly proportional to the height of water impounded in the reservoir. Thus, if a gravity dam is raised to increase the height of stored water, it follows that—in order to ensure safety—the base needs to be thickened in order to accommodate the increased hydrostatic pressure. To raise the height of a gravity dam without increasing the thickness of the base is to court disaster.[30]

By at least the 1890s, some engineers building gravity dams appreciated that water from a reservoir could seep under a dam and exert upward pressure. This phenomenon, known as "uplift," destabilizes a gravity dam because it reduces the structure's effective weight and thereby lessens its ability to resist horizontal water pressure. Uplift can act through bedrock foundations that, in the abstract, are strong enough to bear the weight of the dam but are fractured or fissured and thus susceptible to seepage and water saturation.[31] By the early 1920s, many engineers understood that the deleterious effect of uplift on a gravity dam could be countered in various ways: 1) by excavating foundation "cutoff" trenches that reduce the ability of water to seep under the structure; 2) by grouting the foundation (which involves pressurized injection of wet mortar into drilled holes), thereby filling underground fissures and impeding subsurface water flow; 3) by draining the foundation and the interior of the dam through use of porous pipes, relief wells, and tunnels to remove seepage; 4) by increasing the dam's thickness (and hence weight) to counter the destabilizing effect of water pushing upward.[32]

For reasons related to uplift (and to the use of a steep upstream face that essentially eliminates any vertical component of water pressure to help provide stability), concrete gravity dams require enormous quantities of material. As a result, they can be expensive to build.[33] Notwithstanding, many engineers considered (and still consider) properly designed concrete gravity dams to be reliable—even desirable—structures. And in the early twentieth century, at least part of the attraction of massive concrete dams lay in their imposing downstream facades. More than a few engineers, as well as many politicians, found their monumentality appealing, a symbolic projection of presumed security and civic achievement.[34] Lest it be thought that engineers are unaffected by seeming nontechnical considerations, recall Mulholland's comparison of earth-fill dams to "an old woman's apron," followed by his claim of "a little pardonable pride" in what was his first concrete gravity dam.[35]

"They Paid Attention to What I Said"

In his first day of testimony at the Coroner's Inquest a basic question was posed: "Colonel Mulholland, during the time of the erection of this dam, did you exercise supervision and control over it, and were you there frequently?" Mulholland replied: "They paid attention to what I said. I was frequently on the job and gave directions."[36] This rather tepid response, from the chief engineer himself, suggests that he did not have a particularly direct, hands-on involvement in either the design or construction of St. Francis. At the inquest, the explanation of how the St. Francis Dam design was developed—and how it related to Weid Canyon—was left to his subordinates Edgar Bayley and W. W. Hurlbut. Their testimony,

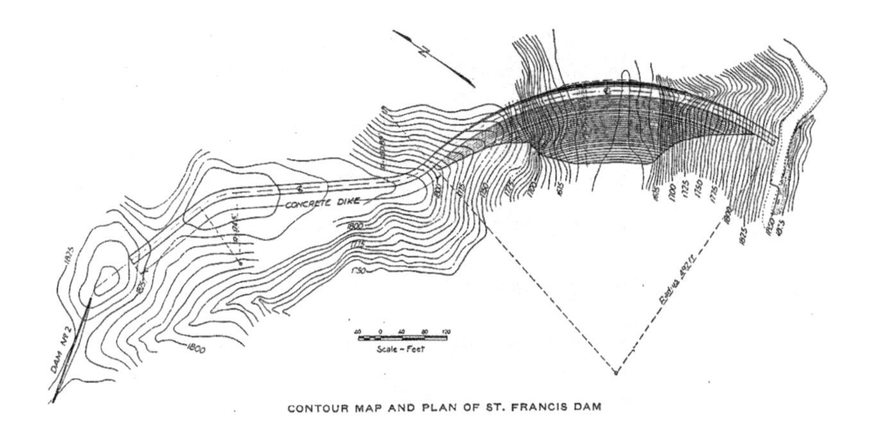

FIGURE 2-14. Plan drawing of the St. Francis Dam, showing the main curved gravity structure on the right and, to the left, the lengthy dike running atop the west abutment. [*Western Construction News*, 1928]

quoted extensively below, underscored Mulholland's detachment from the design process.

In accord with protocol established during construction of the Los Angeles Aqueduct, the initial design for the St. Francis and Weid Canyon dams was apparently delegated to an assistant engineer or draftsman working under the direction of an "office engineer" who reported to Mulholland.[37] Early on at the Coroner's Inquest, William Wilkinson was identified as the assistant engineer charged with developing the St. Francis design, but, surprisingly in retrospect, he was never called to testify.[38] In his place Bayley, the assistant engineer for the Weid Canyon/ Hollywood Dam, testified and described how he assisted Mulholland in developing a preliminary design (at times referred to as the "cross-section profile" or the "cross-sectional transverse profile").[39] Bayley claimed no expertise or experience with concrete gravity dams. Taking care to explicitly deny that he was the dam's designer, he placed responsibility for the two projects on Mulholland:

Q. [By the coroner]: How many of this type [concrete gravity] dams have you designed and constructed?

A. [Bayley]: I have constructed none, have had nothing to do with the construction of any, except being that the Hollywood [Weid Canyon] Dam complied with the profile we had to work by.

Q. Didn't I understand that you are the man that designed the thing, the Hollywood Dam?

A. No, I just testified that I had to do with the design of the cross section profile of the design, with certain limitations. The dam was designed between Mr. Mulholland and myself. Mr. Mulholland set the radius, picked the site, he picked the abutments. We made one or two little changes upstream to get a radial bond.

Q. Mr. Mulholland visited the site?

A. Picked it, considered it suitable for a dam.

Q. And that would be the place to put a dam, [he] said, "I want you to draw me the plans and specifications for a gravity dam"?

A. No, no specifications were written, it was to be done by the department itself, certain dimensions to follow.[40]

Also taking the witness stand at the Coroner's Inquest, office engineer Hurlbut of the Bureau of Water Works and Supply took on the task of explaining (or at least trying to explain) the relationship between the Weid Canyon/Hollywood design and St. Francis:

Q. [By the coroner]: In remodeling the [Hollywood] dam to suit the purposes of the site of the St. Francis Dam, did you change the factors of safety in any way?

A. [Hurlbut]: No sir, the factors were not changed. The lines of stress in all cases lie within the middle third, and are practically the same as that of Hollywood. Some minor changes, but as far as the engineering design of the dam was concerned the lines of stress were entirely within the middle third, and the governing feature of a design of that kind.

Q. Did you make the figures?

A. No sir.

Q. Where were they made?

A. In the drafting room.

Q. Who had charge of it?

A. I had charge of it.

Q. All the work preparatory to building this dam was done in your own department?

A. Yes sir.

Q. You had no advice from the outside or outside your department in designing the Hollywood Dam?

A. I did not have anything to do with the designing of the Hollywood Dam.

Q. So, it was all done in your department?

A. Yes sir.[41]

A short time later, further questions pertaining to the design procedure were posed to Hurlbut:

Q. Were the computations made by one man?

A. The computations, as I have stated, were on the design of the Hollywood Dam, as applied to St. Francis. I said we used the same general design and those computations were made under Mr. Bayley's instructions in the design of that dam. That same general design that was used on the Hollywood Dam was used on the St. Francis, under the Chief Engineer's direction.

Q. There were no new computations?

A. They were not necessary because the dams were practically of the same height.

Q. Did you check those computations or have the[m] checked?

A. They were all checked by individuals in the drafting room. There were several men who worked on the computations.

Q. Can you state who made the original computations?

A. They were made by Mr. Bliss and Mr. Francis.

Q. After you changed the design to suit the different conditions at the St. Francis Dam, were any check computations made?

A. There were no changes in the design made in the design to fit the different conditions. I made the statement that the same design was used for the St. Francis Dam.

Q. Was the radius the same?

A. No sir, the radius was five hundred feet on the St. Francis and five hundred and fifty on the Hollywood.[42]

The Mr. Bliss and Mr. Francis who worked in the BWWS drafting room, like another purported designer, William Wilkinson, were never called to testify at the Coroner's Inquest, and their relationship to the St. Francis or Hollywood designs was never precisely defined. Near the end of his testimony Hurlbut was again questioned on this issue by the water department's attorney, K. K. Scott, and then a new name came up, a draughtsman named Stevens:

Q. You said that Mr. Bliss and Mr. Francis made the computations on the Hollywood design under Mr. Bayley's direction, or did you mean that Mr. Bliss and Mr. Francis made the computations on the St. Francis Dam?

A. Made them on the Hollywood Dam and Mr. Stevens made the drawings, the tracings, and computations in connection with the St. Francis Dam. I said the computations that were made on the Hollywood Dam were made by Mr. Bliss and Mr. Francis.

Q. And those really taken and adapted to the St. Francis Dam by Mr. Stevens, a draftsman?

A. No, those drawings were made following the general analysis which had been made on the gravity type section for the Hollywood Dam under the instructions of the Chief Engineer, to prepare the same plans and adapt them to this site.[43]

The Coroner's Jury was apparently unimpressed by the clarity and forthrightness of the testimony, and a short time later Assistant District Attorney E. J. Dennison made one last attempt to get a straight answer to the basic question.

Q. I want to know if I have this thing straight now. The Coroner asked you this question and I don't think it has been answered definitely: who designed the St. Francis Dam?

A. [Hurlbut]: The St. Francis Dam was designed under the instructions of the Chief Engineer and based on studies which were made for a gravity type dam, which was the Hollywood Dam, that is, the main study is—as I explained—that was made on the basis of a dam two hundred and ten feet high and it was applied to both of the studies in connection with that and were applied to both of these dams.

Q. Now, who designed the St. Francis Dam? Did you design it?

A. I did not.

Q. Did Mr. Mulholland design it?

A. It was designed under his instruction.

Q. Then, am I to understand that Mr. Mulholland designed the St. Francis Dam?

A. It was designed under his instruction.[44]

No one could fault Hurlbut for seeking to protect his boss, but the convoluted, almost tortured, testimony offered by BWWS staff members only confirms Mulholland's distance from the development of the St. Francis design. Yes, he picked out the dam site, gave "instructions" that presumably filtered down to underlings in the drafting room, and was present at the construction site on a more-or-less regular basis. (However, the authors are not aware of any photographs that show Mulholland at the dam site during construction.) But the conjecture that he devoted much time, energy, or careful thought to the St. Francis design is difficult to support. If that were the case, Mulholland would not have relied upon the long-winded and confusing testimony of his subordinates. Instead, he would have taken the lead in explaining to the Coroner's Jury exactly how the St. Francis design was conceived and developed.

"Born with a Stub Toe"

In his testimony, assistant engineer Bayley highlighted how construction of the St. Francis Dam (like that of other BWWS dams) was carried out by city employees working directly under the department's control. Following a precedent established during work on the Los Angeles Aqueduct, no contract for constructing the dam was advertised or put out to bid. Among other things, the decision to forgo a contractor allowed Mulholland and his staff considerable freedom in modifying the design after construction was underway. Because "no specifications were written" and "certain dimensions [were] to follow," the design process was imbued with an air of informality, or at least was much less structured than it would have been had Mulholland engaged an independent contractor. Had it been necessary to renegotiate or adjust a contract as a result of design changes, we would no doubt have a much clearer sense of the exact as-built dimensions of the St. Francis Dam.[45] Nonetheless, there is little doubt that Mulholland decided to raise the height of the dam after construction commenced.

At the Coroner's Inquest—the one time that Mulholland and his staff were publicly questioned about the dam's design and construction—the issue of an enlarged or heightened dam was never raised. That issue also failed to attract the attention of engineers who investigated the disaster in the spring of 1928. (These investigations, and the Coroner's Inquest, are discussed in chapter 5.) In the wake of the collapse, Mulholland and his staff distributed an undated drawing indicating a maximum dam height of 205 feet (extending from the deepest foundations at 1,630 feet above sea level to 1,835 feet at the spillway crest) and a maximum base width of 175 feet (a dimension slightly at odds with a report published in 1926 indicating a maximum thickness of 169 feet).[46] The Governor's Commission included this profile drawing in its report on the disaster, and for many years it was accepted as accurately documenting what would have been a very amply proportioned cross-section.

However, it is now apparent that the drawing published in the Governor's Commission *Report* poses serious problems. For one thing, the drawing fails to accurately record the thirty-inch-diameter outlet pipes that extended through the dam. It shows only four outlet pipes when in fact the dam featured five (the drawing excludes the top pipe, which empties at elevation 1,795 feet).[47] This may not seem to be an egregious error, but it does invite skepticism about the overall accuracy of the drawing. In addition, it highlights the issue of whether or not the drawing in the Governor's Commission *Report* truly documents what was built in San Francisquito Canyon.

Historian Charles Outland was the first to call attention to the heightening of the dam that occurred during the first year of construction, which he detected while researching his book *Man-Made Disaster* in the early 1960s. After studying Board of Public Service Commissioners annual reports from the years 1923 to 1925, he noted that the city had significantly increased the capacity of the reservoir, which was possible only if the dam's height had been raised.[48] Specifically, in July 1923 the city publicized the reservoir capacity at 30,000 acre-feet with the spillway crest at 1,825 feet above sea level;[49] a year later, in July 1924, after "detailed topographic surveys" were completed, a capacity of 32,000 acre-feet at the 1,825-foot spillway level was reported by Mulholland's staff.[50] In this case, the difference in reported capacity can be readily explained by more accurate topographic surveying (the spillway crest elevation does not change; it remains at 1,825 feet). But in 1925 the reported reservoir capacity increases by almost 20 percent, to 38,000 acre-feet. And it is reported that this increased storage capacity derived from "additional surveys and *changes in the plans*" [emphasis added] that created a "crest elevation 1835 feet above sea level."[51] The evidence is clear. Sometime during the first year of construction the dam was raised ten feet in height and the spillway crest was elevated from 1,825 to 1,835 feet above sea level.

Raising the dam's height was not necessarily dangerous but, to fully assure safety in a concrete gravity design, the base width would also need to be increased proportionally. However, photographic evidence uncovered by Outland reveals that no such increase occurred (figure 2-16). While studying construction photographs taken in early 1925 by officials visiting the site for the Santa Clara River Protective Association (and now retained in the archives of the United Water Conservation District in Santa Paula), Outland discerned that the thickness of the dam's base was as much as twenty feet narrower than indicated in the supposed "as-built" cross-sectional profile. As Outland phrased it, "the dam had been born with a stub toe."[52]

Exactly what transpired on-site during construction of the dam will never be known, but there is little doubt that Mulholland and his staff chose to increase the reservoir capacity in a way that was not accurately

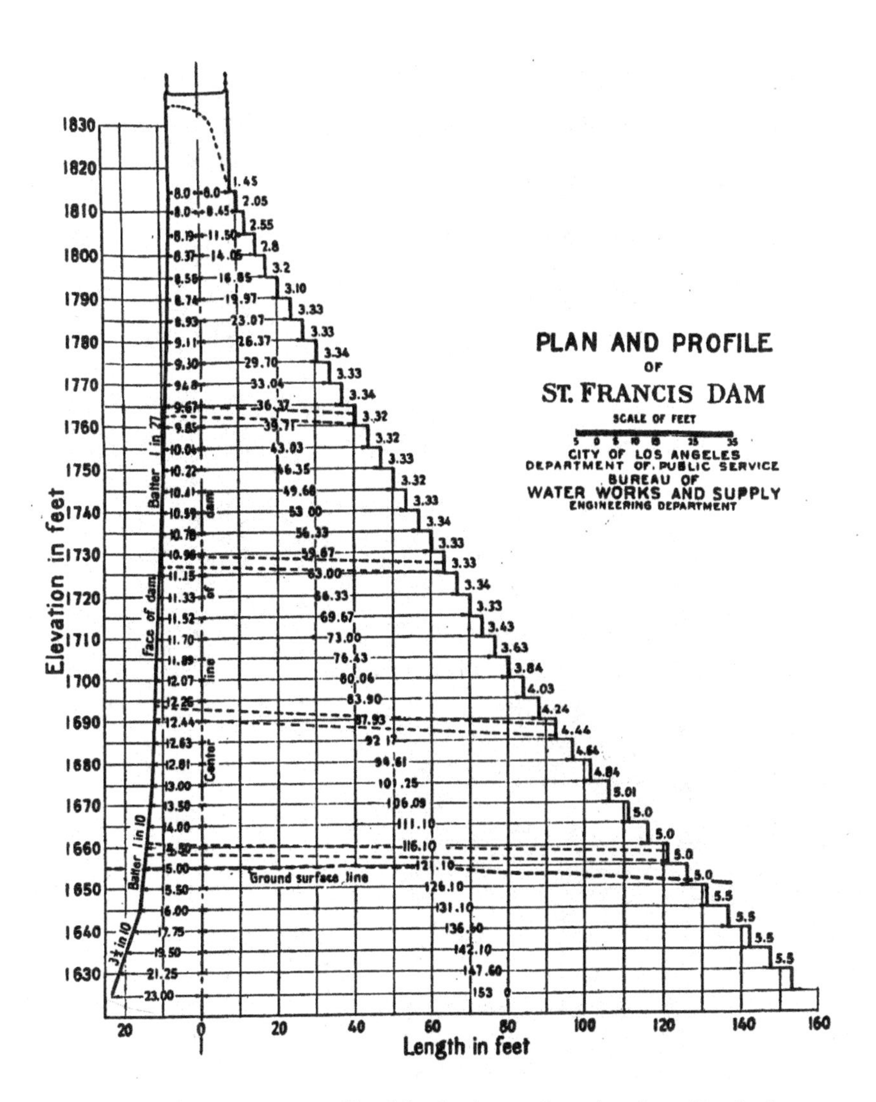

FIGURE 2-15. Cross-section profile of the St. Francis Dam distributed by the Bureau of Water Works and Supply after the collapse. Although investigators considered this an accurate record, there is good reason to believe that it did not depict as-built dimensions, especially the base width. Also, the drawing does not show the fifth outlet pipe at 1,795 feet above sea level. [Governor's Commission *Report*, 1928]

reflected in the drawing provided to and published by the Governor's Commission. Because of ambiguity in the precise dimensions of the dam as built, it is impossible to determine the full extent of instability caused by the increased height.

Uplift and Early Twentieth-Century Dam Design

Mulholland's decision to raise the height of the St. Francis Dam without increasing the base thickness also served to make the structure more vulnerable to the destabilizing effect of uplift. And while he understood the possibility that uplift forces could act upon the dam, he made only a limited effort to mitigate their effect. Specifically, Mulholland placed ten drainage wells across a distance of about 120 feet at the center (and deepest part) of the dam site. Wells of this type help alleviate uplift by providing a way for water to be drained off, and this reduces the destabilizing effect of upward pressure. The ten wells located at the center of the structure constituted the full extent of Mulholland's effort to counter uplift. When questioned at the Coroner's Inquest, Mulholland acknowledged that he addressed the issue of uplift only in the deepest part of the dam foundation because, he claimed, that was where "the rock was fissured."

Q. [By a Juror]: Was this dam under-drained practically for its entire distance?

A. [Mulholland]: No, it was only where the rock was fissured, that is, those igneous rocks are always more or less jointed a little bit, and we find it usually and always expedient to drain them out so there will not be any up-pressure, taking that much pressure of the dam away. So we lead them out. Those drains are provided in every dam I have ever built.

Q. At what intervals were these bleeders put in?

A. About every fifteen or twenty or twenty-five feet.

Q. Practically almost to the top of the dam, as you went along?

A. No, the west end was a homogenous ground. There was no drain necessary in those. It was much tighter. It was about as hard as the other but tighter and more compact. The rocks—the fractured rocks, all the hard rocks in this country are more or less fractured and you can go to the mountains here and look at the granites on every hillside and you will see them fissured and fractured more or less, but they will carry water without doubt, but the prudent thing is to drain them out.

Q. But the points of under drainage [were] put in where the rock was seen to be fractured?

A. Yes.[53]

FIGURE 2-16. Taken in January 1925, this is one of the earliest surviving photographs documenting construction of the St. Francis Dam. At the far left, note how the "stepped construction" does not conform to the design presented in the official drawing. Most important, the photo shows how the downstream "steps" in the lower part of the dam were truncated, narrowing its base width. The precise extent of this truncation is now impossible to determine, but the reduced base width unquestionably affected the stability of the structure. [United Water Conservation District, Santa Paula]

In essence, Mulholland recognized the possibility that uplift would act through the fractured schist. However, while professing that "the prudent thing is to drain them out," he focused his remedial efforts only on the dam's center section. He ignored the fact that, as the reservoir level rose, water would extend up the east canyon wall and then almost certainly seep into the fractured schist foundation forming the east side abutment. The upper canyon walls on both the east and west side abutments were therefore not protected with drainage wells. There was also no pressure grouting of the foundation that could have reduced seepage and subsurface flow; no cutoff trench excavated into bedrock along the length of the foundation; and no drainage or inspection tunnel placed within the dam's interior.[54] So the question arises: Is it reasonable to believe that the structure adhered to contemporary standards of gravity dam design in regard to uplift, in particular the possibility that it could act along the entire length of the foundation—and not just at the center section?

In the 1860s and 1870s uplift was not integrated into gravity dam design protocols,[55] but dam builders soon began to recognize the dangers it posed, and British engineers building Liverpool's Vrynwy Dam (a gravity structure completed in 1892) incorporated drainage wells into its design.[56] Not all civil engineers in the late nineteenth century paid heed, however, prompting Edward Godfrey in 1904 to complain in *Engineering News,* "I find nothing in [books] on dams mentioning this floating tendency of the water which percolates under dams."[57] Four years later Godfrey reiterated his complaint, but in 1910 Charles E. Morrison and Orrin L. Brodie declared in *High Masonry Dam Design* that "present practice requires [that uplift] be considered where a structure of great responsibility is proposed."[58]

Apprehension about uplift intensified following the collapse of a concrete gravity dam in Austin, Pennsylvania, on September 30, 1911. Located about two miles upstream from town, the fifty-foot-high Austin Dam failed catastrophically, taking at least seventy-eight lives in the flood surge.[59] The calamity attracted great public attention and galvanized American dam builders to deal with uplift. John R. Freeman, a prominent advocate of gravity dam technology who also had served as a consulting engineer for the Los Angeles Aqueduct, led this effort.[60] Rushing to Austin, Freeman soon reported in *Engineering News*: "The cause that probably led to the failure of the Austin PA dam [was] the penetration of water-pressure into and underneath the mass of the dam, together with the secondary effect of lessening the stability of the dam against sliding." Freeman implored engineers to understand that "uplift pressures may possibly occur under or within any masonry dam and should always be accounted for."[61] At the time, Freeman was helping oversee construction of New York City's Ashokan (also known as Olive Bridge) Dam and Kensico Dam, two projects that—as *Engineering News* said in describing Kensico Dam in April 1912—countered "upward water pressure" with foundation pressure-grouting and an extensive drainage system.[62]

Arthur P. Davis, chief engineer (later director) of the U.S. Reclamation Service, also believed that the failure of Austin Dam "was caused by an upward pressure on the base of the dam." After visiting the Austin site, Davis became concerned about uplift acting on the service's planned Elephant Butte Dam in southern New Mexico.[63] Soon after, the agency approved a concrete gravity design for Elephant Butte that included extensive grouting, a deep cutoff trench, and a drainage system running the entire length of the dam.[64] For the Reclamation Service's 354-foot-high

FIGURE 2-17. The fifty-foot-high concrete gravity Austin Dam in Pennsylvania failed on September 30, 1911. Uplift acting on the dam's base destabilized the structure, allowing large sections to be pushed downstream. This failure prompted many engineers to sound a warning about uplift acting on gravity dams. [DC Jackson]

FIGURE 2-18. A postcard view of devastated homes and businesses after the Austin Dam disaster in 1911. [DC Jackson]

FIGURE 2-19. Almost eighty people died in the Austin Dam disaster. Here, a recovery team brings a body out of the flood zone. Seventeen years later, there were many similar scenes after the collapse of the St. Francis Dam. [DC Jackson]

concrete gravity Arrowrock Dam, built in 1913–16 near Boise, Idaho, Davis reported that "a line of holes was drilled into the foundation just below the upstream face of the dam to depths of 30 to 40 feet. They were grouted under pressure [and] another line of holes was drilled to serve as drainage holes to relieve any leakage under the dam. These were continued upward into the masonry and emerged into a large tunnel running the entire length of the dam."[65]

In the aftermath of the Austin Dam failure, Freeman, Davis, and the Reclamation Service were hardly alone in their concern about uplift. In 1912, C. L. Harrison addressed it in a prominent article in the *Transactions of the American Society of Civil Engineers*.[66] The next year, *Engineering News* described field tests confirming the existence of uplift pressures, prompting Edward Godfrey to proclaim: "The results of these experiments further emphasize what the author has said before: It is a crime to design a dam without considering upward pressure."[67] Authors

FIGURE 2-20. After the Austin Dam failed, Chief Engineer of the U.S. Reclamation Service Arthur P. Davis took special care that the agency's concrete gravity Elephant Butte Dam in New Mexico (completed in 1916) would be protected against uplift. [DC Jackson]

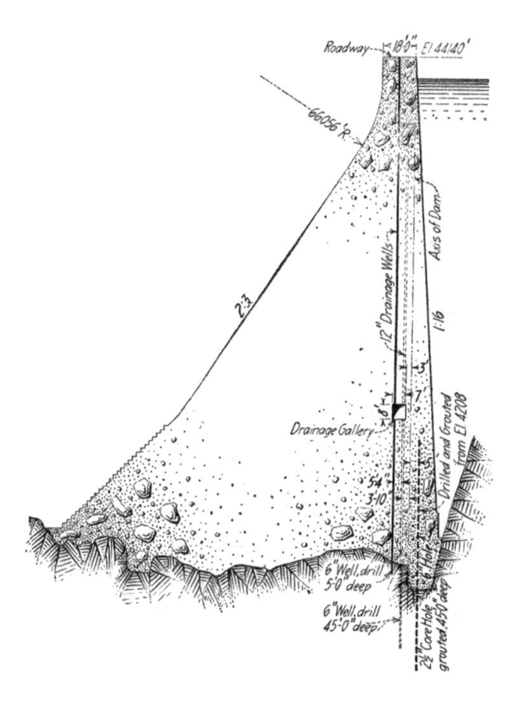

FIGURE 2-21. Cross-section profile of the Elephant Butte design, showing the upstream cutoff trench, the lines of subsurface grouting, and the drainage wells drilled into the foundation. [*Engineering News-Record*, 1917]

of technical books also joined the anti-uplift crusade. Chester W. Smith's *The Construction of Masonry Dams* (1915) included a ten-page section describing how cutoff trenches, foundation grouting, and drainage systems could ameliorate its effects.[68] The 1916 edition of Morrison and Brodie's *High Masonry Dam Design* (retitled *Masonry Dam Design Including High Masonry Dams*) began with an extensive discussion of uplift and described "several ways in which upward pressure may be cared for," including a foundation cutoff trench and "drainage wells and galleries to intercept all entering water."[69] A year later, William Creager's *Masonry Dams* also emphasized the need to counter uplift in the aptly titled chapter "Requirements for Stability of Gravity Dams."[70]

By 1917, American dam engineers' concern about uplift was neither obscure nor unusual, and the placement of drainage wells only in the center section of St. Francis Dam did not reflect standard practice in California for large concrete gravity dams in the mid-1920s. In 1916, for example, Hiram Savage's plans for two municipally owned concrete gravity dams near San Diego called for both grouting and drainage wells along the length of the two structures and for a cutoff trench containing a "continuous 12-[inch] sub drain" to run the length of both dams."[71] In Northern California in 1922, the Scott Dam (also known as Snow Mountain Dam) across the Eel River featured "grouting below the cut-off wall" as well as a network of under-drains to "carry off seepage water...The drains under the dam consist of porous concrete tile...Lines were laid parallel with the axis of the dam and on 15-ft. centers under the entire structure."[72] And in California's Central Valley in the early 1920s, two major curved concrete gravity structures designed by consulting engineer A. J. Wiley—the Don Pedro Dam in Tuolumne County and the Exchequer Dam in Mariposa County—featured both subsurface grouting and a cutoff trench, as well as an extensive drainage system running up both canyon walls.[73]

Perhaps most importantly, San Francisco, the only municipality in California comparable in size and wealth to Los Angeles, began construction in 1919 on a major water supply dam in the Sierra Nevada that reflected careful attention to the danger posed by uplift. The curved concrete gravity Hetch Hetchy Dam (soon renamed O'Shaughnessy Dam) featured an extensive drainage system consisting of 1,600 porous concrete blocks and a cutoff trench running up both canyon walls. The dam reached an initial height of about 340 feet in 1923 (it was extended to 430 feet in 1938) and, as detailed in the *Engineering News-Record* in 1922, "the porous concrete blocks are placed in the bottom of the cutoff trench for its full length, and also in vertical tiers."[74]

FIGURE 2-22. Construction view of the Exchequer Dam near Merced, California. Note the cutoff trench being excavated up the abutment wall and also the line of drainage pipes paralleling the trench. [*Engineering News-Record,* 1925]

FIGURE 2-23. The Exchequer Dam soon after its completion in 1926. This major concrete curved gravity dam was designed by A. J. Wiley, the engineer who later chaired the Governor's Commission that investigated the St. Francis Dam failure. [DC Jackson]

FIGURE 2-24. Don Pedro Dam near Modesto, California, under construction in 1922. It was completed in March 1923, before any construction work began at St. Francis. Like the Exchequer Dam, this curved gravity dam was designed by A. J. Wiley, and it included pressurized grouting of the foundations. Note the discrete sections about fifty feet wide that are separated vertically by expansion/contraction joints, and drainage pipes aligned parallel to the upstream face for the full length of the dam. Mulholland did not incorporate these features at St. Francis. [DC Jackson]

The Hetch Hetchy Dam's drainage system—designed and implemented before 1924—was clearly far more comprehensive than Mulholland's effort to counter uplift in his St. Francis design, and this was noted at the Coroner's Inquest. Specifically, the issue was addressed in testimony from M. H. Slocum, the construction supervisor at Exchequer and Scott dams, who also participated in foundation preparation and early concrete placement at Hetch Hetchy:

Q. [By the coroner]: In your opinion, how could undermining of the [St. Francis Dam foundation] have been prevented?

A. [Slocum]: On other work of such a character with which I have been connected, it has been done by putting in drainage holes, connected up to a drainage gallery which intercepts the water practically at the upstream base, taking away the uplift and letting it run off downstream without any pressure.

Q. [By a juror]: Is it common practice to run the drainage lines you are speaking of pretty well up the sides of the hills?

A. Drainage galleries in Exchequer, Hetch-Hetchy, [and] Snow Mountain run to all intents and purposes to the top of the dams, clear to the top.

Q. Have you ever seen or heard of a dam which you considered to be a safe and properly designed dam, which didn't provide some means of draining up the sides?

A. I have been to a great many dams, and to my memory I can't remember any that haven't had drainage, drainage galleries in them of the gravity type, not of strict arch type, this [St. Francis] was a gravity type.[75]

Prior to the St. Francis disaster, American dam builders fully appreciated the threat posed by uplift, prompting engineer Fred Noetzli to counsel in the June 1927 issue of *Modern Irrigation*: "Conservative engineering requires that gravity dams be designed for uplift."[76] Why Mulholland largely ignored the tocsin sounded by numerous engineers—both in print and in practice—over the dangers of uplift remains a mystery. After all, he was reputedly a voracious, self-schooled devotee of technical information.[77] But labeling his actions mysterious can hardly excuse them. Many American dam engineers of the 1920s and earlier understood the importance of countering uplift with measures that went far beyond the meager steps taken at St. Francis. Mulholland stood well apart from his contemporaries on this crucial issue.

FIGURE 2-25. Like William Mulholland at St. Francis, San Francisco City Engineer M. M. O'Shaughnessy was not subject to state supervision when he set out to build a major dam at Hetch Hetchy Valley for San Francisco's municipal water supply. This photograph, published in *Engineering News-Record* in 1922, illustrates the deep cutoff trench excavated up the Hetch Hetchy canyon's south side abutment. [*Engineering News-Record*, 1922]

FIGURE 2-26. Concerned about uplift, O'Shaughnessy devised drainage blocks to be incorporated into the Hetch Hetchy Dam. About 1,600 of these precast porous blocks were integrated into the dam's structure. Each measured "3 ft. 3 in. square and 3 ft. high" and had a "central cored opening 15 in. square." [*Engineering News-Record*, 1922]

FIGURE 2-27. O'Shaughnessy (left) posing in front of drainage blocks placed along the foundation where it has been excavated for the abutment's cutoff trench. [M. M. O'Shaughnessy Collection, Bancroft Library, University of California, Berkeley]

The Municipal Exemption

Another key issue concerns Mulholland's ability to develop—and considerably enlarge—a design of the scale and importance of St. Francis without any significant review of his work by other engineers or by state officials. Was this common procedure in the 1920s, or was it unusual? The answer to this important question speaks directly to Mulholland's responsibility for the disaster.

The history of California dam safety regulation is commonly told with emphasis solely on the 1929 legislation enacted after the St. Francis tragedy. But the state's prior dam safety law, in place since 1917, was actually quite stringent, at least for certain classes of dam builders, such as state-authorized irrigation districts. The 1917 law stipulated that construction and operation of all sizable dams in the state (those more than ten feet high or impounding more than three million gallons—about ten acre-feet)

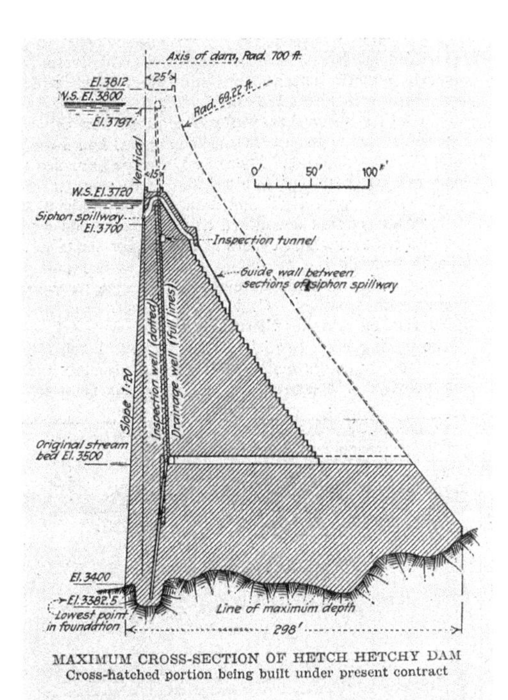

FIGURE 2-28. Cross-section of the Hetch Hetchy Dam showing cutoff trench and interior drainage pipes. Though designed to be more than 400 feet, it was initially built to a height of only 340 feet. But the width of the base was adequate for the full-size dam. [*Engineering News-Record*, 1922]

FIGURE 2-29. Upon initial completion in 1923, Hetch Hetchy Dam was formally named O'Shaughnessy Dam in honor of the chief engineer. This view shows the dam before its height was raised in the 1930s. [DC Jackson]

were to be supervised by the state engineer. Other than federally owned dams, the only exceptions allowed under the law were for structures built or operated under the supervision of the state Railroad Commission, mining dams regulated by the California Debris Commission, and structures owned by a municipality with a department of engineering.[78]

It was the last of these that allowed Mulholland to design and build the St. Francis Dam without any outside review or supervision. In the wake of the disaster it was widely understood that this de facto "municipal exemption" had given Mulholland enormous latitude in carrying out the city's dam-building projects in the 1920s.[79] There is no evidence that Mulholland was ever questioned about how the exemption came to be included in the law—the issue was never raised at the Coroner's Inquest—but it is clear that this regulatory loophole did not come about by happenstance or legislative carelessness. After the St. Francis disaster, San Francisco City Engineer M. M. O'Shaughnessy—who also benefited from the exemption in his construction of the Hetch Hetchy Dam and Aqueduct project—provided a telling explanation of the exemption's origin:

> I had our City Attorney present objections to the State legislative body in Sacramento in 1917, against allowing Mr. McClure [Wilbur F. McClure, then the state engineer] to have anything to do with our dams at Hetch Hetchy, as I did not think, from his previous experience and knowledge, he had the requisite experience to pass on such a subject and I did not care to be subject to his capricious rulings...I did not think that Mr. McClure's previous clerical and engineering experience entitled him to be czar over the plans for our dam.[80]

Mulholland and O'Shaughnessy were engineering rivals of a sort (just as Los Angeles and San Francisco were civic rivals), but one thing they would have shared was an antipathy to "capricious rulings" emanating from a Sacramento-based "czar." Perhaps Mulholland never lobbied the state legislature to incorporate the municipal exemption into the 1917 law. Nonetheless, he well understood—and exercised—the power accorded him as he built the St. Francis Dam free from regulatory constraint or oversight by the state engineer.

In designing, building, and enlarging the dam, Mulholland was not legally required to solicit or undergo outside review. However, by embracing the municipal exemption authorized by state legislators he was by no means prevented from engaging engineers outside the BWWS hierarchy to review the design or the proposed method of construction. The use of consulting engineers for project review was very common by the early twentieth century, and state laws were hardly needed for Mulholland to seek the counsel of his peers. The go-it-alone approach he adopted at St. Francis did not accord with the common practice of contemporary dam builders (and the organizations financing construction) to consult with outside experts.[81] It also did not accord with some of his own previous experience.

For example, in 1906 the city of Los Angeles engaged the engineering services of John R. Freeman, Frederic Stearns, and James D. Schuyler to act as a Board of Consulting Engineers to review plans for the Los Angeles Aqueduct (discussed in chapter 1). Three years later roles were reversed and, when Freeman set out to design Big Bend Dam in Northern California for the Great Western Power Company, he had the company engage Arthur P. Davis and Mulholland to serve as project consultants.[82] In early 1912, Mulholland invited Davis to visit the Lower San Fernando Dam site, and his reason for doing so is instructive, because it could apply with equal force to later projects for which he sought no outside review. "Last Tuesday [I] requested or rather suggested to the Board of Public Service Commissioners," Mulholland told Davis, "that an engineer be employed to examine the proposed San Fernando Dam [site] when it is stripped [down to the foundation] in order to clear them of any charge that might be brought in the future of having proceeded with the work without competent advice."[83] Thus, for the Lower San Fernando Dam, Mulholland advocated a principle that he thereafter largely ignored. Or, as he acknowledged at the Coroner's Inquest: "In general, for the last ten or twelve years, I haven't consulted with anybody, or but very few."[84] In the aftermath of the disaster, Mulholland eventually voiced responsibility for the St. Francis tragedy at the Coroner's Inquest (see chapter 6). But it should be apparent that his failure to seek outside review of the design made any other course of action essentially impossible.

A Possible Alternative

In the early twentieth century a technological alternative to massive embankment and concrete gravity dams became available, an alternative that Mulholland never considered for the St. Francis project but one that, in retrospect, was perhaps better suited to the geological conditions of the upper San Francisquito Canyon. After the Theresa Falls Dam in upstate New York was completed in 1903, flat-slab concrete buttress dams were promoted nationwide by the Ambursen Hydraulic Construction Company of New York. In California, multiple-arch buttress dams were pioneered and promoted by the engineer John S. Eastwood, starting with the Hume Lake Dam in the Sierra Nevada east of Fresno, completed in 1909, and continuing with the Big Bear Valley Dam near San Bernardino in 1911, the Murray and Lake Hodges dams in San Diego County in 1917–18, and the 170-foot-high Littlerock Dam in northern Los Angeles County (about thirty-five miles east of the St. Francis site) in 1922–24.[85] By 1923, multiple-arch dams were a well-known feature of the California hydraulic landscape, and Mulholland was very likely aware of their proliferation.

Buttress dams, and especially multiple-arch designs, were substantially less costly than concrete gravity dams because they required far less material. In addition, because of the substantial spacing between the buttresses (usually between twenty-four and sixty feet), the danger of water seeping through the foundations and pushing up on the base of the dam is eliminated. Uplift, which was a crucial factor in the St. Francis disaster, is the Achilles' heel of concrete gravity dams—the solid base extending the length of the foundation allows for no easy relief of upward pressure. At the time of the St. Francis Dam's construction, therefore, an alternative technology was available that obviated this destabilizing force. So it is worth asking why the opportunity presented by the development of the multiple-arch dam has heretofore never appeared in histories or treatments of the St. Francis Dam disaster.

The reason that Mulholland ignored the possibility of building a multiple-arch buttress dam at St. Francis becomes clear in retrospect. Advocates of gravity dam technology—most prominently John R. Freeman, mentioned above and in chapter 1—criticized multiple-arch dams as an inferior, insubstantial technology that was "not well suited to inspire public confidence" despite any economic or practical advantages it might offer.[86] First at Weid Canyon and then at St. Francis, Mulholland was drawn to concrete designs that presented a monumental façade, structures that presumably reflected the great accomplishments of a great city. In view of their perceived prestige, and of Freeman's endorsement of gravity dams as best suited for great municipal projects, it is easy to understand why Mulholland ignored multiple-arch technology as an option.

But despite Freeman's proselytizing about "public confidence," one major western city did build a large multiple-arch dam for a municipal

FIGURE 2-30. Mountain Dell Dam, near Salt Lake City, following its completion to a height of 150 feet in 1925 (it was built to 110 feet in 1917). City Engineer Sylvester Q. Cannon believed that the multiple-arch buttress design was particularly well suited to the site, precisely because the foundations were composed of "calcerous shale" susceptible to uplift. The use of widely spaced buttresses allowed for water percolating through the foundation to rise harmlessly to the surface without exerting a destabilizing upward force against the base of the dam. [DC Jackson]

reservoir in the early twentieth century, and did so with explicit knowledge of how the technology ameliorated the effect of uplift. Salt Lake City's Mountain Dell Dam was located upon a "calcerous shale not entirely watertight." In 1916 City Engineer Sylvester Q. Cannon commissioned John Eastwood to design a 150-foot-high multiple-arch buttress dam for the site with an understanding that it would provide for "the practical elimination of upward pressure" acting on the structure.[87] Cannon, an 1899 graduate of the Massachusetts Institute of Technology, brought both education and experience to his work in guaranteeing Salt Lake City a secure water supply.[88] Although not as well known as Mulholland, Cannon was a hydraulic engineer highly respected throughout the American West, and his willingness to see multiple-arch technology as a viable option for a major municipal dam is significant—especially because of the way he implemented the design in order to address the effect of uplift acting through less-than-ideal foundations.

FIGURE 2-31. If disinclined to look to Utah as a source for large-scale buttress dam design, Mulholland could have taken a short trip by car, only thirty-five miles, to the Littlerock Dam, east of Palmdale. Here, a 170-foot-high multiple-arch dam designed by California engineer John S. Eastwood (who also designed Mountain Dell) was built by the Littlerock and Palmdale Irrigation Districts in 1922–24. In contrast to Weid Canyon/Mulholland and St. Francis, the Littlerock Dam was built under the authority of the state engineer. [DC Jackson]

The Mountain Dell Dam was built to a height of 110 feet in 1917 and raised to its ultimate height of 150 feet in 1925.[89] Had Mulholland wished, he could have consulted Cannon while planning the St. Francis Dam and explored ways to utilize a multiple-arch buttress design for the site in San Francisquito Canyon. But he focused solely on a massive concrete gravity design—that buttress dams would counter the effect of uplift was of no interest. Thus the alternative offered by the multiple-arch dam seems completely separate from the tragedy associated with Mulholland's ill-fated concrete gravity design. However, as described in chapter 7, the engineering and political tumult that transfixed California in the aftermath of the disaster would have—in less than obvious ways—an enormous impact on the viability and future of multiple-arch dams. In terms of hydraulic technology, it would become one of the most important consequences of decisions Mulholland and his staff made when designing—and enlarging—the St. Francis Dam.

CHAPTER THREE

The Dam: Construction, Operation, Failure

> Changes in the plans for this reservoir have disclosed the fact that at crest elevation 1,835 [feet] above sea level the reservoir will have a capacity of 38,000 acre-feet.
>
> W. W. Hurlbut to William Mulholland, July 1925[1]

> I dug a little drift in on the east side to see if all that shale, what I call shale, was the same, and I dug in about thirty feet, between thirty and forty feet, used a little powder, and it was the same thing, and we filled it up as full as we could by hand and put a grouting machine in.
>
> Stanley Dunham at Coroner's Inquest, March 21, 1928[2]

No great fanfare heralded the start of work on the St. Francis Dam. Instead of press releases and boastful announcements in the *Los Angeles Times*, the first public notice of the city's intent to build the dam appeared quite prosaically in the Board of Public Service Commissioners' annual report for the year ending June 30, 1923. Along with a pre-construction photo of the site, Bureau of Water Works and Supply office engineer W. W. Hurlbut curtly reported to Mulholland: "detail topographic surveys and plans for the San Francisquito Reservoir have been completed together with preliminary designs for a gravity-type masonry dam...The major portion of this reservoir site is owned by the city...totaling approximately 480 acres, all other land in the site belongs to the United States Government."[3]

In August 1923 a city work crew began grading a dirt road up the east side of San Francisquito canyon to replace part of the county road connecting Power Houses No. 1 and No. 2.[4] For many months, BWWS crews apparently did little more than complete the new east side road (necessary because much of the old right-of-way would be flooded by the future reservoir) and continue surveying the inundation zone. In April 1924, BWWS construction engineer Stanley Dunham "took charge of the entire work and started excavation for the foundation of the [St. Francis] dam."[5] But the effort proceeded slowly as construction of the Weid Canyon Dam dominated BWWS activities.[6] Although work at St. Francis was underway in the latter half of 1924, it did not kick into high gear until March 1925, the same month that Weid Canyon was formally dedicated and renamed "Mulholland Dam."

Early Construction

In contrast to the reporting on many major civil engineering projects of the 1920s, no articles detailing the construction of St. Francis Dam ever appeared in the national or regional engineering press (such as *Engineering News-Record, Western Construction News,* or *Southwest Builder and Contractor*).[7] Surviving records relating to its construction, furthermore, are far from thorough, and they largely derive from testimony given at the Coroner's Inquest, from material in the archives of the Los Angeles Department of Water and Power, or from the report by the Governor's Commission formed to investigate the cause of the collapse. Also extant is a large selection of construction photographs, but the earliest of these images date to January 1925. In other words, there is no surviving visual record documenting the first eight months of work at the St. Francis site.

Much of our knowledge of the early work comes from a three-page typewritten report with the prosaic title "Historical Data on Construction of the St. Francis Dam," prepared by an unnamed BWWS employee.[8] Tersely phrased, this memo outlines a basic chronology of construction: the initial use of pressurized water to help clear off loose soil from the canyon walls (often called sluicing or hydraulicking); excavation for a "shutoff wall" into the sand and gravel of the creekbed; construction of a diversion flume; and the erection of the wooden hoisting tower and its replacement by a steel tower.[9] The entries from May 1924 through late January 1925 touch upon crucial aspects of the structure's character and dimensions. They are quoted here in full—because records covering the initial period are so scarce, and also because it was during this period that Mulholland made the crucial decision to raise the dam's height.

FIGURE 3-1. Work underway at St. Francis in January 1925, just before the wooden tower used to hoist and transport mixed concrete was dismantled. This view (along with figure 2-16) represents the earliest surviving image documenting construction of the St. Francis Dam. [United Water Conservation District, Santa Paula]

May 1, 1924. *Hydraulic operation started, sluicing down sides of the canyon at dam site.*

May 19. *Steam shovel began work on the dirt brought down by sluicing.*

June 14. *Trucks started hauling from the* [*steam*] *shovel.*

July 8. *Excavation started on water shutoff trench.*

August 17. *First concrete poured in shutoff trench.*

September 10. *Shutoff dam finished. Elevation 1655.10* [Note: all elevations are given in feet above sea level; when the dam was complete, the crest of the spillway stood at 1,835 feet, meaning that the top of the shutoff dam lay 179.9 feet below the spillway crest of the completed dam.]

> October 1. *Work started pouring* [*concrete*] *behind shutoff wall.*
>
> October 14. *Work started on* [*wooden*] *concrete tower.*
>
> November 9. *Flume for flood waters finished.*
>
> November 10. *Down stream toe forms set for elevation 1630.*
>
> November 12. *First concrete poured from* [*wooden*] tower. *Upstream face of dam raised to elevation 1659.*
>
> December 23. *Set forms of downstream face for elevation 1650.*
>
> December 30. *Shut down pouring to replace broken tower bucket.*
>
> January 12. *Flume over dam taken down.*
>
> January 13. *Outlet pipe set on radial 9 + 25.*
>
> January 24. *Last pouring made from wooden tower. Work started on new steel tower.*
>
> January 26. *Work on new mixing plant and trestle began.*[10]

Despite its brevity, this memo highlights some key issues. First, the depth of excavation for the shutoff trench is not indicated, although the top of the completed shutoff wall is given as 1,655.10 feet. Although it might be surmised that the trench for the wall extended into bedrock, the Governor's Commission *Report* issued after the collapse made clear that it was dug into the sand, gravel, and loose rock in the streambed but stopped before extending down into the schist bedrock. The trench went down only to the 1,638-foot level and was then filled in with "a concrete wall 8 feet thick (narrowed to 5 feet at the top) and about 80 feet long at the bottom and 155 feet at the top." The Governor's Commission described how, once this shutoff wall was in place, the foundation

> was then excavated behind the wall to elevation 1830 across the deepest part of the channel. [As a result] the foundation excavation of the dam was carried 8 feet below the bottom of the [shutoff] wall. The wall was built merely to cut off underflow through the gravel and small freshets that might come from the San Francisquito drainage during early construction stages and convey the waters through a flume past the dam site... the wall itself was [later] incorporated into the upstream face of the dam.[11]

In other words, the structure was called a "shutoff wall" simply because it blocked subsurface flow through the gravel streambed. By this means, it facilitated excavation of sand and gravel lying atop the bedrock foundation downstream from the dam's upstream edge, or heel. Even more significant, the upstream face of St. Francis Dam was actually excavated to notably less depth (eight feet) than the remainder of the structure. This is exactly the opposite of what was done in dams that feature an upstream cutoff wall excavated deeper into the foundation than the rest of the structure (such as the Reclamation Service's Elephant Butte Dam, illustrated in figure 2-21). Why Mulholland and his BWWS subordinates chose to excavate the upstream face only to "tight material" at elevation 1,838 and not down to the schist bedrock at elevation 1,830 is unknown. As a design feature it did nothing to help control seepage or mitigate the effect of uplift once the reservoir began to fill, and it reinforces the view that Mulholland and his subordinates lacked a sophisticated understanding of gravity dam technology.

Second, the memo indicates that excavation of the foundation downstream from the shutoff dam apparently took place over a period of three weeks (September 10–October 1). No precise elevation is given for the depth of excavation along the length of the foundation, but it is reported that downstream form work for the concrete was placed at 1,630 feet, and presumably this represents the elevation of the deepest part of the dam (this corresponds with data provided by the Governor's Commission). Left unstated is the horizontal distance between the upstream face of the shutoff wall and the downstream form work at elevation 1,630. In other words, the memo provides no specific data on the width of the dam at its deepest part.

On a related point, the memo states that "down stream toe forms" for the concrete were set at elevation 1,630 on November 10, and then on December 23 set at elevation 1,650. The drawings of the dam submitted by the BWWS to the Governor's Commission indicate that the width of the dam changed every five feet in elevation (giving the downstream face of the dam its "stepped" appearance). The question therefore arises as to how, whether, and when such steps at elevations 1,635, 1,640, and 1,645 were ever implemented. Of course, this relates directly to design questions raised in chapter 2—particularly by Outland's remark on the dam's "stub toe" profile, and by photographic evidence indicating that the thickness of the dam was not accurately presented in the Governor's Commission drawing.[12] The memo "Historical Data on

Construction" offers no evidence that Mulholland built the lower section of the St. Francis Dam in accord with dimensions indicated in the drawing featured in the Governor's Commission *Report.*

West Side Test Holes and the East Side Tunnel

Other important issues unmentioned in the "Historical Data on Construction" relate to tests Mulholland made to investigate the foundations. At the Coroner's Inquest, construction foreman William Lindsey described some of this early work:

Q. [By the coroner] Were you there all the time the dam was being built?

A. I was there before the dam started, I went up there, Mr. Mulholland sent me there to build the road before they started construction of the dam.

Q. Were you there when the first tests were made at the site of the dam?

A. I was.

Q. Did you see the test made, or did you make them?

A. I had men dig these holes on the west side, at the top of the hill.

Q. What kind of test holes did you dig?

A. Dug a series of holes, I judge four or five, not exactly certain, probably one hundred fifty feet apart from the point of the hill back to the end of the dam, running down probably fourteen or sixteen feet deep.[13]

During the inquest, Mulholland further described the test holes excavated into the west abutment, explaining that he had used them to study how the red conglomerate would react under sustained inundation or exposure to water:

Q. [By a juror]: I wonder if Mr. Mulholland examined the holes [excavated into the red conglomerate], after you put water in to see the effect of the water on the conglomerate?

A. [Mulholland] I ordered the holes in, was very interested in their behavior, put a big hole at the top [of the west abutment], filled it full of water, and it was there about two weeks, and the conglomerate was no softer than it was afterwards; had to bail the hole out, the conglomerate was very tight. We dug holes, had holes dug, I saw the holes, holes dug into that red conglomerate all over, the north flank of the hill.[14]

We have no means of confirming the results of these tests, since there are no known photographs documenting the size or location of these holes in the red conglomerate, or their condition after being "filled with water...[for] about two weeks." But it is reasonable to believe that Mulholland did perform these tests, and did make an effort to explore the suitability of the red conglomerate as a foundation for the dam.

Mulholland also investigated the east side mica schist formation. Here the limitations of the "Historical Data on Construction" memo are again evident—specifically, it does not refer to a large tunnel dug into the lower east abutment. In fact, this sizable excavation goes unmentioned in all post-disaster engineering reports, as well as all historical treatments of the collapse prior to *Heavy Ground*. But the tunnel is clearly described by construction superintendent Stanley Dunham and by carpenter and construction worker Richard Bennett in testimony offered at the Coroner's Inquest. Given the critical role of the east side canyon wall in the failure of the dam, this tunnel—and the extent of its excavation—is of great import. Here are key excerpts from Dunham's testimony:

Q. [from City Attorney Scott]: How deep did you dig in on both sides, on the east side and west side to anchor the dam in?

A. [Dunham]: Until you couldn't dig any more, until solid rock, then I dug a little drift in on the east side to see if all that shale, what I call shale [that is, mica schist], was the same, and I dug in about thirty feet, between thirty and forty feet, used a little powder, and it was the same thing, and we filled it up as full as we could by hand and put a grouting machine in.

Q. You ran a little tunnel into the east side?

A. Yes sir.

Q. About thirty feet.

A. It was a little over thirty.

Q. What was the object of running the tunnel?

A. To see that the rock there, inside, as it was on the outside.

Q. What was the character of the rock?

A. It was the same as on the outside.

Q. Then you covered it up or filled it up?

A. Filled it up with concrete when I built the dam.

A few moments later, the question of using "a little powder" was revisited.

Q. [Scott] Did you do any blasting [during foundation excavation]?

A. [Dunham] No sir, only this little drift [for the tunnel], blasted out a little to get in.[15]

Later in the Coroner's Inquest, the east side tunnel was further discussed by construction worker Bennett:

Q. [By District Attorney Asa Keyes]: Was any tunneling done up there of any kind?

A. [Bennett]: Where they drifted in, one little drift at the edge of the dam, on the right hand side, east side.

Q. How far was that driven in?

A. I wasn't back in. Some said about forty feet.

Q. You saw it?

A. I saw the hole.

Q. What is the size of the hole?

A. Big enough for a man to work running a wheel barrow.[16]

During the early phases of the inquest, when both Dunham and Bennett testified, it was widely assumed that weakness in the red conglomerate of the west abutment precipitated the collapse. (As described later in this chapter, west side leaks were investigated by Mulholland about twelve hours before failure.) So the Coroner's Jury was not particularly curious about how the integrity of the east side abutment might have been disturbed or compromised during construction.[17] But from a historical perspective—especially in view of the role of the east side foundation in the collapse—the testimony presented above is striking. Two employees—one of them the superintendent of construction—gave independent testimony describing the tunnel, and there is little reason to doubt that somewhere near the bottom of the foundations a large opening (big enough for a man pushing a wheelbarrow) extended laterally into the lower east side canyon wall for more than thirty feet. In addition, Dunham acknowledged that he used "a little powder" (that is, explosives) when blasting out the entrance of the tunnel. While Dunham assured the inquest jury that the tunnel was later filled up with concrete, it is reasonable to believe that the integrity of the east side foundation had been damaged by the blasting and excavation of this tunnel. At a minimum, this work would have reduced resistance to subsurface seepage and exacerbated the effect of uplift acting through the lower east abutment.

"Changes in the Plans"

When construction started in the spring of 1924, the St. Francis Dam was to have a spillway crest at 1,825 feet above sea level. This aligned with the Board of Public Service Commissioners' 1923 annual report projecting a reservoir capacity of 30,000 acre-feet.[18] Between July 1923 and June 1924, the BWWS carried out more detailed topographic surveys, and in July 1924 office engineer Hurlbut reported that the proposed reservoir would hold 32,000 acre-feet. Because the spillway level remained unchanged at 1,825 feet, it appears that more precise surveys prompted this revision of the reservoir's capacity.[19]

Sometime over the course of the next nine months, Mulholland acted to increase the size of the reservoir by raising the spillway elevation ten feet.[20] In the 1925 annual report Hurlbut informed Mulholland that "additional surveys and *changes in the plans for this reservoir* have disclosed the fact that at crest elevation 1,835 above sea level the reservoir will have a capacity of 38,000 acre-feet [emphasis added]."[21] The heightening necessary to create the enlarged reservoir would require the construction of a 613-foot-long dike atop the western abutment.[22] But more important, enlarging the storage capacity introduced problems that could—and did—have a profound effect on the stability of the dam's main structure. Raising the dam ten feet did not by itself mean that the dam would be less safe, but Mulholland needed to make sure that the structural base was sufficiently widened to accommodate the increased hydrostatic pressure. As discussed in chapter 2, it appears that no such widening ever occurred. Unfortunately, no photographs exist documenting what was actually done on-site between the start of foundation excavation in May 1924 and the time that concrete placement reached the level of the natural streambed in January 1925. All we know for certain is that, in Hurlbut's words, there were "changes in the plans" and, apparently, they were made rather quickly and without any public airing.

Yet Mulholland's decision to raise the dam was not made on a whim, as the addition of the western dike attests some deliberation. What might have been behind his decision? The answer appears to be related to the political crisis that festered throughout the 1920s: Los Angeles' supposed stealing of water from farmers in the Owens River Valley through the Los Angeles Aqueduct. While *Heavy Ground* is not the forum for a full exploration of the Owens Valley controversy, it is important to acknowledge the political impact

FIGURE 3-2. On November 16, 1924, Owens Valley farmers and ranchers took control of the Alabama Gates on the Los Angeles Aqueduct. For five days, aqueduct water was diverted onto desert land once covered by Owens Lake—and prevented from reaching Los Angeles. This vigilante action greatly angered Mulholland and other city officials, and it may have served as a catalyst for raising the height of St. Francis Dam in mid-construction, thus increasing its capacity. [DC Jackson]

of the sabotage of the aqueduct in May 1924, a dynamite blast ignited about twenty miles north of Haiwee.[23] The damage was quickly repaired (service was interrupted for only a few days), but the point had been made. Los Angeles' water supply was vulnerable to vigilante activism, and the closer water could be brought to the city the easier it would be to mitigate disruptions.

In November 1924, not long after casting of concrete began at St. Francis, the stakes in the controversy were raised when Owens Valley provocateurs staged a forcible takeover of the Alabama Gates, south of

FIGURE 3-3. In late January 1925 the wooden hoisting tower was taken out of commission and replaced by a steel tower that—starting in March 1925—was used for the next fourteen months. Trucks brought sand and stone aggregate to the base of the tower, where it was mixed with water and cement to create concrete. The wet concrete was then lifted up the tower and, via gravity flow, delivered through chutes across the dam site. [Richard Courtney Collection, Huntington Library]

Independence.[24] The activist ranchers did not commit any overt acts of destruction, but they did open the gates so that the full flow of the aqueduct was channeled down onto the bed of Owens Lake. In "wasting" this water, the Owens Valley protestors made certain that no one in the Los Angeles metropole could ever use it for any agricultural, industrial, or domestic purpose.

FIGURE 3-4. Sand and stone used to build the St. Francis Dam were excavated from San Francisquito Creek about a half-mile below the dam. Trucks carried the aggregate from the streambed to the concrete mixing plant located at the downstream face of the dam. [Richard Courtney Collection, Huntington Library]

Mulholland was angered by the Alabama Gates incident, not least because of the sympathetic hearing it engendered for Owens Valley interests across the state.[25] Mulholland's interest in increasing the storage capacity at St. Francis, viewed in this light, was likely spurred, or at least encouraged, by the Owens Valley dispute. An extra 6,000 acre-feet of water at St. Francis, lying far south of Owens Valley, offered the city a cushion in case a disruption of supply—like that at the Alabama Gates in November 1924—were to recur.[26]

Completing the Dam

Overall, the construction of the dam was not complicated. Once the steel hoisting tower became operational in March 1925, work proceeded at a steady pace for the next year. In August 1925 the *Los Angeles Times* reported that "about 160 men are at work on the dam at present, many

FIGURE 3-5. A construction view looking toward the west abutment, late summer 1925. Note that no cutoff trench has been excavated into the foundation. For comparison, see construction views of the Hetch Hetchy/O'Shaughnessy Dam (figures 2-25, 2-26). [Richard Courtney Collection, Huntington Library]

of the families living in the vicinity."[27] Work crews excavated deeply enough along the sloping canyon walls to remove surface soil and reach relatively firm bedrock. Except for some work beneath the dike running

FIGURE 3-6. A detail view looking upstream, showing how concrete was placed atop the west abutment. Note the absence of a cutoff trench. [Richard Courtney Collection, Huntington Library]

along the west ridge, however, there was no effort to excavate a cutoff trench into the bedrock along the upstream face that could help block percolation under the dam. Similarly, there was no grouting—that is, the pressurized injection of cement grout into holes drilled into the foundation—that might have helped seal subsurface fissures. Ten two-inch-diameter drainage pipes were drilled into the foundation of the center (and deepest) section of the dam, but no effort was made to place drainage facilities within the east side or west side abutments as they extended up the canyon walls.

The sand, gravel, and small stones (collectively known to engineers as aggregate) used in the concrete were excavated from the streambed of San Francisquito Creek about half a mile below the dam site. A fleet of seven gasoline-powered trucks delivered this aggregate to the concrete mixing plant at the base of the hoisting tower adjoining the downstream edge of the dam. Cement was brought by truck up San Francisquito Canyon and stored a short distance below the site. From there, it was carried to the mixer via a covered conveyor belt. The concrete was

FIGURE 3-7. Looking across the dam toward the east abutment. The excavation of loose surface material (often called "overburden") had yet to be completed to the top of the structure. [Richard Courtney Collection, Huntington Library]

mixed in one-cubic-yard batches, lifted up the hoisting tower, and then dumped into steel chutes. The wet concrete flowed down the chutes, where it was deposited to form the slowly rising structure. Depending upon where superintendent Dunham wanted work to be done on any

FIGURE 3-8. A construction view showing the east abutment near the end of 1925. [Richard Courtney Collection, Huntington Library]

FIGURE 3-9. Concrete was placed directly atop the east side's broken mica schist, with little foundation preparation. This photograph clearly shows the lack of cutoff trench, grouting, or drainage system along the east canyon wall. [Governor's Commission *Report*]

FIGURE 3-10. The last phase of construction focused on the shallow dike running atop the west abutment. In total, the St. Francis Dam required about 175,000 cubic yards of concrete. [Richard Courtney Collection, Huntington Library]

given day, the chutes could be moved to provide access to various parts of the construction site.[28]

Although some temporary barriers were erected to confine the wet concrete as it hardened, no expansion joints (sometimes referred to as contraction joints) were placed in the structure. St. Francis Dam was cast as a monolithic mass. Later, "temperature cracks"—as Mulholland termed them—formed in the concrete as the structure cooled; these cracks appeared because of the absence of expansion joints. Overall, the quality of the concrete was not particularly high, at least in terms of crushing strength, but concrete in a gravity dam need not be particularly strong. However, the density of the concrete used at St. Francis was low (later tests showed it to be about 141 pounds per cubic foot [pcf], whereas the density of concrete and/or masonry in other gravity dams was generally 145 pcf or higher). This weight deficiency was not the cause of the collapse, but it did act to reduce stability.[29]

With the steel hoisting tower, concrete was cast at a monthly pace of about 12,000 cubic yards until the dam topped out. Approximately 175,000 cubic yards went into the completed structure, and the last batch was placed on May 12, 1926.[30] During the spring and summer of 1926, the cost of the dam was widely reported at $1,250,000.[31] No prominent public ceremonies were held for the dam's dedication, but Mulholland, accompanied by President R. F. del Valle of the Board of Water and Power Commissioners, other board members, and representatives of the Los Angeles City Council, quietly marked its completion on June 5.[32] No record survives of comments made by members of the inspection party. Seven weeks later the *Los Angeles Times* published a short article on the partial filling of the reservoir, but after that the St. Francis Dam largely disappeared from public view.[33] No other articles about the dam would be published until after the collapse.

During the Coroner's Inquest it became known that, in response to reports of possible attack by saboteurs from the Owens Valley, law enforcement personnel from Los Angeles were called to guard the recently completed dam. Dynamite attacks damaged segments of the aqueduct in the late spring and summer of 1927 and, not unreasonably, city officials feared that the dam presented a vulnerable target.[34] Threat of an attack on the dam was never aired in the press prior to the dam's collapse, but the fact that city officials worried about sabotage may explain why little publicity was given to the dam and reservoir after the summer of 1926. In addition, Los Angeles banned recreational use of the reservoir for swimming and boating—for health reasons, there was good reason not to draw attention to a new "lake" north of the city. On occasion employees of the city's power plants and their friends did go out fishing on the reservoir in search of amusement and perhaps an inexpensive dinner, but this was done surreptitiously and in violation of city regulations.[35]

San Francisquito Creek Percolation Test

The primary purpose of the St. Francis reservoir was to store surplus flow from the Los Angeles Aqueduct. But more than thirty-seven square miles of the San Francisquito Creek watershed lay above the dam, and runoff from this area would be captured in the reservoir unless purposely released through the dam outlets. Although the city of Los Angeles publicly projected the St. Francis reservoir as a component of the aqueduct system designed to store water from the Owens Valley, Mulholland understood early on that the city would be well positioned to lay claim to San Francisquito Creek flood flow and divert it out of the Santa Clara River watershed. As engineer Hurlbut reported to Mulholland in the 1923 annual report: "The [St. Francis] reservoir will also derive a supplemental supply from the high mountainous country which it will drain by the natural runoff from a drainage area of 37.5 square miles."[36] In August 1924 a special consulting board on municipal water supply issues projected that 5,240 acre-feet of flow per year from San Francisquito Creek could be impounded for the city's use.[37]

As early as September 1924, the proprietors of the Newhall Ranch (which controlled the Mexican-era Rancho San Francisco and encompassed much of the upper Santa Clara Valley) objected to the St. Francis Dam, claiming that it would "obstruct and cut off all waters arising in the watershed of said canyon [and] thereby the said lands of the Newhall Land and Farming Company will be deprived of their rightful supply of water." Mulholland and the city shrugged off the complaint, ignoring the company's entreaty to "immediately desist and refrain from further construction of said dam."[38] Farmers in the lower Santa Clara Valley also began to express concern, with headlines in local papers such as the *Ventura Free Press* and the *Ventura Post* proclaiming, "Supervisors Oppose Dam Working Harm to Santa Clara," and "Damming of Headwaters of Santa Clara River Menace to Water Users of Valley."[39] Ventura County irrigation interests led by C. C. Teague of the Limoneira Ranch in Santa Paula took action and organized the Santa Clara River Protective Association (later restructured as the Santa Clara Water Conservation District and now known as the United Water Conservation District).[40] Guarding their interest in the stream flow of the entire Santa Clara watershed—which fed the groundwater that many irrigated orange, lemon, and walnut groves depended on—the association embarked on a protracted confrontation with Los Angeles. The goal was to block the city from impounding and diverting water rightfully claimed by downstream landowners.[41]

The dispute revolved around this question: how much of the water passing down San Francisquito Creek in the course of a year was absorbed into the Santa Clara Valley groundwater table? Put another way, how much of San Francisquito Creek's flood flow passed through the Santa Clara Valley and was "wasted" into the Pacific Ocean? Irrigators operating in the valley before the mid-1920s possessed water rights that, in a strictly legal context, Los Angeles could not encroach upon. But city officials asserted that much of San Francisquito Creek's seasonal

FIGURE 3-11. The percolation test of September 1926, in which agricultural interests in the Santa Clara Valley sparred with the city of Los Angeles over plans to divert 5,000 acre-feet of allegedly "unappropriated" floodwater from San Francisquito Creek every year. This view shows the test water flowing at Harry Carey's ranch about six miles below the dam. [United Water Conservation District, Santa Paula]

flood flow was not used for irrigation and, in fact, passed unclaimed into the sea. This floodwater is what Los Angeles sought to store in the St. Francis reservoir and feed into the aqueduct. Because, by definition, surface flood flow reaching the sea could never have been put to beneficial use by irrigators, the city of Los Angeles claimed it was legally available for storage, diversion, and beneficial use by people outside the Santa Clara Valley.

Through 1925 and into 1926 the legal wrangling dragged on, with Santa Clara Valley landowners seeking protective intervention from the state Division of Water Rights.[42] Once the dam was completed in May 1926 and the reservoir began to rise, the question became: how much water legally controlled by Santa Clara Valley farmers should be released into the creekbed to compensate for flow blocked by the dam? In September 1926 a special percolation test overseen by the state Division of Water Rights was undertaken to help answer this question. In the proposed test, water was to be steadily released from the dam over

FIGURE 3-12. For three days, water was released from the reservoir and allowed to flow down the canyon. The artificially induced surge came within a few hundred feet of the main course of the Santa Clara River. However, as shown in this photograph of the highway and railroad crossing of San Francisquito Creek near Castaic Junction, the lowest section of the creekbed remained dry for the duration of the test, vindicating Santa Clara Valley farmers who believed that most of the creek's flow seeped into groundwater. [United Water Conservation District, Santa Paula]

a three-day period and measurements taken along the length of San Francisquito Creek. The objective was to calculate how much of the release seeped into groundwater and how much remained as surface flow reaching the main stem of the Santa Clara River in Saugus. Los Angeles officials hoped for a high surface flow because this would bolster the position that most floodwater in the creek was ultimately wasted into the ocean. Conversely, if the release quickly seeped into the streambed gravels, as valley interests hoped, it would confirm that the creek was an important source of downstream groundwater.

Mulholland was brashly dismissive of his adversaries, predicting that it would take a man on horseback to keep up with the downstream flow as it approached the Santa Clara River and then rushed westward through the valley.[43] His prognostications proved wildly off the mark. From 6:40 a.m. on September 15, 1926, through 5:50 a.m. on September 18, over 770 acre-feet of water gushed through the dam outlets at a rate that varied between 100 and 150 cubic feet per second. It passed sluggishly down the canyon, and after three days no surface flow had reached the Santa Clara River (although it did come within a few hundred feet on the last day of the test). The supporters of valley interests were elated.[44]

With Mulholland's braggadocio punctured, state water rights officials took the test data and continued deliberations on how to handle the city's claims. By March 1928 the dispute remained unresolved, but resentment in the valley had been stirred against both the city and William Mulholland. The confrontation had also brought C. C. Teague into a leadership role with the Protective Association, a position that would bolster his standing in the valley in the tumultuous days after the dam's collapse.

A Silver Anniversary Celebration

For Mulholland, the percolation test and affairs in the Santa Clara Valley were of minor import, as the city's plans to draw water from the Colorado River were accelerating. He later claimed that he had made a habit of visiting the city's dams and reservoirs "at least once [every] 10 days or two weeks," but exactly how often he made the trek to St. Francis is unknown.[45] Photographic evidence of such visits is almost nonexistent; a lone surviving photograph—from the *Los Angeles Times* archives, dating to early 1927—shows him at the dam prior to the collapse.[46]

The proposed Colorado River Aqueduct and advocacy of the Boulder Canyon Project Act (discussed in chapter 6) absorbed much of Mulholland's energy during the latter 1920s, but his twenty-five years of past service to the people of Los Angeles did not go unremarked, and a special banquet was held in his honor. Conjoined with commemoration of the city's "Twenty-Fifth Anniversary of the Acquisition of the Los Angeles Water Works and Supply," the dinner was held at the City Club on the evening of February 12, 1927. The roster of local leaders who came to pay homage underscores Mulholland's political status among the city's elite. Guests included Del Valle, chairman of the Board of Water and Power Commissioners, who served as master of ceremonies and delivered the "Introduction of Mr. Mulholland"; Mayor George Cryer, who conveyed the "Word of Welcome";

FIGURE 3-13. In early 1927, Mulholland (pointing) visited the St. Francis Dam with John R. Haynes (far right) and Reginald Del Valle (second from left), two members of the Board of Water and Power Commissioners. This is the only known (or surviving) photograph of Mulholland at the dam prior to its collapse. [*Los Angeles Times* Photographic Collection, Department of Special Collections, Charles E. Young Research Library, UCLA]

and Harry Chandler, publisher of the *Los Angeles Times*, who offered a toast to "The Chief." Fifty years on from his first labors as a ditch digger, Mulholland had become a dominant figure in the political economy of Southern California. Any complaints lodged by aggrieved citizens in the Santa Clara Valley over San Francisquito Creek water rights barely registered on his radar. Boulder Dam and a tremendous new aqueduct project to ensure Southern California's economic future were what commanded his, and the city's, attention.[47]

Operation of the Dam and Reservoir

Filling of the St. Francis reservoir started in March 1926. In the company of a newspaper reporter, the chief engineer "opened the inlet gate [just

FIGURE 3-14. As the dam neared completion in March 1926, Mulholland's construction crew closed the release valves and the reservoir began to fill. [Richard Courtney Collection, Huntington Library]

FIGURE 3-15. St. Francis Dam and reservoir soon after completion in May 1926. In contrast to the dedication of Weid Canyon/Mulholland Dam in March 1925, no public fanfare accompanied the completion of the monumental concrete gravity structure. [Richard Courtney Collection, Huntington Library]

FIGURE 3-16. In May 1927, the water level came within three feet of the spillway crest, a scene documented by a photographer for the Automobile Club of Southern California. This photo is the best surviving visual record of the (almost) full reservoir. [Automobile Club of Southern California Photo Collection]

below Power House No. 1] and started the formation of a lake that will be three miles long and will contain 12,000,000,000 [gallons] of water for domestic use in Los Angeles." To further emphasize the project's "vital importance as part of the city's water supply system, Mr. Mulholland pointed out [that] while 47,700 acre feet of water previously could be stored in the six principal basins near the city, the new reservoir will increase this storage capacity to 85,700 acre feet." During this initial filling, flow drawn from the aqueduct below Power House No. 1 reached 70 million gallons a day.

During 1926, storage in the reservoir peaked at 13,200 acre-feet in early June (reaching an elevation of 1,779 feet, some 56 feet below the spillway crest) and was held fairly close to this volume before dropping in the fall.[48] In January 1927 levels began to rise, reaching 1,832 feet in May. This was only three feet short of the spillway (and represented storage of over 36,000 acre-feet). At about this time a photographer from the Automobile Club of Southern California brought a model to the dam

FIGURE 3-17. Water in the reservoir was released via pipe outlets extending through the dam's center section. Upon release, water flowed a mile and a half downstream through a concrete-lined channel to the aqueduct below Power House No. 2. [Automobile Club of Southern California Photo Collection]

for some publicity photos. Presumably the idea was to illustrate what an intrepid motorist might experience on a day trip out of Los Angeles. The Automobile Club's photograph of the model gazing out over the reservoir now stands as one of the best visual records of the panoramic view from atop the dam.

Starting in June 1927, the city drew the reservoir down to below 1,820 feet, where it stayed for the remainder of the year. In January 1928 reservoir levels began to rise as seasonal demand in the city dropped, reaching 1,830 feet by early February. On March 5, water came within three inches of the spillway crest (elevation 1,835), where it remained until the evening of March 12. At this time the reservoir was so close to full capacity that wind gusts blew water from the reservoir over the spillway and onto the dam's downstream face. The record of water storage in the St. Francis reservoir from March 1926 through March 1928 does not appear particularly remarkable, but it does show that the reservoir reached full capacity only a week before the collapse. In other words, failure came quickly once full hydrostatic pressure acted upon the dam and its foundation.[49]

In terms of performance, St. Francis Dam generally met the demands placed on it from 1926 through early 1928. The massive structure

FIGURE 3-18. This snapshot shows leakage through a vertical "temperature crack" near the lower west abutment (to the left). The darkened area at the center had a different cause: water intentionally released through the dam's outlet pipes. [DC Jackson]

did not suffer from spalling or deterioration, but it did undergo significant cracking as the concrete cooled. This phenomenon was well known to engineers in the 1920s, and many dams built at the time featured special expansion joints to accommodate shrinkage and expansion that occurred in reaction to temperature changes. (For example, see the photograph of the Don Pedro Dam in figure 2-24.) Mulholland had opted not to build any joints into the St. Francis Dam. But "temperature cracks" formed as the concrete cooled. The flow rate through the temperature cracks in the dam was not egregious, and leakage was reduced by stuffing pliable rope (oakum) into the fissures.[50] Nonetheless, the cracks were a source of unease because they signified that the dam was not operating at optimal efficiency.

Of greater concern were reports of leaks from beneath the dam base and along the wing wall atop the western abutment. As the reservoir approached full capacity in February and early March 1928, these leaks became more noticeable, as confirmed by testimony at the Coroner's Inquest from local farmers and ranchers. For example, Robert Atmore, a rancher who lived near Lake Hughes about fifteen miles north of the dam, traveled up the canyon on March 10 and, looking out at the dam from the east side canyon road, saw leakage through the west abutment:[51]

Q. [By the coroner]: What do you know about the condition of the St. Francis Dam anytime prior to March 13, this year?

A. I was past there on the morning of the tenth.

Q. What did you see as you passed the dam?

A. It was like the dam was leaking very bad on the west side.

Q. Just where on the west side?

A. On the west side of the canyon, the ridge the wing was built on, lots of water coming out of the ground.

Q. Was it coming out of the ground west of the same portion of the dam, west of the abutment there?

A. I couldn't exactly say where it was coming out, but the hillside seemed to be wet, you could see the water running down.

Q. How much water did you see coming out?

A. Seen a stream, from a distance it looked like about the size of a man's leg.

Atmore also stated that he had "passed this place twice before, I think first on the sixth [of March], and then on the eighth, and then I noticed on the tenth there was a great deal more water coming out on the tenth [than] there was previous."

Leakage through the west abutment attracted the most attention because of seepage affecting the dike that ran across the western ridge, and also because the west abutment was readily viewed from the east side county road. But leakage along the east abutment did not pass unnoticed. In his inquest testimony, Henry Ruiz, a San Francisquito Canyon farmer who worked at Power House No. 1 made the point:

Q. [By the coroner]: Were you familiar with conditions at the dam just before the thirteenth of this month?

A. Yes sir, I passed there everyday.

Q. Did you work up close to the dam?

A. I was working in construction camp No. 1.

Q. For the city?

A. Yes.

Q. What did you see there at the time, prior to the time it broke, anything that made you feel uneasy?

A. Yes, leakage on the westerly side there, kind of got on my nerves for awhile, leaking badly...then there was a leakage on the eastern side too, made me feel uneasy."[52]

In other Coroner's Inquest testimony, Chester Smith, a rancher living in the lower San Francisquito Canyon who experienced—and miraculously survived—the flood, described how he had "passed over the road and went around the dam on Saturday, and also passed over the dam on that road on March 12, that is Monday." Smith related observations and conversations from his recent travels through the canyon: "I should judge about a month ago, the water began seeping on the west bank, left wing of the dam, which would be the west wing, and I talked with Tony Harnischfeger about it. He was a good friend of mine, and didn't seem to be alarmed because it was only seeping a little then...and then on Saturday before the dam went out, I went up over the dam and fixed a fence in B Canyon, to take some stock. Harry Nichols [another rancher in the lower canyon] went with me, he had been working for me, and as we got up near to the dam, we looked over and saw considerable water coming out on the west bank."[53] Tony Harnischfeger, the dam keeper who lived downstream, was initially resistant to Smith's entreaties but eventually reacted to the worsening conditions. On Monday morning, March 12, he called Mulholland at his office in Los Angeles.[54]

Mulholland's Last Visit

Precisely what Harnischfeger said to Mulholland on the morning of the dam's collapse is lost to history, but he did express concern that, to his eye, leakage along the west abutment appeared muddy. After taking the call, Mulholland apprised assistant chief engineer Harvey Van Norman of his intent to visit the dam. Soon the two of them headed out to inspect the structure with Mulholland's chauffeur at the wheel of a Marmon sedan.[55] On the first day of the Coroner's Inquest, Mulholland described this late-morning visit:

Q. [From the coroner] What was your observation of the condition of the dam at that time [the morning of March 12]?

A. There was a leak that brought us there. Mr. Van Norman and myself were brought there and it was a leak that had not manifested itself before, a new one. It was running down the slope of a hill and running across an old side hill out where there was a road.

Q. On which side?

A. On the west side. An old construction road running up there and washing the dirt which was thrown from that road, and made dirty water. The keeper [Harnischfeger] telephoned us that the water was dirty and we went up there and the water was not dirty and had not been dirty, but it was cutting [through the dirt road]. It was throwing it to the side of the road, but it was a new leak, because, like all dams, there are little seeps here and there and I will say as to that feature of it that of all the dams I have built and of all the dams I have ever seen, it was the driest dam of its size I ever saw in my life. I have travelled many miles to look at new dams here and there and everywhere...And this year it has gradually filled up to the overflow point and it was the driest dam I ever saw. It was the driest for a massive dam I ever saw. We were there—Mr. Van Norman and I went up—it was about due time for me to go up because I have been in the habit of going up there, and to all the dams we have got, at least once in ten days or two weeks. It is part of my patrol, but Mr. Van Norman and I went up there that morning because Tony Harnischfeger—that is the man who was drowned there, and he was the keeper—

Q. What did Tony tell you?

A. He said the water was muddy and it was a new leak.

Q. What difference does it make if the water was muddy?

A. That is an indication—it is seen as a very bad characteristic of a leak through the earth...It means that the earth is cutting, but there was no earth coming from under the dam. The earth, as I say, came from that side hill cut.

Q. That leak seemed to go through the soft dirt of a road that you had built to get up to the west side of the dam?

A. It was visibly doing so.

Q. Was not all the west side saturated?

A. It was.

A moment later the coroner asked Mulholland the essential question:

Q. When you saw the dam last Monday did it occur to you that it was in danger?

A. No sir, it never occurred to me that it was in danger. I just wanted to see the new leak.[56]

Van Norman's testimony at the Coroner's Inquest, which immediately followed Mulholland's first day on the witness stand, largely corroborates what his boss had to say: "Everything was, as far as I could see, as secure as it had been...The water in every case was crystal clear...I left there with perfect confidence that the dam was perfectly all right." But Van Norman did not at any time describe the St. Francis Dam as the "driest dam" he had ever seen, and he acknowledged that leaks were not confined to the west abutment: "There was some seepage coming from both ends of the dam, some coming from under the very base of the dam."[57] Later in his testimony Van Norman reiterated that the schist on the east side abutment was also experiencing leakage: "There were minute seams in the rock and seepage of water that got into them [on the] east side, there was a little stream of water coming down from approximately two thirds of the way up the hillside, running down in the ground, between the contact of the dam and rock, trickling down the hillside."[58]

In response to a question about why they visited the dam on the morning of March 12, Van Norman offered a slightly different rationale from Mulholland's, one that downplayed any fears about safety:

> A. [Van Norman] Mr. Mulholland—I was in my room next to his—said he was going out to San Francisquito Dam—"It is full and the wind is blowing some, water coming over the spillway. I am going to look at it." I said "I will go along with you."
>
> Q. [From the coroner] It wasn't because it was indicated to you that there was danger of the dam breaking?
>
> A. No, I don't believe there was any indication, to my knowledge, any indication came to me from any of the men up there that there was danger of the dam breaking. I understand such statements have been made—I have been reading the newspapers—I don't know of any such indications coming from my office, and that is as far as I know. The reason Mr. Mulholland and I went to the dam—it is our custom to go to these places at intervals, as a matter of taking care of our work.[59]

When the coroner asked Mulholland, "Did any employees indicate to you they were afraid the dam was unsafe?" his answer—"Absolutely none, no sir"—seemed to be an emphatic denial corroborating Van Norman's testimony. But to a follow-up query—"Did you know any employees who were contemplating moving their families out of the camp?"—Mulholland offered a more ambiguous response: "I knew them both. When I left there Harnischfeger and [Aaron "Jack"] Ely [the assistant dam keeper who also died in the flood], and both of them in the closest kind of contact with Van and I when we were up there laying out some work." The coroner did not press Mulholland further after this rather inchoate response, but evidently Harnischfeger had conveyed to Mulholland some measure of disquiet about living so close below the dam.[60]

No Need for a Paul Revere

In the end, Mulholland and Van Norman stayed at the dam for about an hour and a half before their chauffeur drove them back to Los Angeles, where they had a late lunch about 2:00 p.m. There is no evidence that they took any further action that day or evening related to the dam. Later, in response to the coroner's direct question, "If you had realized the dam was likely to break, what measures would you have taken for the safety of the lives of the people in the canyon?" Mulholland drew upon a trope of American history, replying: "I would have sent a Paul Revere alarm up and down the Valley." And to the follow-up question—"It didn't look to you that there was any occasion for alarming the people in the valley of the canyon?"—he brusquely agreed, "No sir."[61]

Of course, steps could have been taken at midday on March 12 to begin draining the reservoir and thus reduce the hydrostatic pressure acting against the dam. And any reduction in pressure would have acted to increase structural stability and thus lessen the likelihood of failure. At the Coroner's Inquest, Mulholland stated that if all the outlet gates were fully opened, a discharge of about 1,000 cubic feet per second (cfs) was possible. He estimated that the reservoir level could have been lowered at a rate of "about one and a half feet per day—not that much—about a foot a day." In response to the question, "There would be no way to relieve the pressure in a few hours?" Mulholland reiterated, "No, it would come down too slow."[62] When Van Norman took the witness stand, he remained in lockstep with Mulholland on this issue, testifying: "reducing the water level in the reservoir...was impossible to accomplish. From the time we were up there [on the morning of March 12] until the day the dam failed [that is, that evening at midnight], the great quantity of water there, and the capacity of the channel, and the facilities for releasing it were so limited, it couldn't be reduced but a very little in that period."[63]

Such assertions have been accepted at face value for more than eighty years. However, as detailed in Appendix A, if a 1,000 (cfs) discharge rate and a reservoir surface area of 615 acres are assumed, the reservoir could have been lowered one foot in less than seven and a half hours (or lowered three feet in less than twenty-four hours). In truth, Mulholland's claims were far off the mark and, at the very least, extremely misleading. Given conditions at the dam at midday on March 12, it is by no means certain that an emergency effort to draw down the reservoir as rapidly as possible would have averted collapse. But, at the Coroner's Inquest, Mulholland bluffed his way past the issue of what action he could have taken to reduce the reservoir level—and, by extension, reduce the possibility of failure.[64]

With hindsight, it is easy to conclude that disaster might have been averted if only *something* had been done at midday on March 12. But if Mulholland and Van Norman had taken substantive action at that time, the resulting discharge would have called public attention to conditions at the dam and fueled concern that something was seriously wrong. And once opened, the Pandora's Box of public trepidation would have been impossible to close. In addition, one of the reasons the St. Francis reservoir was filled to the brim was that all city reservoirs farther downstream were also full. This meant that any release from St. Francis would not be diverted into the aqueduct at Power House No. 2. Instead, the sustained discharge of 1,000 cfs would pass downstream along San Francisquito Creek and, eventually, into the Santa Clara River at Saugus. At a time when the city was battling with Santa Clara Valley interests over water rights to the creek, any large voluntary release below Power House No. 2 could only complicate the city's case for its rights to unappropriated floodwaters. Yes, if Mulholland and Van Norman had been absolutely certain that the dam was in imminent danger of collapse, they would have taken action to warn city employees and citizenry downstream. But anything short of certainty was subject to a different calculus—one giving great weight to the cost of "crying wolf" while discounting, if not denying, the possibility of tragic failure.[65]

Orders to Stay at the Dam?

In their testimony at the Coroner's Inquest, Mulholland and Van Norman both insisted that they departed San Francisquito Canyon free of any concerns about the dam's safety. Nonetheless, there is evidence—absent in prior histories of the disaster—that Mulholland was more worried by his March 12 visit than either he or Van Norman acknowledged on the witness stand.

This information did not come to light at the Coroner's Inquest or during investigations made in the immediate aftermath of the flood. It surfaced in the fall of 1928 while the city was engaged in settling death and damage claims (the issue of reparations is discussed in detail in chapter 5). Leona Johnson, who had died in the flood, presented an unusual problem because of uncertainty about her marital status. Had she been divorced from her first husband? And were she and Tony Harnischfeger legally married? This was of interest to the city because Henry F. Johnson, her first husband, was insisting that Leona was still legally his wife. Seeking to invalidate claims based on their marital status, the city sought out court records and interviewed people who had knowledge of her personal life.

Leona's marital status is not germane to the dam's stability, but the investigation of her case turned up information relevant to Mulholland's concerns about conditions at the site. This information was revealed in a statement by Archibald (Archie) Eley, a friend of Tony Harnischfeger, about a phone call he received from Leona the evening of March 12. Eley was a retired chief of the Los Angeles Fire Department who, often in the company of Harnischfeger, enjoyed taking camping and prospecting trips into the Southern California backcountry.[66] The two men had planned such a trip for Sunday, March 18, but Leona called Eley to let him know that Tony was going to have to stay in San Francisquito Canyon. Here is part of the transcript that the city's claims investigator recorded as "Statement, Etc. of Archie J. Eley. Oct. 26, 1928":

> I personally talked to Leona at 6:30 p.m. on March 12, 1928, at which time she said to me on the phone (long distance) to my office in Universal City "Is this Chief Eley?" Answer, "Yes" "This is Leona, Tony asked me to phone you and tell you that he could not go to Mohave with you next Sunday as he had orders to stay on the job as the dam was leaking bad and he had orders to report three times a day." I have heard Leona tell Tony that if the dam burst or anything else happened she would stick right with him.[67]

Eley was never called to testify at the Coroner's Inquest and his name does not appear anywhere else in the saga of the St. Francis Dam disaster. His statement was recorded six months after the flood and perhaps his memory was playing tricks on him. Or perhaps he accurately recalled what Leona told him about why Harnischfeger's plans needed to be changed (to the authors of *Heavy Ground*, the specificity of Eley's statement gives it considerable authority). Nonetheless, it was never mentioned publicly that Tony "had orders to stay on the job" and "report three times a day" because the "dam was leaking bad." But viewed in this light, Mulholland's testimony at the Coroner's Inquest appears less than truthful. He was likely more concerned about conditions at the dam than his testimony was intended to reveal.

Final Hours

After Mulholland and Van Norman departed for Los Angeles in the early afternoon on March 12, life went on in the upper San Francisquito Creek watershed.[68] Windblown water continued to overtop the spillway crest and, in combination with rain and the leakage that prompted Harnischfeger's call to Mulholland, fed a modest flow down the concrete conduit between the dam and Power House No. 2. This flow was not needed for the city's water supply, and during the afternoon a small crew at Power House No. 2 worked to close the adit connecting San Francisquito Creek to the aqueduct. Later, there would be suspicion that this closure showed that city officials were anticipating the dam's failure.[69] In truth, this action was taken because there was a surplus of water in reservoirs closer to the city, and the city could make no profitable use of any flow within the creek. Water that was passing through Power House No. 2 to generate electricity was exceeding demand, and some of this flow was already being released from the aqueduct into the creek about half a mile downstream at Drinkwater Canyon. In March 1928, Los Angeles was suffering from too much water, not from drought.[70]

To outward appearances, nothing seemed to be happening at the dam site in the hours following Mulholland's visit. But underground, conditions were changing. Water under pressure from the full reservoir continued to percolate into the mica schist foundations and slowly began to push the concrete structure upward. Compelling evidence of this upward displacement is provided by the Stevens water gauge, maintained atop the center section of the dam to record the reservoir level. Data from this gauge miraculously survived the flood. Although unreliable for a precise minute-by-minute record, the gauge nonetheless appears to show a steady lowering in the reservoir level starting around 8:00 p.m. on the night of March 12 and a further drop starting about half an hour before the collapse. Several early investigations of the dam failure took this as a sign of accelerating leakage through the west side abutment. However, the gauge was in fact documenting a gradually rising dam (not a lowering reservoir), which was being pushed upward by water pressure acting through the foundation.[71]

Once the reservoir began impounding water in March 1926, water started seeping into the fractured schist of the east side formation. But the effect of this seepage had been constrained by the height of the reservoir—which stayed at least three feet below the spillway level for almost two years after filling commenced. The reservoir did not approach the spillway crest until about a week before the collapse, and it was only under the pressure of the full reservoir that structural stability was affected by conditions within the foundations. As pressure slowly intensified along the west abutment, water seeped through the interface between the bottom of the concrete dam and the foundation surface at an increasing rate. Leakage also passed through the east canyon wall, but even more important was the unseen subsurface water that saturated the fractured mica schist. This infiltration—now under pressure from a full reservoir—threatened to reactivate the ancient landslide that, millions of years earlier, had formed the east side of the upper San Francisquito Creek watershed. As geologist Willis described it:

> The old slide had ceased moving and presumably would not have renewed its activity [if not] disturbed by the [foundation] excavation...and if its base had not become saturated when the reservoir was filled... Whatever openings the water found it entered, and it penetrated along all the capillary spaces. The base of the [ancient] slide became wet and lost the friction by which it held back the superincumbent weight. Moving forward, even through fractions of an inch only, the slide pushed its weight against the arched dam.[72]

During the afternoon and evening of March 12, pressure within the mica schist formation continued to build, exacerbated by the lack of cutoff trenching and by the absence of drainage wells extending up the east canyon wall. Now the effect of raising the dam ten feet in height without broadening the base would be felt. So long as the reservoir was kept a few feet below the spillway crest, the design's dimensions (and weight) were sufficient to maintain stability. But the hydrostatic pressure exerted by the full reservoir exceeded the limit that the gravity dam could safely withstand.[73] Like straws accumulating on the proverbial camel's back, pressure generated by the reservoir when it rose to 1,835 feet produced instability that did not exist at lower levels.

Collapse

Exactly what transpired as midnight approached on March 12 can never be known with certainty. No surveillance cameras were pointed toward the dam, as one might expect in the twenty-first century. There are no surviving eyewitness accounts and no transcripts of phone calls that

FIGURE 3-19. The east abutment above the dam, from a photo taken during construction. The road running past the dam is visible in the bottom foreground. At top center is the Southern California Edison transmission line pole, which recorded the exact time that power was cut—and at which the structure failed: 11:57:30 p.m., March 12, 1928. [Richard Courtney Collection, Huntington Library]

might have been made by dam keeper Tony Harnischfeger or Leona Johnson as they reacted to unusual sounds emanating from the dam. We are left with testimony of a few Power House No. 1 employees who drove past the dam in the half-hour before failure and with the physical evidence the deluge left behind. From this, a convincing—if necessarily conjectural—account of how failure was initiated, and how it progressed to encompass both sides of the concrete structure, can be deduced.[74]

In the minutes before the collapse at least four people drove up San Francisquito Canyon from Power House No. 2 and past the dam on their way to Power House No. 1. None of these witnesses observed anything alarming and none reported any unusual water flow in the creek and channel near Power House No. 2.[75] Around 11:30 p.m. Dean Keagy drove by the dam, and he later testified that as far as he could tell everything seemed fine. However, Keagy did report seeing lights in the canyon below and thought they might have been from some type of camp. (In response to a question from the coroner asking if he had "seen any light on the dam or anybody on the dam," Keagy replied, "No, the only light was down in the canyon in a sort of a camp.")[76] Exactly what these lights were remains unknown, but quite possibly they were lamps carried by Tony Harnischfeger or Leona Johnson (reportedly, her body was found near the dam and well *upstream* from the cabin she and Tony shared), intent on investigating sounds of high pressure flow emanating from the dam's east foundation.[77] It is also possible that Harnischfeger's assistants Jack Ely and Earl Pike were members of a party carrying lanterns up toward the dam.

Near the time Keagy passed Power House No. 2, Katherine Spann and Helmer Steen dropped off their friend Harley Berry at his bungalow near the power plant. They were returning from a trip to Newhall to pick up provisions. Berry's wife invited them in for a late night snack but Spann and Steen demurred, staying only five to ten minutes. They wanted to get to their homes up at Power House No. 1. When passing by the dam and reservoir a short time after 11:30 p.m., they noticed nothing out of the ordinary, except perhaps that everything seemed especially calm. In her testimony at the Coroner's Inquest, Spann, who served as the nurse at Power House No. 1, recalled telling Steen, "It is quite spooky tonight, terribly quiet."[78]

The last person known to have seen the dam intact was Ace Hopewell, a city employee who was riding his motorcycle up to Power House No. 1. He saw the structure about 11:50 p.m. or perhaps a bit later, and observed nothing unusual as he drove along the road above the east abutment. By this time pressurized water was jetting through the lower east side canyon wall, but this would not have been visible from high up on the east side canyon. And any sound produced by such an outflow would have been inaudible over the roar of his motorcycle engine. About a mile past the dam Hopewell stopped to smoke a cigarette. At this time he heard a low rumbling coming up the canyon from the dam site. He shrugged it off and continued on to Power House No. 1.[79] The rumbling Hopewell heard was the sound of the dam collapsing amid a large east side landslide.[80]

FIGURE 3-20. Aerial view showing the massive east side slide, taken soon after the failure. [DC Jackson]

The first sign of failure came near the bottom of the east side canyon wall. Here a chunk of the dam (shown as block no. 35 in the diagram [fig. 3-22] developed by Charles Lee for his June 1928 article in *Western Construction News*) became dislodged by water pressure acting through the fractured mica schist. Geologist J. David Rogers explains that this would have constituted a "high velocity orifice outflow" driven by water from the reservoir pressing on and through the lower depths of the east side canyon.[81] As this high-pressure release gradually expanded after 11:30 p.m., it produced sounds that likely attracted the attention of Harnischfeger and others in the cabins below the dam.

Significantly, the site of initial failure lay close to where, in the early stages of construction, the thirty- to forty-foot-long tunnel (big enough for a man pushing a wheel barrow) had been excavated into the east side mica schist formation. Perhaps this was only coincidental. But, as geologist Willis pointed out, excavation of any type into the ancient land-

FIGURE 3-21. Detail view of the east side canyon wall following the landslide. The large piece of concrete in the foreground (usually referred to as block 5) was originally located near the top of the structure along the east canyon wall. The slide pushed the block down below the surviving center section, where it stayed for the duration of the flood outflow. [DC Jackson]

slide of mica schist would facilitate the action of seepage and saturation. And as noted earlier, superintendent of construction Stanley Dunham testified at the Coroner's Inquest that some blasting for the east side tunnel had been carried out during the excavation. In disturbing the mica schist formation, the tunnel would have weakened the foundation underlying block 35. Notably, block 35 was the concrete fragment that, after the flood surge was spent, came to rest farthest downstream from the dam, a distance of more than half a mile (see fig. 3-23).[82]

By about 11:55 p.m., near the location of block 35, water was jetting out from the lower reaches of the east side abutment. This precipitated erosion of the dam's lower east side foundation and weakened the structure's resistance to the pressure of the reactivated ancient landslide. The jet of water quickly became a huge blowout, undermining the structure adjacent to and above block 35. As Rogers describes it, "a massive chain

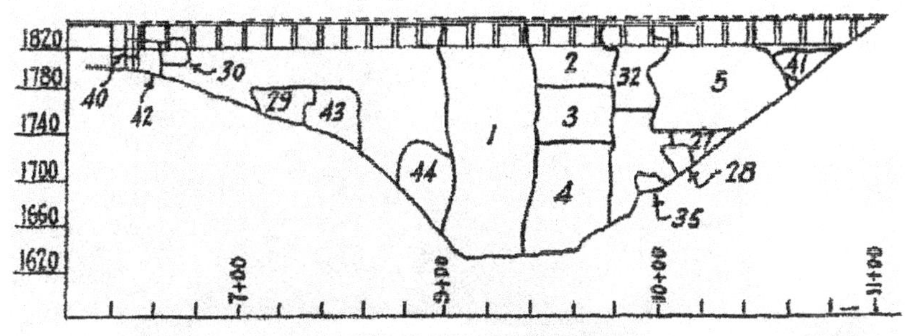

FIGURE 3-22. Drawing of the downstream side of the dam, showing the placement of surviving blocks within the original structure. Not all remnants could be identified, but because the dimensions of the stepped downstream face varied depending on the elevation in the original structure, it was possible to link many of the surviving pieces to a precise location. Failure of the dam started along the lower east abutment, near block 35. [*Western Construction News*, 1928]

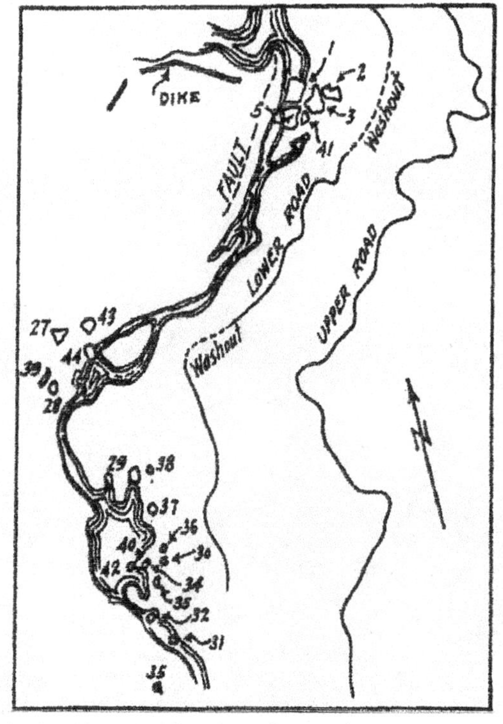

FIGURE 3-23. Drawing of the debris field below the dam, showing the location of various blocks referenced in figure 3-22. Note that block 35 is the one that came to rest farthest downstream, evidence that it was dislodged in the earliest moments of the collapse. Engineer Charles Lee created both drawings for his article in *Western Construction News*, June 1928. Lee's analysis indicated that the dam's collapse was initiated by failure at the east abutment. [*Western Construction News*, 1928]

FIGURE 3-24. Detail view of block 5, with broken mica schist deposited on top of the block. [Richard Courtney Collection, Huntington Library]

FIGURE 3-25. Block 5 is visible in the lower-left foreground. Toppled to the right are blocks 2, 3, and 4. These large remnants sheared off from the surviving center section (block 1) late in the flood. [Richard Courtney Collection, Huntington Library]

reaction failure" quickly ensued as the large section of the dam above block 35 (and encompassing block 32 at the dam's crest) was pushed downstream, removing any support for the saturated mica schist that made up the east side canyon wall.[83] At 11:57:30 p.m. an immense slide of this schist descended down the east side abutment, bringing with it the Southern California Edison transmission line pole and immediately cutting off power along this line. This slide, estimated at a minimum of 500,000 cubic yards of disintegrated mica schist, marked the beginning of calamitous failure, but the full force of the flood was momentarily held in check as the slide filled in the lower east side of the canyon and restrained the deluge.[84] This delay was reflected in testimony from the survivors at Power House No. 2, who described how a modest flow preceded the full force of the flood for a short time. It was their prompt reaction to the preliminary flooding—creating "a misty haze over everything," in the words of Lillian Curtis—that allowed a lucky few to escape before the main torrent hit.[85]

The respite was brief. Seepage and erosion quickly ate away at the earthen barrier and soon the reservoir broke through the east side gap. As water rushed through unimpeded, it created a strong current that pushed against the water gauge stilling pipe used to record the reservoir level, as mentioned above. This pipe was attached to the surviving center section of the dam (block 1), and its upper section was later found to be bent toward the east side gap, thus revealing the direction of outflow at the beginning of the flood. In short, when the reservoir first began to empty, the rush of water came through the canyon's east side.[86]

As the east side collapsed in a catastrophic landslide, the large concrete remnants that Charles Lee designated as block 5 and block 27 were carried from atop the east side of the dam down to the bottom of the canyon. Here they came to rest behind the surviving center section that was later labeled block 1, protected from the full force of the deluge. As block 5 slid down and across the canyon it slammed into block 1, knocking off a part of the downstream face. With the floodwater pouring through the east gap, the center section of the dam rocked back on its foundations. Evidence of this dramatic tipping motion comes from a wooden ladder that lodged in the crack formed near the bottom of the center section's upstream face. The ladder was caught in the opening when the dam rocked back into place, and there it remained, as photographed by post-flood investigators. The huge center section survived the east side slide and subsequent onslaught of rushing water, but later surveys revealed that it had rotated several inches downstream

FIGURE 3-26. The surviving center section of the dam, perched atop part of the mica schist foundation. After the disaster, some engineering investigators celebrated the section's survival as clear evidence of the superiority of concrete gravity dam technology. But the section came perilously close to failure. [DC Jackson]

FIGURE 3-27. Detail view of the diagonal crack in the base of the surviving center section (block 1), illustrating a wooden ladder caught in the crack when the center section tipped back—and almost failed—in the midst of the flood. This photograph, featuring C. C. Teague, was published by C. E. Grunsky in his May 1928 *Western Construction News* article on the disaster. [St. Francis Dam Disaster Papers, Huntington Library]

FIGURE 3-28. The concrete structure atop the west abutment collapsed after the east side failure. Cracks formed by the displacement of the center section helped to destabilize the west side of the dam and precipitate its washout. This view shows the west abutment and highlights the geologic "contact plane" between the mica schist and the red conglomerate-sandstone. In the first weeks after the disaster, some investigators theorized that the contact plane, or fault, played a major role in the collapse, but more careful analysis showed otherwise. [DC Jackson]

and toward the east. Relatively late in the failure sequence, the foundation underlying the three large remnants labeled blocks 2, 3, and 4 was undermined, causing them to break free from block 1 and topple into the east side gap.[87]

The wrenching forces that acted on the center section also worked to separate it from the portion of the dam atop the west side abutment. Some uplift was acting on the west side due to seepage, and this force was exacerbated by movement of the center section, which worked to separate the concrete base from the foundation. With the west side of the dam detached from the center section, water infiltrated the growing divide between the foundation and the concrete base, destabilizing the west side structure and precipitating its collapse.[88]

The destruction of the dam's west side came at a point when the reservoir had been significantly diminished because of outflow through the east

FIGURE 3-29. Looking downstream toward Power House No. 2, the landscape littered with the array of concrete blocks and remnants left by the receding flood. The outline of the road that ascended the east canyon wall, visible at center left, was traversed by motorcyclist Ace Hopewell about ten minutes before the dam collapsed. [St. Francis Dam Disaster Papers, Huntington Library]

side gap. Nonetheless, there was still sufficient flow to carry some large concrete fragments from the west side down the canyon about a quarter of a mile. (Block 29, one of the largest surviving fragments, was originally located above and west of the contact line between the red conglomerate and mica schist formations.)[89] Erosion, or lack thereof, along the lower west side canyon below the dam—and the intact survival of a hairpin turn on the roadway extending up the west side canyon directly below the dam—indicates that the west side did not fail until the reservoir had dropped in elevation at least forty feet. As discussed in chapter 6, some early investigations (including the one by the Governor's Commission) sought to lay blame for the collapse on the west side red conglomerate, where Mulholland had been called by Harnischfeger to investigate leakage on the morning of March 12. But this proved to be a red herring. The mica schist formation—Lippincott's "heavy ground"—constituted the site of cataclysmic failure. Destruction of the dam's west side came later and was secondary to the east side collapse.

FIGURE 3-30. Block 29 weighed over 10,000 tons. The largest of the surviving concrete remnants, it was carried over a quarter mile downstream. [DC Jackson]

FIGURE 3-31. A large number of concrete blocks were transported downstream by the flood surge. Taken after reconstruction of the Edison power line down San Francisquito Canyon, this photo illustrates the stepped faces on surviving remnants, clues to their original position in the dam. [DC Jackson]

With the east and west sides of the dam gone, the reservoir emptied in about forty-five minutes, with most outflow discharged in the first twenty minutes. At its peak, the flood reached more than 1 million cubic feet per second, washing up high on the canyon banks below the dam, obliterating the cabins occupied by Tony Harnischfeger and his assistants, and surging toward Power House No. 2.[90] The dam was now history. The disaster was only beginning.

CHAPTER FOUR

Disaster Unleashed

> I could hear it coming, hear the trees breaking, and could hear a big pole snapping, could hear the wires on the electric poles going.
>
> Chester Smith at Coroner's Inquest, March 22, 1928[1]

> With the dead stacked in piles, the French & Skillin Chapel in Fillmore today took on a grim aspect as the silent workers hurried back and forth bringing in their gruesome burdens. Many of the dead remain as yet unidentified, although the bodies are being washed and every possible opportunity is being given to friends and relatives who fearfully pass by glimpsing at the distorted faces.
>
> *Santa Paula Chronicle*, March 13, 1928[2]

To the outside world, the first sign of trouble came moments before midnight on March 12, when a voltage drop disrupted the regional power grid. Detected at 11:57:30 p.m. at the city's San Francisquito Power House No. 1, the same drop registered at two Bureau of Power and Light switchboards in Los Angeles and at the Southern California Edison substation in Saugus about nine miles below the dam. At Saugus, the pressure drop destroyed an oil switch and shorted out Edison's power line from Lancaster, while at the Los Angeles stations it was weaker, lasting only two seconds. The anomaly briefly caught the attention of Power House No. 1 operator Henry Silvey and relief operator H. L. Tate. "We got what we call a nibble or a fish bite," recalled Tate, but it elicited no concern as "everything was clear and there was no indication of any trouble."[3]

Unbeknown to Silvey and Tate, the nibble reflected much more than a minor glitch: Edison's 60,000-volt Borel-Lancaster electric power line,

FIGURE 4-1. Looking down San Francisquito Canyon, with surviving center section of dam in foreground. The tremendous size of the flood surge is evident in the scouring of the canyon walls. It is estimated that at peak flow, the flood's discharge approached 1 million cubic feet per second. [DC Jackson]

strung through San Francisquito Canyon, had just been washed away. Originating at the Borel hydroelectric power plant in Kern County, the line connected through Lancaster on its way to the Saugus substation. As the dam broke, the power line collapsed down the east abutment, engulfed in a mass of water-saturated schist.[4] The force of the flood as it left the reservoir was so powerful that huge concrete chunks of the dam—the largest weighing several thousand tons—were carried as much as half a mile downstream.

First Victims

The torrent quickly claimed its first victims: dam keeper Tony Harnischfeger, his companion, Leona Johnson, and his six-year-old son, Coder. Years earlier Harnischfeger had served Mulholland as watchman at Jawbone Siphon during construction of the Los Angeles Aqueduct. After the St. Francis Dam was completed, he moved to a cottage a few

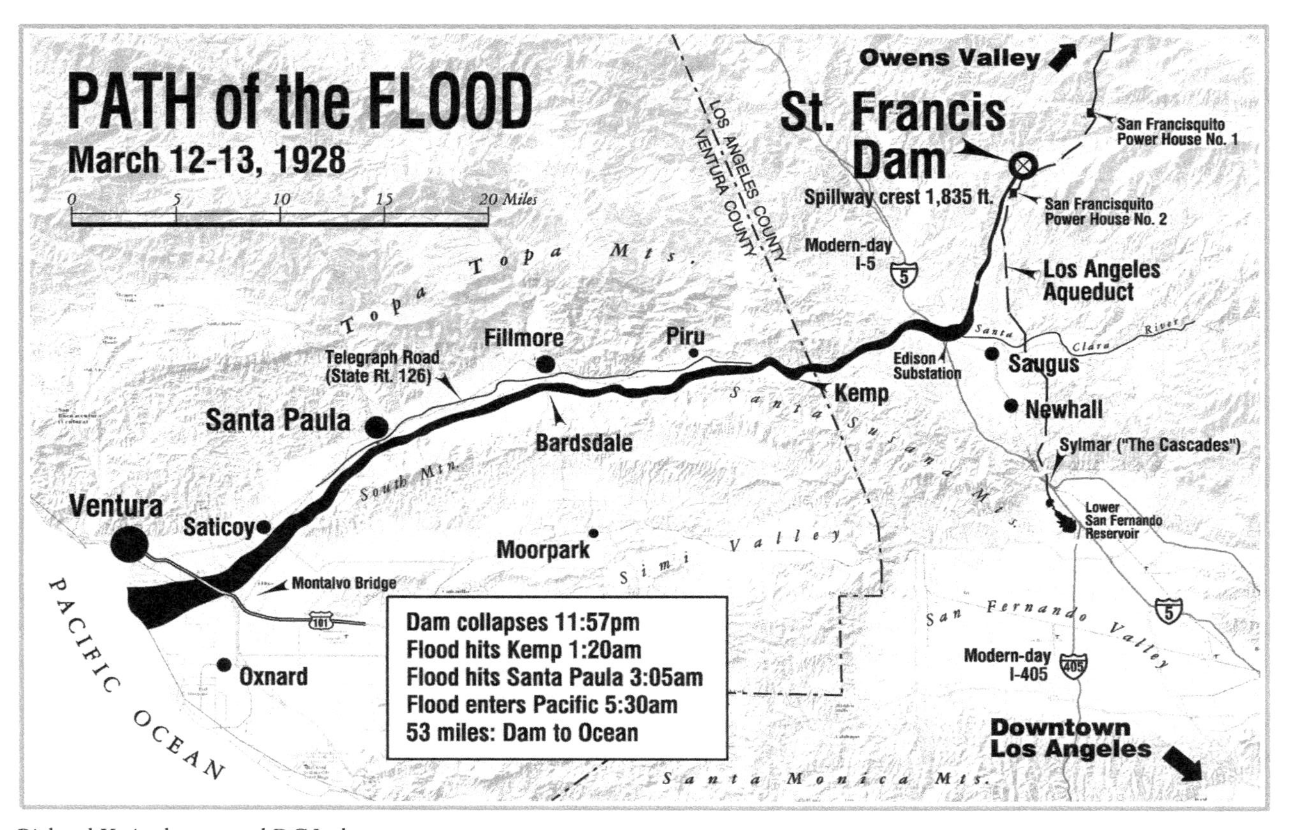

Richard K. Anderson and DC Jackson

hundred yards downstream in San Francisquito Canyon. Within a minute of the dam's rupture a cascade of water demolished his house, probably with his son inside. Tony's and Coder's bodies were never found, but Leona's was reportedly recovered well upstream from the family cottage, evidence that she (and perhaps Tony as well) had ventured up toward the dam as pressurized water began to surge from the east side abutment.[5] Also killed in the initial surge were two assistant dam keepers, Aaron "Jack" Ely and Earl Pike, who lived close to the Harnischfeger cabin, along with their wives and children.[6]

In the upper San Francisquito Canyon the flood traveled at almost twenty miles per hour, and less than five minutes separated the obliteration of the Harnischfeger, Ely, and Pike cottages and the destruction of Power House No. 2.[7] At 12:02:30 a.m. generation at Power House No. 2 abruptly ceased. "It went down in a heap," remembered operator Tate at Power House No. 1. "There was just a dead short." He and co-worker Silvey could not explain it. "Did you ever see anything like that?" Silvey wondered. "No," replied Tate, "I never did either; it has got me stumped."[8] Only later would the two men realize that the "dead short" marked the end of Power House No. 2 and the drowning of most who lived nearby.[9]

E. H. Thomas, who worked at the surge chamber on the hill above the power plant, was the first to encounter the carnage at Power House No. 2. Awakened by his mother, who shared his living quarters, he recalled that "the lights went out, the house just started to tremble, the windows and doors [were] just trembling." When the shaking finally stopped, he grabbed a flashlight and scrambled down the hill, soon to discover that "the power house had gone." Checking his watch he noted the time—12:15 a.m. The first wave of the flood had overwhelmed the power house about twelve minutes earlier. A reporter visiting later that day described the scene: "Near the power station were almost a score of comfortable houses for the families of the workmen. Neither stick stone nor nail is left of them to mark the spot. The location is as barren as the most arid desert floor."[10]

Built just a few years before, the settlement of city employees and their families was not a collection of itinerant workers devoid of personal connections. This is clear in testimony from D. C. Mathews, who spoke of the power house community at the Coroner's Inquest. Mathews did general labor for the city's Bureau of Power and Light, but unlike his brother Carl—who lived with his family near Power House No. 2—he resided in Newhall. His testimony helped bring to life the human collective destroyed by the flood:

FIGURE 4-2. San Francisquito Power House No. 2 before the flood. The house on the left (sometimes referred to as the clubhouse) was part of the residential community built for Bureau of Power and Light employees and their families. [Southern California Edison Photographs and Negatives, Huntington Library]

> [Harry Burns' house] was the last city house down the canyon from the power house. There were four of the city houses below the power house in the canyon, anywheres from a hundred and fifty to three hundred yards down the canyon, and maybe a quarter mile below. [Between Harry Burns' house and the power house were] the house that Harry Mathis lived in and the house that John Seddon used to live in—he had moved away the fore part of January—and Harry B. and another was a shack just below the power house where Maxie Browski and Earl Kerr slept, and there was the clubhouse nearly opposite the power house, and then above that the first house was Mr. Hughes and to the left of Mr. Hughes house was a little canyon that led off the main canyon,

FIGURE 4-3. William Weinland and his family lived in the workers' compound at San Francisquito Power House No. 2. He, wife Ina, and son Lloyd were all killed within minutes of the dam's collapse. Weinland's father took this photograph outside the family bungalow near the power house a few months before the disaster. Here, in a classic scene of Americana, a mother poses with her son, daughter-in-law, and grandson during a family visit. [William H. Weinland Photograph Collection, Huntington Library]

FIGURE 4-4. Interior view of Power House No. 2 around 1927, with plant operator William Weinland posing in front of the two 14,000 kW generators. The flood unleashed by the dam's collapse would destroy the entire reinforced concrete structure, leaving behind only the two generators. [William H. Weinland Photograph Collection, Huntington Library]

FIGURE 4-5. The site of Power House No. 2 after the flood. Nothing above the surface survived except for the turbine-generators. The height of the deluge that engulfed the power house is evident in the scoured terrain extending up the canyon walls. [DC Jackson]

where there were two old houses that were put up during construction of the power house...Mr. Homer Coe's family lived in one of those, and the next house was where Ray Rising lived with his family and the next house was where Lyman Curtis lived. He was the road boss, and the next house in that same canyon above Lyman Curtis was where Lew Burns and his family lived and in that main canyon just above Mr. Hughes was where Mr. Weinland, one of the operators lived, and the next house above Mr. Weinland was where my brother lived. He was a relief operator, and above his house is where Mr. [Farrell] Hopp lived, and his family, and a few yards above that was a new school house built a couple of years ago. That was all the city houses, except some up at the dam...Tony [Harnischfeger's] house and Jack Ely and the other man [Earl Pike] that worked with him, lived, were

FIGURE 4-6. Detail view looking down at the destroyed power house, showing the outline of the building's rectangular foundation. The turbine-generator on the left (with water spout) was in operation at the time of the flood and suffered more damage than its companion. [DC Jackson]

> right up close to the dam...Practically all of them were in the main channel of the canyon, except those four that led off in that little canyon to the left of Hughes' house. That was Coe's, Rising's, Curtis' and Lew Burns'...They were protected in a certain way. In case of excess water they had a better chance to get out of there than anybody on account of being out of the main canyon.[11]

While Mathews' description makes clear that more than a few lives were lost at Power House No. 2, he omitted some other unfortunate victims. On the night of March 12 Ina Weinland's sister, Olive Imus, and her eight-month-old baby boy, Myron, were visiting from their home in Oxnard. They, like their hosts, perished in the flood.[12]

The ubiquity and seeming blandness of "drowning," typically given as the cause of death, should not obscure the physical torment that victims endured. Their suffering is tersely conveyed in testimony offered at the

Los Angeles County Coroner's Inquest. The autopsy of Julia Rising, who lived with her husband near Power House No. 2, was entered into the public record by the county's assistant autopsy surgeon, Frank Webb:

> The body was a female of the white race, aged twenty-nine years and five months, height five foot five inches, estimated weight about a hundred seventy five pounds, dark brown hair and light complexion. Further examination showed numerous superficial scratches and bruises scattered over the body, the face, and the limbs. There was a deep gash three inches long from the center of the left leg. There was a laceration diagonally of the right forehead. The lungs were red and inflamed and contained water. The trachea contained mud and silt. The other organs were normal except for marked congestion. The stomach contained a considerable amount of silt. From these findings the deduction was made that the death was due to drowning.[13]

Of more than sixty people residing in the power house compound, only three survived the flood: Lillian Curtis, her young son, Daniel, and their neighbor, Ray Rising. All three had lived in a side canyon to the west of the power house and, in Mathews' view, this is what saved them. Lillian Curtis grabbed her son and ran up the canyon wall just ahead of the rising water. (Mathews believed that her escape proved possible because of "the way the water slopped over those houses back at my brother's place, I have an idea it washed them up on the hill there.") Rising grabbed onto a floating roof and, before the impromptu raft splintered apart, miraculously jumped free when it washed up against a hillside.[14] As the tumult descended San Francisquito Canyon, the odds of cheating death were slim. But a few survived to bear witness.

"The Devil Was on Me"

About two miles below Power House No. 2 in the twisting canyon, the flood hit the cattle ranch of Chester Smith. As he later testified at the Coroner's Inquest, Smith had been among those concerned about "water...seeping on the [dam's] west bank" as the reservoir filled to capacity. On the night of March 12, he slept in his barn with the door open so that, if the dam failed, he could make a fast getaway to higher ground.

Living in a nearby ranch house were an employee, Hugh Nichols, and his wife. Another family—Alva Kennedy, his wife, and their two young children—had recently arrived at the ranch and were encamped even closer to San Francisquito Creek.[15] Sleeping fitfully, Smith was awakened by his dog and strange noises. "I could hear it coming, hear the trees breaking, and could hear a big pole snapping, could hear the wires on the electric poles going." Running toward the house where Nichols and his wife were sleeping, he rousted them with the cry, "The dam is broke!" Together they "ran a hundred and fifty feet to the hill, and as we ran up the hill, the water was right behind us." They escaped, but the Kennedy family did not. "People have asked me why I didn't run and notify the others," lamented Smith. "It was impossible."[16]

"Impossible" also conveys the plight of Henry and Rosario Ruiz a short distance downstream from Smith. With four of their children, they lived on a ranch that had been in the family for many years. A fifth child, Rose, lived with her husband, James Erratchuo, and their infant son on an adjacent parcel. During the dam's construction, Erratchuo had worked for the city doing odd jobs and, like Chester Smith, had felt trepidation about the dam's safety. "Looked like to me it wasn't built strong enough, didn't look good to me," he later recounted at the Coroner's Inquest. Coroner Frank Nance pressed him further: "But in spite of that you didn't take any precautions to move out of the canyon?" "That was my home," replied Erratchuo, "and...I thought everything would be OK, didn't realize how it [the dam] was going to go." On the night of March 12, Erratchuo "went to bed as usual, feeling...perfectly safe." As the water approached, a loud roar filled his ears. Running to the front of the house with his wife, he kicked out the door, only to be overwhelmed by the maelstrom: "The devil was on me." Grabbing for his wife's arm, he lost hold as barbed wire from uprooted fences tore at him. Swallowing water and mud, he gasped for air. Then somehow his feet struck solid ground and the danger passed. He was safe. But his wife, son, father-in-law, and mother-in-law and his wife's four siblings were gone. James Erratchuo was the lone survivor of the Ruiz clan of nine that slept in the canyon that night.[17]

Below the Ruiz Ranch the flood ravaged the Hunick family. At home that night were Bert W. Hunick, a beekeeper, who shared a room with his father, and in their own rooms were his sister and his uncle. "I heard a roar and I jumped up," recalled Hunick. He shouted to his father, "What is that, a tornado?" He answered his own question, "No, the big dam is busted." Calling out to his sister and uncle he "could hear no answer or anything," with the water making "such a crash." What followed was chaotic terror:

FIGURE 4-7. The upper San Francisquito Canyon is quite narrow, but about six miles below the dam site the valley begins to widen out. This photograph was taken in September 1926 during the "percolation test" discussed in chapter 3. An almost idyllic scene, this view shows what the lower canyon looked like near Harey Carey's ranch before the flood. [United Water Conservation District, Santa Paula]

> The water was in the house and…I grabbed a piece of board that floated against my knee and knocked the window out and throwed my foot up to give the screen a kick and my first intention was to jump out and I started to make the jump and was throwed back, but I got up again and got to the window again and then the house was struck and lurching and jumping and then I was going to try the window again and all at once it looked like something give way, part of the house, I guess, and I could see the outlines of the hills and I jumped right off into the water about to the waist and turned around and hollered and grabbed a piece of the house and thought I could hold it, thought the folks were in there, but it jerked away, leaving me with a little piece of board in my hand.

Hunick ran up the hillside, yelling for his relatives. But he "could hear nothing, only the roars of the flood."[18] Later he learned that his father, though knocked unconscious, had floated to safety on a mattress. His uncle and sister were not so lucky. Her body was later identified at the Newhall morgue and his was recovered over twenty miles downstream, trapped amid the debris in Bardsdale.[19]

Caught next were the Price Ranch—where five members of the Price and Halen families died—and motion-picture actor Harry Carey's Indian Trading Post, a popular tourist attraction in the lower San Francisquito Canyon.[20] The store and nearby "Indian village" disappeared into the flood, killing Clinton Harter and his wife, Marian, who served as attendants. Harry Carey's own residence lay on high ground to the west, well back from the flood path, and escaped damage. Carey was not home that night—he, his wife, and two children were in New York City awaiting the premiere of the movie *The Trail of '98*. But his good fortune was limited, as a reporter described the bleak scene: "His trading post which lured many a tourist is gone. In its place is a stream of rushing murky water and beyond a waste of sand...A huge iron safe from the post was found a half-mile downstream. The cash register was near by." The morning after the flood, "small sticks with white handkerchiefs, or a piece of rag waving from their tops," signaled to ambulance crews and pack-horse teams where lifeless bodies lay in the sandy waste of the trading post site.[21]

The flood laid waste to the settlement at Power House No. 2 and the ranches of San Francisquito Canyon with equal ferocity. As it plunged through the narrow valley, there was little opportunity for escape. In the words of a workman helping in the cleanup: "The poor folks never had a beggar's chance." The day after the flood, the *Los Angeles Times* characterized the devastation in equally stark terms: "Scores of dead horses, cattle and fowl lay half-buried in the sand...It was in this section, below the power house where the farmers plied hoe and plow to productive acres of their little homesteads. Nothing remains but water-scarred rocks and sand and gravel."[22] A reporter for the *New York Times* offered a similarly bleak description of Harry Carey's trading post and the Price Ranch the day after the flood. Here stood Mrs. Russell Halen, "huddled in a vivid red sweater, wringing her hands." She could do little more than point to "where [her] grandmother's home had stood, beside a cottonwood tree. It was nothing but a tablelike surface of yellow sand and even the cottonwood tree had been stripped of its bark by the wild water."[23]

FIGURE 4-8. Lower San Francisquito Canyon after the flood. A thick layer of rock and sand (much of it from the landslide of fractured mica schist from the east side) has obliterated the once-edenic landscape of the lower canyon. [DC Jackson]

Newhall Ranch and Castaic Junction

Below Harry Carey's Indian Trading Post and more than seven miles downstream from the broken dam, the flood entered the northern reaches of the 40,000-acre Newhall Ranch. Founded as Rancho San Francisco during the era of Mexican rule, since the 1870s the ranch—which extended down the Santa Clara Valley and into eastern Ventura County—had been controlled by Henry Mayo Newhall and his heirs through the Newhall Land and Farming Company. Where San Francisquito Creek entered the ranch, the canyon broadened, and the flood became wider and shallower. Its velocity may have fallen below thirteen

FIGURE 4-9. The Southern California Edison substation at Saugus after the flood. Located on the south side of present-day Magic Mountain Parkway in Santa Clarita, the substation lay at the southern edge of the flood as it passed from San Francisquito Creek into the main Santa Clara River Valley. Edison personnel at the substation were aware of the high water (note the debris caught in the steel framing), but no clear warning was given to the outside world. [Southern California Edison Photographs and Negatives, Huntington Library]

miles an hour, but the water continued to move with enormous destructive force.[24]

Today, San Francisquito Creek flows on a relatively straight course in its lower stretch, joining the main thread of the Santa Clara River to the east of an imposing knoll (known to the Newhall family as Round Mountain) on the north bank. But prior to the flood, the creek took a westward turn in its lower stretch and joined the main river downstream from Round Mountain. Here, a Southern Pacific trestle and a small highway bridge crossed San Francisquito Creek. To the south, a larger Southern Pacific bridge crossed the main stem of the Santa Clara River, as did a large truss bridge that carried highway traffic along the famous Ridge Route heading from the San Fernando Valley to Bakersfield and points north.[25]

Below the boundary of the Newhall Ranch, 350 acres of farmland, designated San Francisquito Fields Nos. 1 and 2, were covered with a thick layer of sand, gravel, and debris. (As the flood slowed, the broken mica schist carried by the water began to settle out, leaving a rocky and sandy residue down the length of the Santa Clara Valley.) The cascade pushed south over the Santa Clara River channel, inundating both the ranch's 200-acre Santa Clara Field and Southern California Edison's Saugus substation (located on the south bank of the river along modern-day Magic Mountain Parkway), where several small buildings were demolished and the transformer yard became littered with debris. To the west, the surge also carried away twelve steel towers supporting Edison's trunk line, which descended Castaic Creek Canyon from the Big Creek hydroelectric plants near Fresno.[26]

Downstream from the Saugus substation, the flood destroyed the two large railroad and highway bridges crossing the main stem of the Santa Clara River. It also swamped the flat stretch of land lying downstream and west of Round Mountain—for about twenty minutes, this high ground was an island. As with their larger counterparts crossing the main river, the railroad and highway bridges over San Francisquito Creek northwest of Round Mountain were torn from their foundations. In this area the Newhall Ranch was especially hard hit, with hundreds of acres in Rye Field, Reservoir Field, Grapevine Field, and other tracts both eroded and heavily layered with sediment.[27]

Heading westward down the Santa Clara Valley toward Piru, the flood immersed the western outskirts of Saugus (now part of Santa Clarita and commonly known as Valencia) and Castaic Junction, a rural transportation hub connecting highways north and south (the Ridge Route) as well as to the west (Telegraph Road, modern-day Henry Mayo Drive, or California Route 126).[28] At Castaic Junction, Robert McIntyre's restaurant, gasoline station, and tourist camp, including eight automobiles and their sleeping owners, were swept away. McIntyre's son George survived (his father and brother perished), and he later reported that he had "heard several automobiles passing his place [on the Ridge Route] going toward Saugus and Newhall shortly before the flood struck...[He had seen] seven pairs of automobile headlights topping a rise before dropping down to the bridge to cross the [Santa Clara] river, the lights never reappeared on the road."[29] Some drivers outraced the flood. Joe Sockel, a poultry buyer from Los Angeles on a late-night run up to Porterville in the San Joaquin Valley, later told a reporter for the *Los Angeles Times*:

FIGURE 4-10. The Southern Pacific's railway bridge crossed the Santa Clara River only a few hundred yards north of Edison's Saugus substation, but the force of the flood swept the steel truss off its foundations and carried the structure more than a quarter mile downstream. The damaged truss came to rest close to where modern-day Interstate 5 crosses the river. "Beneath its girders," wrote a reporter for the *Los Angeles Times*, "the searchers found the body of a baby girl with a rag doll clutched tightly to the breast." [DC Jackson]

> When I reached Castaic I saw a great wall of water, seemingly about six blocks away...The noise was terrific, and I was almost paralyzed with fright. Men whom I afterward learned were Edison company employees, came tearing down the highway toward me and shouted a warning to turn back...In some manner I made the turn in the highway at lightning speed and dashed for Saugus, I pushed my truck to its full speed, thirty-five miles an hour. With that terrible roar behind me and death threatening at every moment, the truck seemed to just creep along...I headed for the hills near Newhall and there in comparative safety, held vigil through the night.[30]

FIGURE 4-11. The Ridge Route crossing of the Santa Clara River in Saugus, from a postcard view of around 1926. This highway was the equivalent of modern-day Interstate 5, climbing over the Tehachapi Mountains ("the Grapevine") to Bakersfield and points north. [DC Jackson]

The Southern Pacific section headquarters and station at Castaic Junction lay a short distance west of the Ridge Route near McIntyre's roadhouse and, although it was not widely reported in the daily press, were essentially wiped out. Two months after the flood the journal *Railway Engineering & Maintenance* reported: "among the 400 persons reported killed or missing [in the flood]...were 18 men, women and children at the section headquarters." The victims included six members of the Martinez family, five members of the Alvarez family, and five members of the Alvarado family.[31] While much of the Southern Pacific's forty-five-mile-long Montalvo-to-Saugus rail line was undermined or torn up, "serious damage was confined to the 9½ miles of the line below where the flood first struck the tracks [at Saugus]. In this district three miles of track was completely washed away and 1,150 feet of trestle bridges, and a 150-ft steel truss bridge were destroyed. In addition [the flood] destroyed section houses, corrals, station buildings, telegraph and telephone lines and fences." Overall, the railroad's property damage was estimated at $250,000 to $275,000.[32]

FIGURE 4-12. The flood washed away the Ridge Route's massive steel truss, leaving behind only the reinforced concrete piers and the north girder approach span. The bridge was later rebuilt about 1,000 feet upstream along what is now the Old Road crossing of the Santa Clara River. The old piers still survive (as of 2014) and are visible north of Feedmill Road near the back entrance to Six Flags Magic Mountain. [DC Jackson]

FIGURE 4-13. The crumpled remains of the steel highway truss that carried motorists across the Santa Clara River. At a distance nine miles below the dam site, the flood still held great destructive power. [DC Jackson]

FIGURE 4-14. As the flood coursed down the Santa Clara Valley, it wreaked havoc on the valley's agricultural fields. This view shows part of the Newhall Ranch near Castaic Junction. The trees are bent to the left (downstream), reflecting the flood's direction as it ripped through a once-fecund orchard. [DC Jackson]

Coursing down the valley, and paralleling modern-day Route 126, the flood carried away the railroad and highway crossings over Castaic Creek and left behind deep gullies and strata of rock and sand.[33] One of the tenant families in this area, the Riveras, met a tragic fate. As the flood reached Castaic, thirteen-year-old Louis heard the rumbling of the approaching flood and went outside to investigate. Returning to warn his father, he was rebuffed: "Oh, it's nothing but the rain." Fearing the worst, Louis grabbed his eight-year-old brother and nine-year-old sister and ran outside toward higher ground. Looking back, he saw his parents caught as they came out the door. He also saw one of his two older brothers desperately trying to start the family car, "only to be swept away as he sat in the machine." Louis and his two younger siblings were the only members of the Rivera family to survive. Through the remainder of the night and for much of the next day the three hid in the hills above the flood zone until a search party found them late on Tuesday.[34]

Below Castaic Junction the Newhall Ranch's holdings extended down the valley for about ten more miles. Near the Los Angeles–Ventura county line several tracts, including Pecan Orchards Nos. 1 and 2 and

FIGURE 4-15. About 1:20 a.m. (a little less than 90 minutes after the dam's collapse), the flood inundated the Southern California Edison camp near the railroad siding at Kemp. Edison had rented the field from the Newhall Ranch as a temporary domicile for a 150-man crew building a transmission line down the valley. This photograph of an overturned roadster only hints at the terror that engulfed the camp. [Automobile Club of Southern California]

Blue Cut Field (where tenant Joe Gottardi lost his wife and daughters, as well as his house), suffered extensive damage. Beyond the human toll and dead livestock, the company estimated that some 1,720 acres of its agricultural land were "totally destroyed."[35]

Kemp and Bardsdale

As the flood crossed into Ventura County, the brunt of the catastrophe fell not on travelers, townspeople, or farmers, but on 150 Southern California Edison employees encamped at a railroad siding known as Kemp. Located almost seventeen miles below the dam and within a half-mile of the county line, the camp occupied part of the Newhall Ranch's County Line Field, which had been leased to Edison for a few weeks.[36] On a broad field adjacent to the river channel, the temporary tent community housed

FIGURE 4-16. Carnage at Kemp. [DC Jackson]

a large crew engaged in building steel pylons and stringing a power transmission line down the valley from Saugus to Saticoy.

In the early morning hours, one person awake at Kemp was the company's night watchman, Edward Locke. Not long after 1:00 a.m., a muffled sound, gradually building in volume, broke the stillness. Locke began moving through the camp, stirring the sleeping men. His efforts helped save some, but many were trapped inside their canvas tents when the water hit at 1:20 a.m. The Blue Cut promontory ridge, which extended out from the valley's north side, acted as a partial dam, blocking the flow and creating a deadly, churning backwash. Near the tents, men had parked their automobiles for the night. When dawn came on March 13, the wrecked cars lay strewn in disarray across the valley, and eighty-four Edison employees, including watchman Locke, were dead.[37] The death toll at Kemp was rivaled only by the loss of life at Power House No. 2.[38]

FIGURE 4-17. More than half of the men at Kemp died in the flood, and the bodies of many of the eighty-four victims were discovered miles downstream near Fillmore, Santa Paula, and Oxnard. The unidentified man perched pensively atop a truck mangled by the flood was likely one of the fortunate survivors from the camp. [DC Jackson]

FIGURE 4-18. At Kemp, clearing autos and other wreckage took many days of hard work. [DC Jackson]

Three miles below Kemp, the main buildings of the historic Camulos Ranch (renowned as a supposed inspiration for *Ramona*, the novel about Mexican-era California) miraculously survived intact, the edge of the flood passing a few yards to the south. A mile and a half below Camulos the small town of Piru also weathered the flood with little damage because most of the town was perched on bench land along the valley's north side. About eight miles west of Piru, the commercial center of Fillmore was similarly spared because of its siting above the valley floor. But Bardsdale, the low-lying orchard enclave opposite Fillmore to the south, was not so fortunate. Flooded at about 2:25 a.m., it was the last sizable community caught unaware, before widespread warnings roused people from their beds.

By the time the flood reached Bardsdale it had slowed to about eleven miles per hour. It did not so much crush houses as dislodge them from their foundations, forcefully propelling them into adjacent orchards. The *Ventura County News* described Bardsdale, in the wake of the flood, as "one of the real garden spots of Ventura County," which now presented "a picture of desolation, with hundreds of acres of orange and walnut groves uprooted or buried beneath great masses of debris." Among the community's fatalities were Ida Kelley and her two young daughters, Dolores and Phyllis, and George Cummings, his wife, and their twelve-year-old daughter, May. George Basolo, a member of a prominent Bardsdale family, got "his bride of but a few months to a point of safety" before he and his brother-in-law, Clifford Corvin, "headed back to their homes to see what might be saved." But to no avail: "The coupe in which they were traveling got caught by the current and turned over in the stream." Corvin survived by clinging to a treetop until the flood crest passed. Basolo, apparently struck on the head by floating debris, was rendered unconscious and drowned, his body coming to rest in a nearby orchard.[39]

A letter Frank Chandler wrote to his mother in Tennessee vividly described the cacophonous confusion in the Fillmore-Bardsdale district:

> The whole town was wild as there were no lights, no gas and lots of telephone wires down. They were shooting, ringing bells, blowing sirens, yelling and everything to wake everyone...very few people had warning, as lots of them had no telephones, and it came so fast that Central did not have time to notify many that did have [phones]. And although some of them had warning,

FIGURE 4-19. Bardsdale (the agricultural community opposite Fillmore on the south side of the Santa Clara River) was the last settlement to be hit by the flood before warnings were widely sounded. Although its force had diminished, several Bardsdale residents died, and many houses were severely damaged. [DC Jackson]

> they thought it was false and did not move out as we had not had any rain and 90 percent of the people did not know there was a dam that would turn water this way even though it did break.[40]

The Tocsin Sounds

Other victims would die after the flood devastated Bardsdale, but—some two hours after Edison's Borel-Lancaster power line went down at the dam site—news of the collapse was at last beginning to spread through the lower Santa Clara Valley. But troubling questions linger: Why the delay in sending out a general alarm? It seems more than plausible that warnings could have reached the Edison workers at Kemp or the ranches at Bardsdale: both were inundated long after the dam's collapse and after the flood had swept through Saugus. Why wasn't the sudden break in transmission along Edison's Borel-Lancaster line perceived as a potential sign of the dam's failure? Shouldn't the cessation of generation at Power House No. 2—and the failure to reach the power house staff by phone—have immediately signaled that something was seriously amiss? And, most obviously, shouldn't the inundation of Edison's Saugus substation have sparked widespread warnings to the lower valley? In hindsight, the red flags raised by these events are easy to interpret.[41] However, in the turmoil of the moment, apparently no one wanted to believe that the cataclysmic failure of the St. Francis Dam was the cause of the unfolding events. Or at least they did not want to believe it until receiving eye-witness confirmation that the reservoir was empty.[42]

Sometime around 12:30 a.m. the Edison dispatcher at Saugus telephoned the Bureau of Power and Light dispatcher in Los Angeles with word that "someone reported lots of water" in the vicinity of Saugus. Both dispatchers were aware of the earlier disruptions in electric power transmission from San Francisquito Canyon, and the report of "lots of water" heightened their concern. They now "began to suspect trouble at the St. Francis Dam or the Los Angeles Aqueduct." While their suspicions were not enough to sound a general alarm, the bureau dispatcher nonetheless contacted the operator at Power House No. 1, ordering: "Send someone down the Canyon to make an inspection." In response, Power House No. 1 chief operator Oscar Spainhower and a highline patrolman motored down the dirt road toward the dam. About thirty minutes later Spainhower returned with the news: "The St. Francis reservoir [is] empty." Assistant operator Henry Silvey quickly conveyed the information to headquarters. "As our phone line was out, I got busy immediately on the carrier wave [wireless radio] system we had at the plant and soon got our Los Angeles dispatcher on the line. This was exactly at 1:09 a.m."[43] The Los Angeles dispatcher immediately made an emergency call to Mulholland's lieutenant, Harvey Van Norman, who directed that telephone warnings be sent to the Santa Clara Valley.

Responsibility for spreading the alarm now fell to people in the lower valley. The first to act were Pacific Telephone and Telegraph Company operators, who, at about 1:15 a.m., began ringing up and notifying other operators in the valley "of the impending flood waters." At 1:20 a.m., the Los Angeles County Sheriff's Office at Pelton got through by telephone to the Ventura County Sheriff's Office, warning: "Dam...broke... tell all the people to get out of the river bottom." Notices also went out to police departments in the region, the Red Cross, and the Southern Pacific Railway Company. Soon telephone operators, sheriff's deputies,

FIGURE 4-20. A 1930 photograph of the Bureau of Power and Light load dispatch center in Los Angeles. From this communication hub, belated word of the flood finally got out to telephone operators and authorities in the western Santa Clara Valley. [Los Angeles Department of Water and Power]

police officers, and residents were sounding the tocsin of the approaching danger.[44] Of course, once the flood had passed Castaic Junction there was no way for phone calls to connect westward down the Santa Clara Valley. Word had to spread eastward up the valley from western Ventura County, which was connected to Los Angeles via phone lines that extended over the Santa Susana Mountains through Moorpark and Camarillo.[45]

Had the flood come in the middle of the day, disseminating the news would likely have been easier. Still, many frantic calls made in the dead of night did get through, including one to Ventura County deputy sheriff Edwin Hearne at about 1:20 a.m. while he was on duty at the county jail. The emergency call from the Los Angeles Sheriff's Office minced no words: "Notify the people right away from Moorpark, Santa Paula and Fillmore...get the people out...the dam [has] been dynamited." (Early on, violent sabotage was a common—and eventually discounted—rumor.)[46] Hearne and his partner, Martin Jensen, sped first to Santa Paula, about eleven miles east of Ventura up Telegraph Road. Here they "met with the constable and told him to notify all the bigger places." They also "went on down the street blowing [their] siren." They then headed east to Fillmore where, about 2:20 a.m., they could find "no one in sight." They started ringing the fire bell to rouse the town. At 2:25, the bridge connecting Fillmore to Bardsdale gave way, accompanied by "crashes and roar of the water and noises and shrieks."

"I couldn't see anything," Hearne later recalled, "there [were] no electric lights on that night, not anything of that kind." At daybreak he could finally comprehend what had occurred. The town proper of Fillmore was spared because it hugged the north side of the valley. But Bardsdale lay low on the river's south side, directly in the flood path.[47]

Raymond Ransdell and fellow officer Carl Wallace followed a course similar to that taken by Hearne and Jensen. Upon receiving "orders to proceed eastward as far as we could and give warning of the danger," they headed to Santa Paula, reaching it at approximately 1:40 a.m. After alerting townsfolk, they too set out for Fillmore, arriving about 2:10. With still no flood in sight, they continued three-quarters of a mile farther east along Telegraph Road (now Route 126) until they encountered water covering the road. Forced back toward Fillmore, they raced to awaken residents until "a great body of water" stopped them in their tracks.[48]

No law officer gained greater recognition for heroic service that night than Thornton Edwards, a California highway patrolman assigned to Santa Paula. Alerted by telephone a little after 1:30 a.m. (or about an hour and forty minutes before the flood hit Santa Paula), he first warned his immediate neighbors and got his wife and son to higher ground. He then sped off on his motorcycle to rouse others in the slumbering valley, stopping at every third house and imploring residents to awaken their neighbors. When this tactic seemed too slow, he circled through the town with siren blaring, yelling to anyone who came outdoors to warn others of the danger. He was soon joined by Santa Paula motorcycle officer Stanley Baker. Others followed their example, including someone alert enough to trigger a fire alarm; the firemen who came to the station were then assigned to direct people to higher ground, away from the Santa Clara River.[49]

Edwards and Baker also carried news of impending danger to farmers and dairy ranchers in the nearby countryside. Edwards covered the low-lying areas west of town while Baker alerted farmers along the east-west thoroughfare of Telegraph Road. With his "siren shrieking," Edwards then returned to the lower district of Santa Paula, driving

FIGURE 4-21. West of Fillmore and Bardsdale, warning of the approaching flood began to spread. Damage to orchard groves in this district proved especially devastating. [DC Jackson]

> up and down the streets in the now devastated area until at last, speeding up River Street in the dark, a three foot rush of water caught him and put an end to his heroic ride. [Edwards] was upset, but managed to free himself from his motor cycle and saved his life. The motorcycle was not recovered until the water receded.[50]

In the coming days he was dubbed "Flood Revere" and a "motorized Paul Revere." The California Division of Highways later honored Edwards with a citation for bravery and the California Highway Patrol awarded him a medal for exemplary service.[51]

Also playing a vital role that night were the telephone operators who notified deputy sheriffs, patrolmen, medical personnel, civic officials, and myriad residents of the impending flood.[52] And once the high water passed, their work continued for many days in support of rescue and recovery efforts. One hero of the wires was Louise Gipe, night operator in Santa Paula, who arrived at her office as usual at midnight only to have the electricity fail forty-five minutes later. Scrambling for candles

FIGURE 4-22. Louisa Gipe posing at her switchboard in Santa Paula. She and other telephone operators helped sound the alarm throughout the lower Santa Clara Valley. For days afterward Gipe and her colleagues helped facilitate rescue and recovery efforts; they also helped to get information to anxious friends and loved ones residing beyond the confines of the valley. [Los Angeles Public Library Photo Collection]

and matches, she remained at her post. Although the long-distance lines to the east were dead, at 1:30 a.m. the chief operator of the Pacific Telephone and Telegraph Company reached her and warned of the water headed down the valley. Gipe then called Thornton Edwards and sent him off on his mission. She was later joined at her post by Bertha Clark to help with rescue and recovery communications. Throughout the morning and into the afternoon, the two women exchanged information with fellow operators about who was safe, who was missing, and whose loved ones had been identified in the flood zone and in the morgues. Other operators who logged long hours included Carrie Johnson and Myron Dressel in Fillmore, Ora Hill in Oxnard, Mabel Bradley in Moorpark, and Althea Marks in Saticoy. Among those at the phones in Fillmore was Ethel Basolo, who had left the telephone company several months earlier but returned to help at the exchange. Even after learning that her brother-in-law, George Basolo, had drowned near Bardsdale, she remained at her post.[53]

Santa Paula to the Sea

The flood swept through Bardsdale at 2:25 a.m. and after about twenty minutes had dropped to levels posing little danger. The warnings of officers Ransdell, Wallace, Hearne, and Jensen helped residents in the valley below Fillmore get out of harm's way and, from this point on, the human toll was significantly diminished. Nonetheless, the surge—still moving at about eleven miles per hour—could inflict serious damage.[54] Carrying an enormous amount of debris, it was described by Charles Outland as "50% water, 25% mud, and 25% miscellaneous trash."[55]

Outland, who was a senior at Santa Paula High School, witnessed the flood firsthand when it arrived shortly after 3:00 a.m.[56] Most Santa Paula residents had long since been warned, with many taking refuge on the elevated terrain near Main Street or in the hills north of town. Many years later, the renowned Southern California architect Zelma Wilson recalled hearing the police megaphones and evacuating with her family to higher ground:

> They were calling on everyone to get dressed as quickly as possible and go north to the hills because there was a flood coming through Santa Paula. They didn't know the size or nature of the disaster, but we should hurry...So we dressed as quickly as possible. As a matter of fact, I don't think we even dressed. We just kept our nightgowns on and put on our slippers and bathrobe and blankets. My mother and sister—I don't remember my father being there—and I rushed up to the hills. I remember being terrified, excited, thrilled,

> and all of the things that children are. I was around nine or ten years old. My sister was twelve...We went as high as we could until we came to many groups of people who were sitting on the side of the hill. We sat amongst them. There was lots of excitement, and the kids were screaming and hollering and were looking for the water. [Then] we heard a roaring, but it was something that you couldn't identify as being a flood. It was just a terrible roaring sound, and it wasn't that close to us. It was about a mile away. We stayed on the side of the hill, and the children were playing games or huddled around their mothers. Finally, the police came and said that there had been a flood in the south end of town in the riverbed, covering about half a mile swath through Santa Clara Valley, and that we were all safe.[57]

Wilson soon learned of the damage on the south side of town (her elementary school had been flooded and classes were suspended), and she later recalled an eventual "outbreak of typhoid." Though still a young girl she was aware that "many of the people in town were Mexican-Americans and were the ones who were killed because they were the poorest people in town and lived in riverbeds."[58]

A small crowd had gathered on the Willard Bridge in Santa Paula to witness the arrival of the floodwaters, but fortunately they were dispersed by patrolman Edwards in time to avert tragedy. Debris quickly piled up behind the truss bridge and all but one of the spans buckled, subsumed into the flow. As the bridge collapsed, the twelve-inch-diameter gas main supported by the structure ruptured and an explosion boomed across the town.

Much of Santa Paula, although covered by water, did not experience the full force of the surge at the center of the river channel. Among the damaged homes, many were simply lifted off their foundations and floated a few hundred yards before being set down when the flood passed. As Outland recalled, "one bungalow left Seventh Street, managed to float through the logjam [of other floating houses] on the [high school] grounds, and came to rest in the intersection of Olive and Ventura streets, four blocks away." The deep water had passed by 3:30 a.m., and within an hour sustained flow was confined to the normal channel.[59]

Santa Paula was the most populous community affected by the flood. Ten days later, the *Fillmore Herald* reported that Santa Paula's

FIGURE 4-23. Much of Santa Paula lies high above the riverbed, and the city was spared widespread destruction. But the flood ripped away most of the Willard Bridge, which provided passage to the south side of the Santa Clara River. [DC Jackson]

FIGURE 4-24. A low-lying section of Santa Paula after the inundation. [DC Jackson]

FIGURE 4-25. Most of Santa Paula's residents lived on high ground and were unaffected by the flood, but the city's "Mexican" community (as they were referred to in contemporary sources) lived closer to the river and suffered disproportionately. This photo, taken the day after the flood, shows a community in distress. [St. Francis Dam Disaster Papers, Huntington Library]

FIGURE 4-26. The day after: a Santa Paula family left homeless by the flood. [St. Francis Dam Disaster Papers, Huntington Library]

FIGURE 4-27. A building in lower Santa Paula that survived the flood; the height of the muddy inundation is marked on the walls. [St. Francis Dam Disaster Papers, Huntington Library]

FIGURE 4-28. A wooden house on Seventh Street in Santa Paula that was floated off its foundation. Taken the day after the flood, the photo shows a sheet of water still covering the paved street. [DC Jackson]

FIGURE 4-29. Santa Paula High School (in background) easily withstood the flood, but several nearby wooden bungalows were floated off their foundations and on to the schoolyard. [DC Jackson]

"damage zone" contained 273 homes with "135 all gone, 87 badly damaged...and 51 flooded." In addition, "163 families have been left without food, clothing or shelter for a total of 768 persons."[60] Fortunately, human casualties were modest, given the physical destruction. On April 6 the *Santa Paula Chronicle* reported that only sixteen residents of Santa Paula were known to have died in the flood.[61]

At approximately 4:15 a.m. the floodwaters, now slowed to about six miles per hour, reached the agricultural community of Saticoy, long empty of residents thanks to warnings through the night.[62] At the Saticoy Bridge—where sleeping hobos had been awakened in their makeshift camp—the rising waters submerged the bridge's flooring, but the structure held fast. In the days after the disaster it provided the sole auto route across the river channel in Ventura County.

Below Saticoy the natural floodplain of the Santa Clara River flattens, and the surge widened to two miles or more. Farmers and dairy ranchers had received timely notice to vacate the rural flats, and they retreated to higher ground, where they watched the last of the flood's debris-laden journey. At Montalvo two bridges—one carrying highway traffic near the site of present-day California Route 101 and the other a part of the Southern Pacific coast line—were the last two major structures lying between the flood and the ocean. Highway patrolmen blocked trains and automobiles and sightseers from crossing the

FIGURE 4-30. Below Santa Paula the terrain flattens out, and the floodwaters spread to a width of two miles or more. This photo of the highway between Ventura and Santa Paula was taken the day after the disaster. [*Los Angeles Times* Photographic Archive, Department of Special Collections, Charles E. Young Research Library, UCLA]

bridges, although one Pickwick Stage bus driver reportedly cursed mightily about the delay.[63] When the flood arrived about 5:00 a.m., several spans of the Montalvo highway truss bridge collapsed but, aside from damage to an approach embankment, the railroad bridge survived intact. A half-mile south of the Montalvo Bridge, the El Rio Maintenance Station of the California Highway Department stood in the flood path. Apprised of the dam failure about 2:00 a.m., foreman L. B. Prosper "rounded up his crew and stationed men at either side of the Santa Clara River [Montalvo] Bridge to warn motorists of the coming danger." While patrolling the bridge Prosper heard the onrushing water and

> fearing for the safety of his wife in the maintenance cottage, he rushed to her aid and they drove east in their car [to high ground] just as a four foot wall of water struck the El Rio Maintenance Yard...the back fence was toppled over and mud and debris was deposited everywhere. Road equipment was buried hub deep in

FIGURE 4-31. After five hours the flood finally approached the Pacific Ocean. Located a few miles from the ocean (and close to where modern-day Route 101 crosses the Santa Clara River), the Montalvo Bridge was the last major structure damaged by the surge. Thanks to highway patrol officers, passage across the span had been blocked, and no one died when several steel trusses were knocked from their piers. [DC Jackson]

> mud both in the garage and out in the yard. The muddy waters filled the cottage cellar and coming to within six inches of the windows leaked through the closed doors, flooding the interior of the house. Damage at the Maintenance Yard was estimated at $2,000.00.[64]

At the El Rio yard no lives were lost and the damage was slight (although filled with water, Prosper's cottage remained on its foundation). The once raging torrent had devolved into a lumbering cascade of great width but modest force. At 5:25 a.m. the flood's front edge passed into the Pacific Ocean. Perhaps with a bit of hyperbole, the *Fillmore Herald* described the sea as "brown from mud for a distance of six or seven miles out in the ocean and for a width of several miles, while the water in the muddy area was thick with driftwood and all kinds of debris."[65] By the time the once pristine water stored behind the St. Francis Dam reached the sea, it was a squalid concoction of mud, brush, trees and branches, fence posts, wire, structural debris, animal carcasses, and at least a few human remains. At dawn's light the flood itself may have passed into history, but not so the human tragedy.[66]

Fast Cars, Inoculations, and the Red Cross

Among survivors found on the morning of March 13, several were stranded and shivering on sandbars, clad in torn and mud-caked nightclothes.[67] Some were completely naked and many, like Sisto Luna, had only narrowly escaped death. Luna and her three children, the youngest less than a month old, were in their home on the outskirts of Santa Paula when the flood carried it away. Clinging to a feather mattress, they floated two miles until caught in a tree branch, where a rescue squad found them the next morning. Nearby, another woman learned that her husband, after reportedly swimming and floating for nearly a mile, had safely reached dry ground with their six-month-old child still in his grasp. These exhausted and frightened survivors were among at least fifty-six persons aided by rescue workers near Santa Paula the morning after the flood.[68]

Early on March 13, hundreds of law enforcement officials were dispatched from Los Angeles to the Santa Clara Valley. Officers traveling in "fast cars" via Newhall Pass to San Francisquito Canyon, Saugus, and Castaic Junction aided rescue work in the flood zone closest to the dam. To the west in the lower valley, forty-three members of the Los Angeles Police Department's Detective Bureau were deployed to Saticoy, site of the only surviving highway bridge across the Santa Clara River.[69] By daybreak "more than 600 deputy sheriffs, police officers, and State motor-vehicle officers had arrived on the scene and Sheriff Traeger and his chief deputies Eugene Biscailuz and William Bright, began the organization that was to carry on the work of rescue and maintenance of order" in the Los Angeles County sector of the flood zone. Speed was of the essence, and Biscailuz advised the press that "every available deputy sheriff who worked today will be pressed into service to renew the search tomorrow morning at daybreak."[70] All available employees from the Bureau of Water Works and Supply and the Bureau of Power and Light were also pressed into duty, and "by daylight practically all of the automotive equipment of both bureaus had either departed for the devastated area or were waiting for instructions. The warehouses of both bureaus had already sent out such material as telephone wire, axes, rope and equipment to pull the debris apart as soon as it was light enough to work."[71]

Among those joining the early volunteers, city employees, and law enforcement officers were state health personnel bringing medical supplies and sanitation equipment. The threats to health were many, with "scores of dead horses, cattle and fowl [lying] half-buried in the sand,"

FIGURE 4-32. The morning after the flood, a truck transports dazed Edison employees, survivors of the tragedy at Kemp. A lingering question is why Edison employees at the Saugus substation (hit by the flood at about 12:30 a.m.) were unable to warn their brethren at Kemp before the camp was swamped at 1:20 a.m. [Southern California Edison Photographs and Negatives, Huntington Library]

as well as stray dogs. As the *Los Angeles Times* reported three days after the flood, "State health officials arrived in the stricken area with gallons of disinfectants which were spread over dead animals in the river. Scores of stray dogs, preying on [animal] bodies in the river bed, were shot yesterday near Piru." The decision to shoot abandoned dogs was no doubt justifiable on public health grounds, but it underscored the fear that gripped the valley.[72]

Under the direction of Dr. Walter M. Dickie, director of the California Department of Health, with assistance from Santa Paula physician D. Henry Wyatt, a massive typhus inoculation program was underway within forty-eight hours of the flood's passing.[73] But this program also sparked agitation, at least among the valley's Mexican community. As the *Los Angeles Times* reported on March 18, "Last night more than 250 people who were touched by the flood waters when the septic tanks [in Santa Paula] went out had been inoculated against typh[oid]. Several Mexican families who revolted against the inoculation procedure were advised that it was necessary in order to 'have eats.' The 'strike' was soon called off."[74]

Red Cross chapters from Los Angeles, Ventura, and Kern counties quickly dispatched personnel to the scene, not only providing first aid

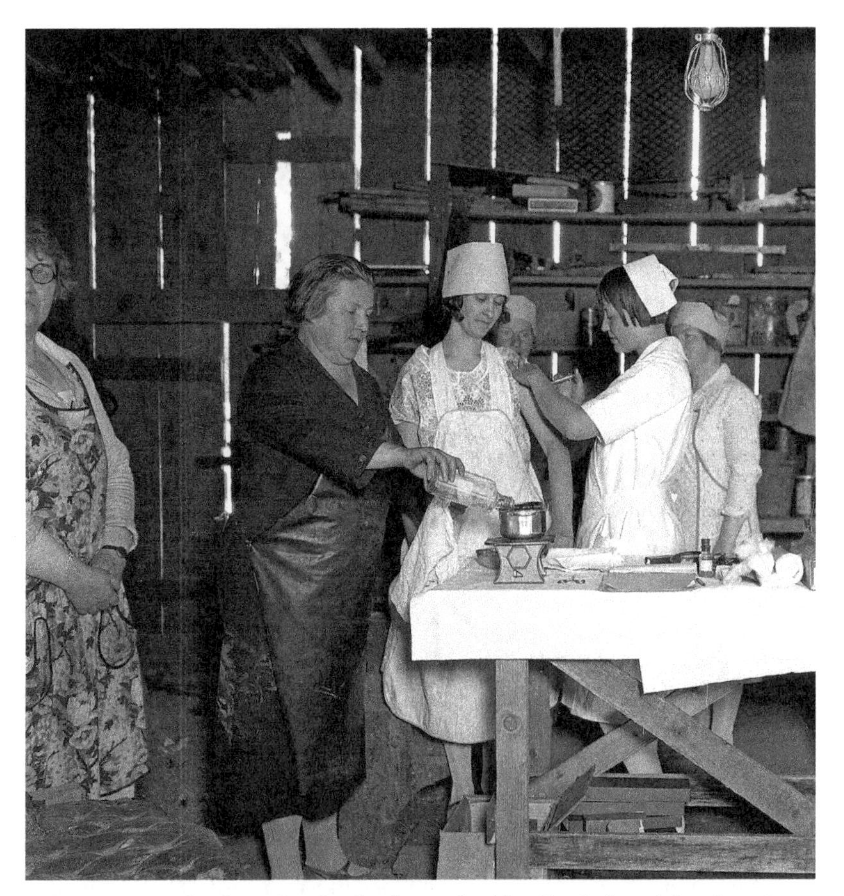

FIGURE 4-33. In the days after the flood, state health officials feared that decaying animal carcasses and untreated sewage would contaminate the valley's water supply. This anxiety over public health gave urgency to the effort by the Red Cross to inoculate survivors and rescue crews against typhoid. [*Los Angeles Times* Photographic Archive, Department of Special Collections, Charles E. Young Research Library, UCLA]

but also setting up temporary housing and kitchens, supplying clothing, and assisting in the recovery and identification of bodies. A Red Cross field headquarters was established at Newhall on March 13 and soon J. W. Richardson, the assistant national director of disaster relief, arrived to oversee the organization's work.[75] Red Cross staff members were joined by volunteers from other organizations, including the Salvation Army, the Catholic Welfare Bureau, and the American Legion. As will be detailed in chapter 5, donations poured in from an array of private groups and individuals to support the relief effort.[76]

FIGURE 4-34. Hundreds of people helping in rescue and recovery operations were fed round-the-clock by Red Cross volunteers. [*Los Angeles Times* Photographic Archive, Department of Special Collections, Charles E. Young Research Library, UCLA]

An "emergency canteen" was in place in Santa Paula on the morning of March 13 and by the next day the Red Cross was meeting the needs of hundreds of people left homeless. Within a week after the flood,

> 300 refugees were moved into a tent city located in the eastern section of Santa Paula. The little city has its own commissary, kitchens, outdoor shower baths, and other conveniences. The tent city [is to] be operated and care for the flood refugees until rehabilitation work is completed.[77]

The Red Cross also provided nourishment for the growing army of rescue and recovery workers. The canteen operated on a twenty-four-hour schedule, "using in all [a staff of] about 80 Americans and 12 Mexican women and 6 men per day. Teachers of grammar schools took charge

FIGURE 4-35. Santa Paula had the greatest number of evacuees forced from their homes by the flood. On the outskirts of town, the Red Cross erected a substantial tent camp to sustain families as they awaited the rebuilding of permanent housing. [*Los Angeles Times* Photographic Archive, Department of Special Collections, Charles E. Young Research Library, UCLA]

of work from 6pm until 6am." The meals provided "consisted of all food served to laborers in devastated areas, refugees in camp, workers at headquarters, policeman and guards, nurses and patients in Clinic Hospital, sick and well refugees housed with friends, children in supervised camp on school grounds, and [the] day nursery." In addition, the canteen had "bottled water sent to all crews along [the] river bed" and to districts where water facilities or pumping plants were inoperable. This effort continued for ten days before conditions in the valley returned to a level approaching normalcy.[78]

The flood wreaked havoc on the valley's transportation and communication infrastructure. Roads, highways, and bridges were washed out; tracks and railroad bridges of the Southern Pacific line near Castaic Junction were upended; telephone and power lines from Saugus to Santa Paula were in shambles. Hundreds of linemen from the Pacific Telephone and Telegraph Company and Southern California Edison canvassed the area, replacing downed wires and poles and restoring service as quickly as possible.[79] And, as directed by Governor C. C. Young—who had rushed to Santa Paula on the evening of March 13—the state Department of Public Works quickly authorized emergency funds to repair highways and to assist in relief work.[80]

For days after the flood, curiosity seekers converged on the disaster zone by automobile and impeded the movement of both rescue workers and supply shipments. Recognizing that there were legitimate reasons to visit the valley, the Los Angeles sheriff's office started "issuing passes through lines only to those seeking some trace of lost relatives or to survivors to return to their former homes in the district affected."[81] The first weekend after the flood (March 17–18), "thousands of sightseers were turned back on all highways leading to the devastated area...Hundreds of motorists, failing to heed the warnings of the officials, attempted to break through the cordon thrown around the section by police and traffic officials."[82] To discourage gawkers, police in Ventura County set up roadblocks at Camarillo and Moorpark to turn back most automobiles headed from Los Angeles to Santa Paula, Saticoy, and other parts of the lower valley.

Motorists with valid reasons to enter the flood zone were to be allowed access, but persuading officers to issue a pass could prove difficult. For example, Mrs. J. R. Schwartz, a resident of Santa Paula, had been returning from a trip to Phoenix when the dam broke. Her two daughters had remained in Santa Paula, and she was unable to get through by phone. In the company of two friends from Banning, she headed off by auto to find out if her children were safe. The *Banning Herald* reported on her odyssey:

> Arriving at the boundary of the district that is being policed, the Banning people [the Schwartz party] were told that they could not cross. They were asked to go to certain towns to view the bodies in an endeavor to find the Schwartz children. It was after great difficulty that they were permitted to cross and at three o'clock in the morning the party arrived at the Santa Paula home. Their home is on high ground on the river bank and the children were not [even] awakened by the roar of the flood."[83]

Mrs. Schwartz's fears proved unfounded, but many others were dealt a different hand.

Gathering the Dead

The emotionally charged task of gathering the dead began immediately, with fear of disease fueling the search. The goal was to transport recovered corpses to morgues for identification and notification of next of kin, a task made easier by ambulances and by "relief cars" brought to the scene by the Automobile Club of Southern California.[84] Within two hours of sunrise hundreds of deputy sheriffs, police officers, and local volunteers were scouring the area from San Francisquito Canyon westward to the lowlands of the Oxnard Plain. As the *Los Angeles Times* reported on March 14:

> Bodies of victims lined the banks [of San Francisquito Canyon]...Squads of deputy sheriffs and police officers had great difficulty in bringing the bodies to places where they could be placed in ambulances...Banks were caving before the still-running waters. Deep sand and mud hampered efforts to retrieve the bodies. All along through the canyon observers saw men bringing forth victims in improvised stretchers. Here was a small baby whose funeral cortege was two brawny hip-booted men. Its shroud a rug, probably from its former home, slung between poles to support the body. Behind the cortege in many instances came men carrying hastily made canvas and rug stretchers bearing, perhaps, the father or the mother. All the victims were badly battered.[85]

Among those rushing to help in Ventura County was Walter Stephens, who had recently moved to Southern California from Blount County, Tennessee. At the time of the disaster he was living in Los Angeles, but fellow migrants and friends from back East had settled in the Santa Clara Valley. "I received the news of the flood just at daylight by radio," he recalled in a March 28 letter to the local newspaper in Marysville, Tennessee. "Knowing a number of Blount County families [who] lived in the flooded district or near it (the Hitches, Davises, Waters, and others of my old friends from Old Blount), I was soon in my car and on my way to the scene taking a load of men to help in the rescue." Among other tasks, he wrote, "it was my sad lot to help in [recovering the] bodies" of two relatives, Mary and Grace Stephens, a mother and daughter, who drowned in the flood at Fillmore. "Oh, you don't know what a terrible sight it is, an awful tragedy which you may be glad you did not witness."[86]

FIGURE 4-36. Throughout the fifty-three-mile-long flood zone, crews searched for the dead. [DC Jackson]

FIGURE 4-37. In the upper stretch of the flood zone, bodies that lay atop wide stretches of barren sand residue were quickly recovered. But below Saugus the task became more difficult, as uprooted orchards often entombed the bodies. [DC Jackson]

FIGURE 4-38. A body is discovered within the dense, entangled debris. [St. Francis Dam Disaster Papers, Huntington Library]

FIGURE 4-39. Shorn of their nightclothes by the flood, many victims were found naked. [DC Jackson]

FIGURE 4-40. A body carried by hand from the riverbed. [DC Jackson]

FIGURE 4-41. Once recovered, victims were transported to the nearest morgue. [DC Jackson]

FIGURE 4-42. The search for bodies could become very personal. Joe Gottardi was a Newhall Ranch tenant who farmed the Blue Cut tract near Camulos Ranch. His wife, three sons, and two daughters were lost in the flood; the bodies of all but daughter Lenore were identified and buried in the cemetery at Piru. This photo, taken a week and a half after the flood, shows Joe digging intently in the debris in search of four-year-old Lenore. There is no evidence that her body was ever found. [SCVHistory.com]

Ventura County Coroner Reardon cautioned that "persons who may have occasion to handle bodies...[should wear] rubber gloves as the danger of infection has now become very great. Rubber gloves are provided for this purpose at all of the different flood relief headquarters."[87] Digging into the piles of debris left by the flood was grueling work and, as the days passed, fatigue took its toll. A later report described how American Legion volunteers carried out the search in Ventura County, where

> groups of men stayed out for periods varying from four to eight hours and were then relieved so they could snatch a cup of coffee, some food, a short nap, and be back again in the harness, giving relief to some other group in another sector. Back and forth across

> the devastated area they pushed through the mud and mire...There was present always, the ghastly expectancy of happening upon a lifeless human body with its face buried in the mud.[88]

How rescue and recovery teams dealt with exhaustion and psychological stress is largely unrecorded. However, after days spent searching for bodies, the crew at Bardsdale apparently required strong fortification to keep their spirits up. Or at least they did in the eyes of Ventura County Sheriff Robert Clark, who purchased three pints of whiskey from Santa Paula druggist Lester Tozier to help soothe their nerves. In 1928, Prohibition was the law of the land and only a prescription from a medical doctor could have made such a purchase legal. But Sheriff Clark took this extraordinary step on his own authority, later averring that the whiskey was "delivered to me for use in the Bardsdale District by men searching for bodies under conditions where artificial stimulation was necessary to carry on the work."[89]

As the cleanup continued, heavy machinery helped in clearing huge debris mounds that defied the labors of men armed with shovels and picks. Stretching for about "50 square miles from Castaic to the sea... there stood mountains of debris—tangled masses of willow and oak, of wrenched bridges and vanity dressers, of automobiles and farm machinery, of orange trees and sage brush, of human beings and animals, piled into heaps sometimes twenty feet high and five hundred feet long."[90] Mechanizing the work expedited the recovery effort, but machinery complicated the delicate search for human remains. In fact, the less-than-careful use of heavy dredges and clamshell buckets eventually prompted Ventura County civic leader C. C. Teague to complain to Los Angeles authorities:

> There should be one man stationed at each dredger to constantly watch the pile to see that any bodies are uncovered before the clam shell takes the next load from the pile. In a number of cases, by reasons of not having this inspection, the dredger has badly mutilated bodies before they were discovered...I would recommend that you authorize the putting on of two local men who will be stationed to continually watch for bodies as the dredger does the work.[91]

Early on, before mechanized equipment was available, there was little fear that bodies of the victims would be inadvertently mutilated by overzealous searchers. Tremendous clumps of debris—uprooted orange, lemon, pecan, walnut, oak, and willow trees embedded in huge deposits of sand and silt—posed the greater problem. Some bodies were quickly discovered on the surface of the floodplain, but unearthing many others required arduous digging. Heavy equipment changed the character of the clearing and recovery process, but there always remained a need for careful on-the-ground inspection.[92]

By nightfall on March 13 some sixty bodies had been brought to the improvised morgue at Newhall. In happier times the small building had served the community as a Masonic lodge and a dance hall, but it was now filled with "rough boards, mounted on chairs, benches, or anything available [as] the resting places for the silent, sheet-covered rows of the dead."[93] The anxiety of those compelled to come to the morgue was captured by a *Los Angeles Times* reporter:

> Across the street from the Sheriff's substation in Newhall hundreds of friends and members of families gathered about the morgue and receiving hospital, each within a few yards of the other, hoping for the best, but concentrating more on the morgue than the hospital, looking for friends or members of the family lost in the catastrophe. Few injured were found. There were mostly lifeless bodies.[94]

The *World* newspaper described the Newhall morgue as a dance hall pavilion where "dingy decorations left over from Saturday night's festivities still adorned the bare interior. Streamers and lanterns hung from the rafters and above the door a huge faded floral sign bade the searchers an ironic 'Welcome.'" Here, "nearly fifty bodies lay in rows on tilted boards," awaiting apprehensive visitors:

> In a single file the quiet line moved slowly through the hall between the rows of sheeted figures, each of which had one foot extended from under the shroud. The only sound was the shuffling of feet on the wooden floor and the occasional gasp as someone in the line recognized the face of a relative or friend. Then a tag

FIGURE 4-43. The makeshift morgues in Newhall, Santa Paula, Fillmore, Moorpark, Oxnard, and Ventura were scenes of wrenching heartbreak, as family members and loved ones trudged through, fearing the worst while also hoping to bring closure to their torment. In Newhall, the former Gully/Swall General Store (known in the 1920s as the Hap-A-Lan dance hall) on Railroad Avenue at Market Street was converted into an ad hoc mortuary. [*Los Angeles Times* Photographic Collection, Department of Special Archive, Charles E. Young Research Library, UCLA]

> would be attached to the protruding ankle and the line would resume its slow march.[95]

Makeshift morgues were also set up at Moorpark, Fillmore, Santa Paula, Oxnard, and Ventura as victims were discovered and exhumed in the lower sections of the watershed. As at Newhall, the magnitude of the disaster quickly overwhelmed morticians in the western valley's small towns. One of the first post-flood editions of the *Santa Paula Chronicle* ran the headline "Fillmore Funeral Chapel Stacked with Bodies of Victims of the Flood," reporting that:

> With the dead stacked in piles, the French & Skillin Chapel in Fillmore today took on a grim aspect as the silent work-

FIGURE 4-44. Fearful friends and family members gather outside the Newhall morgue, with most turning their faces away from the *Los Angeles Times* photographer. Not long after the flood, the wooden frame building housing the Newhall morgue was torn down. [*Los Angeles Times* Photographic Archive, Department of Special Collections, Charles E. Young Research Library, UCLA]

> ers hurried back and forth bringing in their gruesome burdens. Many of the dead remain as yet unidentified, although the bodies are being washed and every possible opportunity is being given to friends and relatives who fearfully pass by glimpsing at the distorted faces.[96]

When bodies could be identified, the task of notifying far-distant next-of-kin often fell to morgue personnel, sometimes through heartbreakingly terse telegrams. The afternoon following the flood, this is how the Weinland family in Banning learned of their son's death:

> Power House Washed Out Completely by Break of Dam Above Wm Wineland and Family Drowned Have Identified Body William
>
> [signed] Dr Sarah L Murray[97]

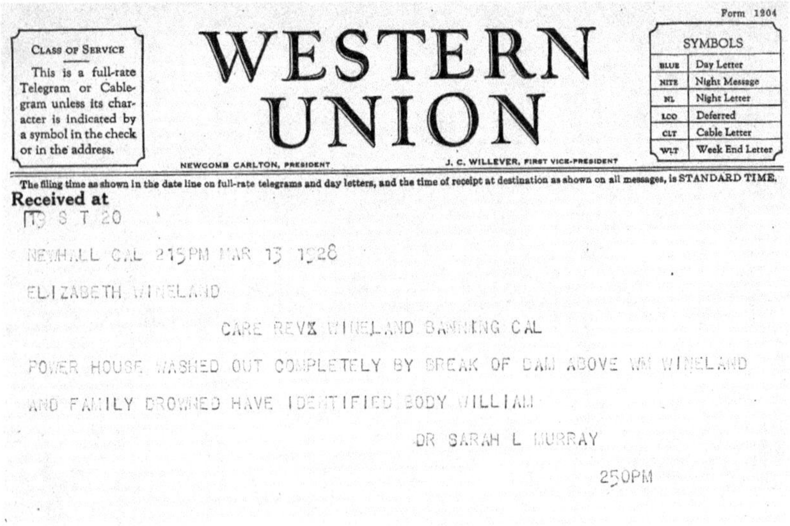

Form 1204

CLASS OF SERVICE

This is a full-rate Telegram or Cablegram unless its character is indicated by a symbol in the check or in the address.

WESTERN UNION

SYMBOLS	
BLUE	Day Letter
NITE	Night Message
NL	Night Letter
LCO	Deferred
CLT	Cable Letter
WLT	Week End Letter

NEWCOMB CARLTON, PRESIDENT J. C. WILLEVER, FIRST VICE-PRESIDENT

The filing time as shown in the date line on full-rate telegrams and day letters, and the time of receipt at destination as shown on all messages, is STANDARD TIME.

Received at

[illegible] S T 20

NEWHALL CAL 215PM MAR 13 1928

ELIZABETH WINELAND

CARE REVX WINELAND BANNING CAL

POWER HOUSE WASHED OUT COMPLETELY BY BREAK OF DAM ABOVE WM WINELAND AND FAMILY DROWNED HAVE IDENTIFIED BODY WILLIAM

DR SARAH L MURRAY

250PM

FIGURE 4-45. Within hours after the flood, this terse telegram was sent to Elizabeth Weinland in Banning, California, with the news that her son William's body had been identified at the Newhall morgue, and that his wife and child were deemed "drowned." Elizabeth and her husband, William, preserved the telegram and later donated it to the Huntington Library along with other papers related to the disaster. [William H. Weinland Photograph Collection, Huntington Library]

Families waited for news, any news, about the fate of missing loved ones. Gladys Harnischfeger, the former wife of dam keeper Tony Harnischfeger, rushed to the Santa Clara Valley in hopes of, at the very least, recovering the body of her son, Coder, who had been living with Tony in a cottage below the dam. Almost two weeks passed and she still held vigil at the Newhall morgue, where a reporter for the *Los Angeles Record* captured her anguish: "We'll never stop looking. They [morgue attendants] have promised not to bury him unless I am there. I must find my little boy." She paid heartfelt tribute to the volunteers at the morgue and in the field: "Nothing was too dirty for them to undertake in the salvaging of human bodies from the frightful wreckage...they have been a great comfort to me in these long, weary, watching days." There is no record that her son's body was ever found.[98]

Gladys Harnischfeger was hardly alone in her plight. She was simply among the most visible of those who grieved and prayed for word of a loved one. After one body was found in the surf at the mouth of the Santa Clara River, the *Fillmore Herald* reported that "the American Legion of Ventura requisitioned eight small craft to

FIGURE 4-46. Interior of the Newhall morgue. The shrouded bodies lay on slanted boards, awaiting identification. [*Los Angeles Times* Photographic Archive, Department of Special Collections, Charles E. Young Research Library, UCLA]

patrol the water" in hopes of recovering more who might "eventually be floated back to the beach."[99] And, a week and a half after the flood, members of the Weinland and Imus families in Banning came to the Santa Clara Valley hoping to retrieve the body of their young cousin, who had lived with his parents at San Francisquito Power House No. 2. Their bleak and ultimately fruitless journey was described in the *Banning Record:* "Yesterday [March 21] Elmer and Earl Imus made a search through the several morgues in the district seeking to identify the remains of little Lloyd Weinland. Lloyd is the only member of the [family] who has not been found and identified."[100] As with Coder Harnischfeger, there is no record that Lloyd Weinland's body was ever recovered.

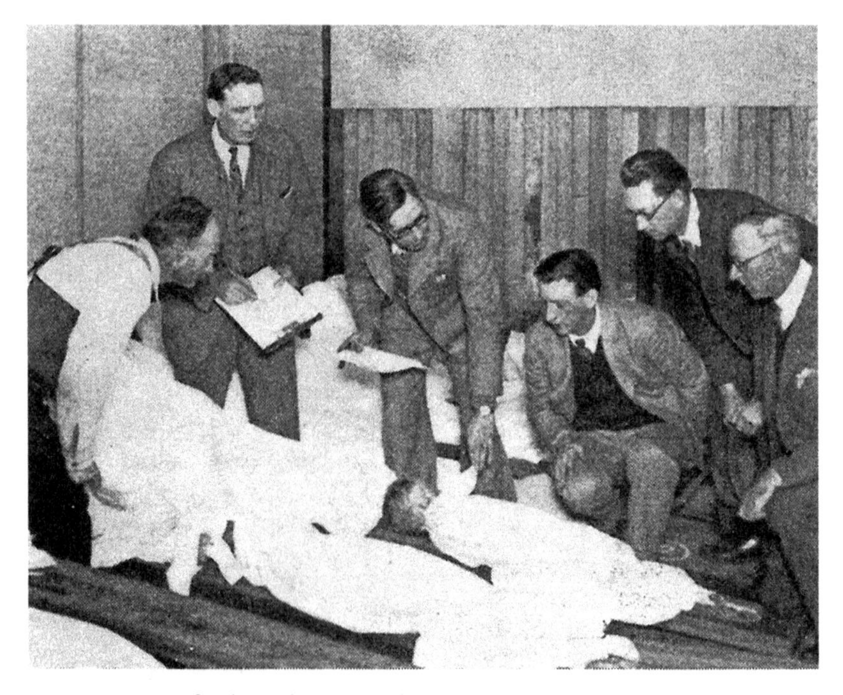

FIGURE 4-47. Officials confer over the body of a young child in the Newhall morgue. [*Literary Digest*, 1928]

Inquests and Victim Identification

Civil authorities convene coroner's inquests to investigate possible crimes when citizens die in unnatural circumstances. These inquests can also aid in the identification of victims and help establish reliable statistics relating to disasters both large and small. In the case of the St. Francis Dam disaster the proximate cause of death for victims was not in dispute: they died because a man-made tsunami killed them. Perhaps death came from water-filled lungs or perhaps it came from a knock on the head delivered by a floating fencepost. It didn't matter. The precise cause was moot in the various coroner's inquests, replaced by other issues.

The flood had originated in Los Angeles County, and there was no doubt that it was caused by the collapse of a large dam built by the city of Los Angeles. Thus Los Angeles County Coroner Frank Nance—along with the city attorney, the district attorney, and myriad civic leaders—was interested not so much in how individual victims died but rather in why Mulholland's dam failed in the first place (see chapter 6 for details).[101] In contrast to proceedings in Los Angeles, the Ventura County inquests focused on ascertaining the identity—and by extension the number—of bodies scattered for some thirty-five miles along the Santa Clara River floodplain. Mindful of the great distance traversed by the flood, Ventura County authorities ordered separate coroner's inquests at Fillmore and Moorpark to the east and at Santa Paula, Oxnard, and Ventura to the west. These were solemn affairs, with the *Moorpark Enterprise* admonishing citizens "who may be attracted solely out of morbid curiosity" to "remain away as their presence will only hamper the work of the officials who already have a great demand on their time."[102]

Ventura County District Attorney James Hollingsworth clearly understood the differing challenges faced by officials in Ventura and Los Angeles Counties. At the Fillmore inquest on March 15, he declared: "There are a great many bodies being held at various places in this county...[and] there is no opportunity at this time to go into any extended investigation to determine why this particular dam broke...[Our purpose] is to identify as many of these bodies as possible so they may be released to their friends and representatives." As for "those that are left unidentified or unclaimed...Every body has been photographed, tagged, and catalogued, and the photographs when developed will be filed...so that anyone who comes along subsequently will have an opportunity to examine these photographs for the purpose of making identification."[103]

As is typical with destructive river flooding, most victims found at a specific location had not lived there. For example, only two of the victims found near Fillmore and covered by the Fillmore inquest were residents of the town. Ten of the thirty-three victims whose identity could be established in Fillmore were residents of Los Angeles County: four from Power House No. 2 (over twenty-five miles upstream from Fillmore), two from Harry Carey's ranch, two from the Newhall Ranch near Saugus, and two from Castaic Junction. Most of the rest could be traced to eastern Ventura County: six workmen from the Edison camp at Kemp, seven residents of Piru, and one from Bardsdale. The identity of thirteen others remained undetermined, with six entered into the record simply as "Mexican."[104] The proceedings at Santa Paula and at the three other inquests held in Ventura County followed the template established at Fillmore.[105]

Ventura County Coroner Oliver Reardon understood the difficulties posed by rapid corporeal deterioration in the water-saturated debris field. After only a few days, victim identification was difficult, if not impossible, absent distinguishing jewelry, clothing, tattoos, or body features. On April 10 Corrine Cowden's body was discovered in a "big drift of debris on the Baker Ranch" about a mile and a half below Bardsdale,

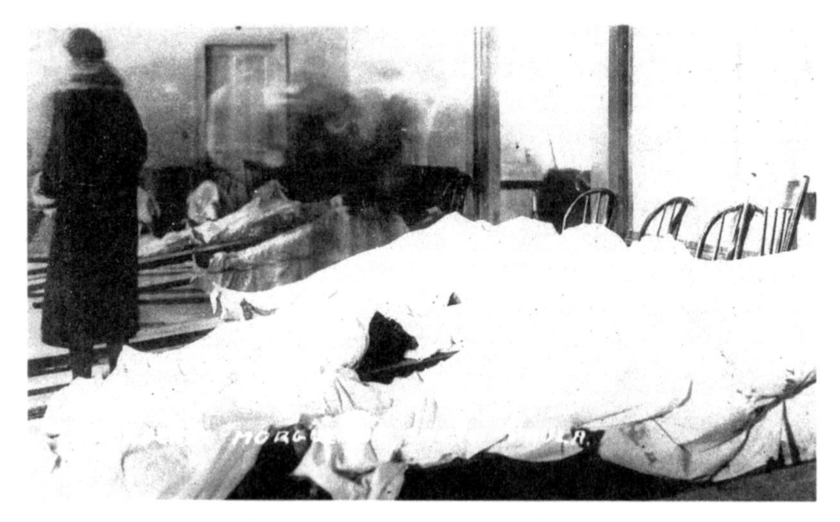

FIGURE 4-48. In an era before DNA technology, visual identification was the primary means of attaching names to victims. Morgue attendants needed to quickly create photographic records to aid identification, since bodies rapidly deteriorated in the absence of refrigeration. This dimly lit photograph captures the somber interior of the makeshift morgue at Santa Paula. [DC Jackson]

and identification was possible only because of "a ring and a necklace she was wearing at the time."[106] Ina Weinland had lived with her husband William at Power House No. 2 and, while his body was discovered the morning after the flood and quickly identified, a full week passed before her body was recovered.[107] And even then, as the *Banning Record* reported: "Identification [of Ina Weinland] was not fully settled upon for some time. However, the remains have been accepted as those of the departed...burial will take place at Banning tomorrow."[108]

Before DNA analysis, and at a time when it was impossible to preserve remains for more than a few days, photography of the dead was an essential component of morgue work. As directed by the district attorney, photographers fastidiously documented each victim brought to the morgues. On March 23 the *Ventura Free Press* related how a body "found in the stream of the Santa Clara River, partially lodged in the mud" would "never be identified," in the view of Coroner Reardon, because of "decomposition having gone too far." Nonetheless, "a [p]icture for the files will be taken at once and the body will be buried with the other unidentified."[109]

Through April and into the summer, bodies continued to be unearthed, some identifiable and others not. For example, under the headline "Three More Bodies Are Recovered," the April 6 *Santa Paula*

Chronicle reported that the remains of Harold Kelley of Bardsdale and Eddie Ritchie, an Edison employee staying at Kemp, had been discovered; a third "unidentified body...was taken to the Reardon undertaking parlors at Ventura."[110] The next day the *Ventura Free Press* reported two other unidentified fatalities, one a "young boy between age 14 and 17. Dark brown hair...stocky build. He wore a pair of blue dungaree trousers and brown belt with brass buckle. His teeth were rather large. He was found yesterday afternoon in Sespe and will be buried this afternoon." The other body was "of a man at least 45 years of age," with brown hair, weighing 145 to 155 pounds and distinctive because "the left side of the body was paralyzed during life...No clothing was on the body when it was found. The body was buried yesterday."[111] And when at the end of March "the sea returned [a corpse] to the sands on the Silver Strand beach at Oxnard," county authorities deemed identification impossible and the body was quickly buried.[112]

The desire to help grieving families reach closure inspired many of the crews and officials working in the flood zone. Sadly, there were instances in which the search for the missing encountered resistance from landowners who did not wish their property disturbed. The most dramatic instance was a confrontation between Jesus Torres and absentee landowner Richard Storke. The families of Torres and Hipolito Cerna resided on a small tract in Santa Paula's lower agricultural district. When the flood hit, several of their relatives (and Hipolito himself) died as the water poured through the citrus groves. Some of their bodies were quickly recovered, and Torres believed the others were buried in debris on an adjacent orchard owned by Storke. A Red Cross official described their quest: "Senor Torres and Senor [Pedro] Cerna believe that the bodies of their families were washed below their two acres of land and desire to have the land cleared and searched for them."[113] The cleanup crews were more than willing to clear Storke's land but he refused to allow access. The impasse lingered into the summer when officials in Ventura County wrote Storke at his home in Santa Barbara, asking him to reconsider:

> In settling Death claims resulting from the St. Francis Dam disaster, we have met with great deal of bitterness on the part of certain Mexicans [in Santa Paula]...They state that they believe that there are still bodies of their dear ones in the debris in the river bottom. We understand that Mr. McNab, Chairman of the debris removal committee, spoke to you in regard to this matter, but

> that at the time you preferred that the clam shells not be allowed to go on your property for the purpose of removing the debris to ascertain whether or not there were any bodies in it. You will understand that people are somewhat sensitive about a matter like this particularly the Mexicans who have great family affection and are particularly anxious to recover the bodies of their dear ones. If you could reconsider your decision to allow this work to proceed in the river bottom where the debris is, I am sure that such action would have a very beneficial effect.[114]

Storke promptly replied, but offered nothing to placate Torres or Cerna. While acknowledging that there was "a Mexican family living just off the ranch that was very seriously affected," Storke protested that "it would be impossible to identify any remains of his family. They were in bed at the time of the disaster, and were devoid of any identifying marks, if found." For Storke, it was all a matter of good business:

> I purchased from J. B. Taylor two years ago some 170 acres of the river bed, in order that I might have the protection the willows gave to my orchard [from seasonal flooding]. I am unwilling that these willows shall be torn loose from the lands, in fruitless effort to uncover dead after they have been decaying for some four months...I have some 300 acres in the river bed and these dead [bodies], if they be on my lands, are as likely to be in one place as another. It would hardly be desired to dig up some three hundred acres, to discover if by any possibility any of them may be found therein.[115]

He concluded: "Regretting that we cannot see the conditions from the same point of view," and signed off. Storke closed the door to compromise, leaving the Torres and Cerna families with little more than bitter regret. Officials in Ventura County were also disappointed with Storke's intransigence but they felt stymied. As C. C. Teague explained to J. W. Richardson of the Red Cross:

> The City [crews employed by Los Angeles] started in at work turning over the piles of debris on the Storke Ranch with their big clam shell outfits and were stopped by Mr. Storke...The City then withdrew their heavy machinery as their legal department [was] of the opinion that they had no right and could not force Mr. Storke to allow them to proceed with this work on his property. Probably if the Coroner actually knew the bodies were there he could force some sort of action, but the attorneys are of the opinion that where it is only a matter of surmise they would not have the legal right to enter the property and search for bodies. Storke is a very peculiar man and one that no one can influence; so I doubt if anything more can be done in the matter.[116]

How Many Died?

Storke's intransigence presents an extreme case, but it helps highlight the difficulties of precisely enumerating the flood victims. Amid the confusion of the immediate aftermath, newspapers were inclined to trumpet inflated numbers of both fatalities and people believed missing. On March 14 the *Los Angeles Times* was relatively conservative in headlining "200 Dead, 300 Missing...in St Francis Dam Disaster," while the *Chicago Tribune* blared, "Flood Deaths Near 1,000," and an Associated Press wire story claimed, "A rapidly mounting death toll...Tuesday night showed 275 had lost their lives, while upwards of 700 persons were reported missing."[117] By the next day, reports of the numbers were more restrained, with the *Los Angeles Times* informing its readers that the "death list of disaster now stands at 192 victims; 169 missing but many may be found alive." More brashly, the *Chicago Tribune* claimed that "the known dead in the disaster [had] reached a total of 305. Three hundred men, women and children are still missing."[118]

Of the two, the March 15 assessment by the Los Angeles newspaper better reflected what was being discovered in the flood zone. The Ventura County coroner inquests, undertaken during March 15–16, compiled a total of 126 deaths, a relatively small number compared to earlier estimates.[119] The same pattern held true for the Los Angeles County inquest, which, on March 21, revealed that only sixty-nine victims had been recovered in the territory east of the Los Angeles–Ventura county line.[120] While these numbers did not reflect presumed victims reported as missing, they nonetheless pointed toward a death count far lower than initial expectations.

Because the Ventura County inquests were held within three days of the flood, many victims in the lower valley were yet to be discovered. On Monday, March 19, the *Los Angeles Times* updated its totals:

> Late last night complete figures show 273 bodies recovered. Of this number 227 have been identified while forty-six await identification in the morgues at Newhall, Santa Paula and Fillmore. The number of missing, now believed dead, stands at 177, making the total dead and missing in the disaster 450, according to last night's check.[121]

A week later the Governor's Commission investigating the disaster stated, "The record of known dead at this time [March 25] is 236 and 200 are still missing."[122] Because the findings of the Governor's Commission were so broadly distributed (see chapter 6 for details), a consensus took hold that the death toll stood somewhere in the mid-four-hundred range. Reinforcing this view, on March 26 the *Los Angeles Times* reported that the flood had "brought death to approximately 450 people."[123]

Nonetheless, tabulations made in May 1928—after several weeks of arduous recovery efforts—could not substantiate the estimates broadcast in late March. On May 12 the *Los Angeles Times* reported a toll of only 378, with 297 recovered bodies and 81 "still listed as missing."[124] These numbers had, a week before, been corroborated in part by a report by the city of Los Angeles indicating that 223 bodies had been identified in various morgues, with another 161 people designated as "missing victims"—a total of 384 fatalities.[125] At the end of May, a "List of Missing and Dead St. Francis Dam Disaster" was drafted by the Department of Water and Power, with 314 victims listed by name as either "identified dead" or as people known to be missing and another 32 "unidentified" victims taken to morgues in Ventura County.[126] A year later, in July 1929, the Death and Disability Claims subcommittee established by Los Angeles and Ventura County authorities reported a total of 370 fatalities: 306 dead or known missing and another 64 bodies recovered but never identified.[127]

If nothing else, the welter of reports and press accounts highlights the difficulty of establishing a definitive tally. However, the recent work of California State University–Northridge anthropologist Ann Stansell—which made extensive use of claims records held in the Los Angeles Department of Water and Power Archives—is generally aligned with the July 1929 accounting.[128] While Ms. Stansell's careful research has provided as much clarity as possible in denoting the number of fatalities (about 400 total, with 240 identified bodies, 68 unidentified bodies, and 195 reported missing), there will always be unresolved questions. For example, how many people were swept into the sea or buried deep under sand and debris, never to be recovered or exhumed? And how many drifters or itinerant workers—largely invisible to county authorities—were caught unawares while sleeping in camps near the riverbed? This latter question is particularly relevant to the Hispanic ("Mexican") community that resided along the Santa Clara River floodplain near Santa Paula. In the days immediately following the flood, there was concern that some refugees were avoiding authorities because "as they were principally Mexicans they would hesitate, for reasons of their own, to come to [Red Cross] headquarters and make their presence known to the authorities; presumably because they might be unlawfully in the U.S."[129]

That some four hundred people died is far from inconsiderable. However, given the length of the fifty-three-mile flood zone and the tremendous energy released as over 12 billion gallons of water descended 1,700 feet from the reservoir to the sea, the death toll could have been far higher. Most of the victims resided in (or were staying in) the upper half of the valley; below Bardsdale, the flood warnings sounded by patrolman Edwards, his police and sheriff department colleagues, and the valley's switchboard operators, proved remarkably effective. The St. Francis Dam disaster was shocking enough in terms of death and destruction, but at least the lower Santa Clara Valley was spared even greater grief.

Funerals

A desire to provide a proper interment for loved ones energized Jesus Torres and the Cerna family in their quest to clear the Storke orchard, and they were hardly alone. Funerals brought the community and family members together to mourn those lost, particularly at the end of the first week following the flood, when scores of victims were buried in services held in Santa Clara Valley, elsewhere in Southern California, and beyond.

Perhaps the most heart-wrenching funeral was held on Sunday, March 18, when eight members of the Ruiz family were laid to rest in the family cemetery about five miles downstream from the dam. Caught in the early stages of the deluge, everyone in the extended family staying on the ranch that night died except son-in-law James Erratchuo. The

FIGURE 4-49. Funerals marked the end of the beginning for the St. Francis disaster. On the Sunday following the flood, burial of the Ruiz family was held in San Francisquito Canyon. After services in Newhall, mourners accompanied the eight bodies to the family cemetery plot on the canyon's western slope. [*Los Angeles Times* Photographic Archive, Department of Special Collections, Charles E. Young Research Library, UCLA]

Ruiz clan had roots in the Santa Clara Valley going back decades, and hundreds of friends and associates attended the ceremony. In the morning, Catholic services were held in the San Fernando home of Henry Ruiz's brother, with Father Sienes of San Fernando officiating.[130] The bodies were then taken to San Francisquito Canyon for burial. The *Los Angeles Times* provided detailed coverage of the interment:

> The long funeral procession of nearly 200 automobiles left Newhall early yesterday and progressed, part of the way on a reconstructed one-way road, to a point above the ruins of the Harry Carey trading post. Horses and wagons then bore the coffins across the stream and the half mile of desolation to the cemetery...The service was simplicity itself. More than 500 friends, neighbors mostly, assembled under the shoulder of the mountain, to lay flowers on the one large grave for the eight dead, to bow their heads in prayer, and lift up their voices in

FIGURE 4-50. Son-in-law James Erratchuo grieves as his wife and infant son (who shared a casket) and six other members of the Ruiz family are laid to rest. [*Los Angeles Times* Photographic Archive, Department of Special Collections, Charles E. Young Research Library, UCLA]

> two simple hymns...Those buried were Henry Ruiz and Rosario, his wife. Their daughter, Rose, and her year-old baby, Roland. The other children were Mary, Martin, Raymond and Susy. There were only seven coffins for the baby was buried in its mother's arms.[131]

Later the same day at the Newhall Community Church, the Reverend W. H. Evans (known locally as "the Shepherd of the Hills") held "a memorial service for all the dead, calling on his congregation, all of whom had suffered some loss, to put their trust in God." At these services a more worldly plea was offered by H. Clay Needham, who "urged the people of Newhall to wait in patience [for] the coming of facts in relation to the breaking of the dam, when blame may be placed on where it belongs."[132]

To the west in Ventura County funeral rites at Fillmore and Bardsdale were held for nine flood victims that same Sunday, and in Santa Paula services were held for "A. C. Joseph and Robert McIntyre, who resided at Castaic, where they conducted the service station at the highway junction...Floral tributes from many friends of the deceased gave the caskets an appearance of huge banks of flowers."[133] The next day in Santa Paula, a larger funeral service was held "with every minister in the city, representing practically all religious denominations," participating in the memorial.

> On a sunbathed hillside at the Santa Paula Cemetery approximately 2000 persons gathered to hear the simple services. Songs welled as the caskets were lowered into the graves surrounded by huge banks of flowers...Thirteen were laid to rest, including members of the Lima, Torres, Gutierrez, Samaniejo, and Perez families.[134]

During the afternoon service, flags in Santa Paula were flown at half-staff and businesses closed their doors. The Santa Paula Cemetery later became the resting place for many of the unidentified dead. For these and other flood victims the city of Los Angeles paid burial costs.[135]

Funerals for flood victims were not confined to the Santa Clara Valley, of course. Ida Parker, Leona Johnson, Lyman Curtis and his two daughters, and others were interred at Forest Lawn Memorial Park in Glendale, and Ray Rising's wife, Julia, and three daughters were buried at the Oakwood Cemetery in Chatsworth, west of San Fernando.[136] Others were laid to rest in hometowns far distant from the disaster zone. The *Los Angeles Times* reported that "the body of O. R. Westbrook has been sent to Hutchison, [Kansas]. Edward Vickery was buried at Exeter [south of Fresno]. Funeral arrangements are being made at Porterville [also south of Fresno] for Gerald Kimball. Oscar J. Doty was buried there Sunday."[137] Services for William Weinland and his sister-in-law, Olive Imus, and her child were held in Banning, about a hundred miles southeast of San Francisquito Canyon, on March 19. At the time the services were held the bodies of Weinland's wife and son had not been recovered. (Her body was soon found and interred with her husband, but the son's body was

FIGURE 4-51. The Monday after the flood a large funeral service was held in Santa Paula. The Santa Paula Cemetery became the resting place for many victims whose bodies were never identified. [DC Jackson]

never recovered.) The rites were described by the *Banning Record* as "one of the saddest funerals ever held in Banning... friends of the family filled the church and crowded out to the sidewalks. Many of them were Indians from the Morongo Reservation where William was born, and where his father has labored for many years as the Moravian missionary."[138]

The collapse of the St. Francis Dam was a tragedy with long tentacles, affecting people across the state and nation who had never heard of San Francisquito Canyon or understood its place in the hydraulic infrastructure of Southern California.[139]

No Turning Back

The search for lost bodies continued after the first wave of funerals. But by the second week after the flood, recovery of the Santa Clara Valley began to take hold. As the days passed, the focus gradually shifted from tragic past to promising future and the pace of rebuilding quickened. Even during the Ruiz funeral in San Francisquito Canyon the momentum was evident, with the *Los Angeles Times* observing:

> Around the bend of the canyon, where the waters clashed, workmen with the aid of tractors were tearing apart tree trunks and rubbish in a search for where the old Ruiz ranch house once stood...hymns [sung at the service were] punctuated by blasts from the tractors below, [as] they sang "Nearer my God to Thee" and "Abide with Me."[140]

San Francisquito Canyon would not stagnate, a mausoleum frozen in tragedy. The blast of tractors in the midst of hymns for the dead served as a harbinger of renewal. Up at Power House No. 2, more than three hundred employees from the Department of Water and Power were engaged in a massive effort to restore the washed-out aqueduct and bring the hydroelectric generating plant back online. The latter task would not be accomplished for several months, but the 412-foot-long section of the Los Angeles Aqueduct running between the power house and the tunnel to Dry Canyon was back in service by March 24, less than two weeks after its obliteration. The St. Francis Dam was gone, never to be rebuilt, but the aqueduct—the all-important aqueduct—was operational once more. This may have been of little concern to the people down below in the Santa Clara Valley, but restoration of the aqueduct signaled that the economic life of Southern California was moving on.[141]

Starting the morning of the flood, the California Division of Highways worked at a frenetic pace to restore traffic across the Santa Clara River and allow north-south travel on the interregional Ridge Route. The next goal was to reopen the highway down the valley from Saugus and Castaic Junction to Piru, Fillmore, Santa Paula, and points west. The task was daunting.[142] In many places near Castaic Junction two feet or more of silt and mud covered the roads; the three-span girder carrying the highway across San Francisquito Creek had been swept away, leaving behind a treacherous gully; and the once imposing steel-truss highway bridge across the Santa Clara River lay half a mile downstream from its concrete abutments, a crumpled mass of metal. Undeterred, highway personnel devised solutions that would quickly restore traffic

FIGURE 4-52. Restoration of San Francisquito Power House No. 2 began immediately after the flood, and the full plant was back online by the end of the year. Here, repair work on the two turbine-generator units is underway about a month after the flood. [DC Jackson]

flow, while more permanent structures would be built in the months ahead. Thus the San Francisquito Creek crossing was realigned upstream where a makeshift right-of-way provided temporary access across the streambed. Rather than rebuild the main Santa Clara Bridge with a completely new truss structure, state highway officials decided to replace the lost span with a hastily erected temporary wooden trestle and, at a later date, build a completely new bridge upstream from the washed-out crossing.[143]

The bridges and roads were back in service by March 19, with the *Los Angeles Times* reporting that "crews of workmen yesterday succeeded in clearing the Piru-Castaic highway of debris...the road is again open for travel from Santa Paula to Saugus." Not everything was back to normal (passage along the highway "was under the strict control of traffic police"), but the tide was turning.[144] About a week later, crews restored train service "over the branch line of the Southern Pacific from Saugus to Santa Paula," offering yet one more sign of progress.[145] And on March 23 word came that Los Angeles city officials had engaged the services of nine large contractors affiliated with the Associated General Contractors of America (AGC) to carry out a massive cleanup of the Santa Clara Valley. Thousands of men working with heavy mechanized equipment would soon be on the job. The *Los Angeles Times* announced:

FIGURE 4-53. By late March, Associated General Contractors crews were clearing and burning the vast expanses of uprooted trees and debris that covered the Santa Clara Valley. Mechanized cranes and tractors speeded their work, and by the end of the summer the cleanup phase was largely completed. [*Los Angeles Times* Photographic Archive, Department of Special Collections, Charles E. Young Research Library, UCLA]

FIGURE 4-54. Near Saugus, huge deposits of silt covered much of the landscape, requiring extensive excavation. This snapshot shows roadway clearing underway near Castaic Junction. [DC Jackson]

FIGURE 4-55. The heavily eroded streambed of the Santa Clara River in Saugus. Restoration work is underway and tents for workers are visible at left. [DC Jackson]

FIGURE 4-56. The temporary bridge built to carry the Ridge Route across the Santa Clara River in Saugus. This wooden trestle was later replaced by a steel bridge built a few hundred yards upstream. Round Mountain is visible at upper right. [*California Highways and Public Works*, 1928]

> At day-break today a well-equipped, organized and financed army of workmen, under the direction of trained leaders, will move from four strategic points on a ten-mile front into the flood devastated area between Piru and Santa Paula, and the battle to remove the debris and wreckage caused by the collapse of the St. Francis Dam and to restore the Santa Clara Valley to a land of fertility will be under way.[146]

The contract between Los Angeles and the AGC represented more than a bureaucratic move to bring additional machines and personnel into the valley. It also marked the city's acceptance of responsibility for the devastation brought by the failure of Mulholland's dam. In hindsight it might appear obvious that the city would bear legal and financial as well as moral responsibility for the flood. But exactly how the city and leaders in the Santa Clara Valley addressed the issue of reparations is central to the history of the St. Francis Dam disaster.

CHAPTER FIVE

Civic Responsibility and Reparations

> Ventura County has borne the brunt of the greatest disaster of its kind in the history of the State. The St. Francis Dam was the creature of the city of Los Angeles...[The city] toyed with the lives and property of the citizens of Ventura County. The tragic loss of life and the tremendous destruction of property in this county can never be adequately compensated.
>
> Ventura County District Attorney James C. Hollingsworth, March 16, 1928[1]

> Conspicuously absent from the picture are the harpies and jackals who occasionally congregate at the scene of a disaster, the shyster lawyers who take advantage of distress for the benefit of their own pocketbooks, who induce the filing of extravagant suits...Such [lawyers] as have appeared have been driven out by the vigilance of the committees in charge.
>
> *Los Angeles Times* editorial, April 7, 1928[2]

In the wake of the dam failure, money and goods poured into the Santa Clara Valley to help survivors. Private efforts were notable, often less for the size of individual contributions than for the vast array of people willing to aid their fellow citizens. By March 14 the *Los Angeles Times* had set up a relief fund, kicking in an initial contribution of $1,000 to support the Red Cross.[3] Within a week the *Times* fund had raised over $78,000 from several hundred citizens, organizations, and businesses. Contributors included the Los Angeles Stock Exchange ($1,000); businessman and electric power entrepreneur William G. Kerckhoff

($1,000); "Friday evening collection at Temple Israel of Hollywood" ($180); "Gross receipts from a benefit picture show, held in Lone Pine Hall, Lone Pine, Cal." ($112.50); a multitude of service organizations, including the "Trona Chapter of the Amalgamated Order of Billy Goats" ($70); "Proceeds of a bridge game at the home of Mrs. Caroline Spencer, Pasadena" ($21); and the Union Church of the Deaf ($9). Most gifts were in the $5 to $50 range and some were smaller still, like the $2 cash donation credited to a "90-year-old Civil War widow." The *Times* was happy to accept them all and the total proved considerable. In May the fund topped out at more than $94,000.[4]

Not to be upstaged, the *Los Angeles Examiner* also set up a relief fund on March 14, donating $1,000 to prime the pump. The *Examiner* appears to have had a special connection to Hollywood, with film mogul Louis B. Mayer of Metro-Goldwyn-Mayer ($150), producer Joseph M. Schenck of United Artists ($100), and leading man Douglas Fairbanks ($100) quickly pitching in. By the next day theater impresario Sid Grauman and cowboy star William S. Hart also stepped up, offering gifts of $150 and $100 respectively. By March 17 the *Examiner* fund had reached $10,000, with the Pacific Mutual Life Insurance Company and William Wrigley Jr. (of chewing gum and Chicago Cubs fame) each contributing $1,000.[5]

Celebrities did more than simply contributing to relief funds sponsored by newspapers, and some volunteered for a special Hollywood fund-raiser at the Metropolitan Theater on March 21. The coterie of stars included Charlie Chaplin, Gloria Swanson, Tom Mix, and Irving Berlin. Fans flocked to the sold-out gala and proceeds came to almost $10,000.[6] Three days later, a special boxing exhibition at San Francisco's State Armory, featuring an appearance by famed pugilist Jack Dempsey, generated another $10,000.[7]

Fund-raising initiatives spanned the cultural spectrum, including a dance sponsored by the Independent Order of Foresters in Los Angeles, a performance of the play "The Captive" at the Mayan Theater, a golf exhibition at the Lakeside Country Club featuring Walter Hagen, and a Los Angeles Philharmonic concert. The Philharmonic presumably attracted a genteel crowd, while a different sort of clientele likely attended a 300-point billiards match between Edward Horemans of Belgium and American Welker Cochran. Held on Sunday afternoon, March 25, at Jones Billiard Parlor, the event was free, and money for flood victims was to be raised "by passing the hat among the spectators." To help induce a good turnout, Horemans and Cochran also agreed to give an "exhibition of trick shots" after the main match.[8]

Exactly how much collected at the Jones Billiard Parlor made its way to the afflicted is unknown, but by early May donations to the Red Cross had exceeded $200,000. Significantly, all of this money was to go directly to relief work, providing clothes, food, shelter, and general sustenance for flood refugees. A paragon of progressive efficiency, the Red Cross expressed pride that it did not siphon off contributions to support a top-heavy administrative structure. As D. C. McWatters, head of Red Cross operations in Los Angeles, advised donors: "These funds will be used exclusively for the relief and rehabilitation work which the Red Cross is doing in the devastated area, without deduction of a single penny for expenses, because the entire cost of administering the fund is paid by the National Red Cross."[9]

"A Tragic Fact and Not a Theory"

Private contributions, however, could not begin to cover the cost of the massive cleanup and claims for property damage, injury, and death. In the days following the flood, Los Angeles officials struggled with both the enormity of the carnage and the issue of liability. The city had built, owned, and operated the dam; it seemed only common sense that the city would bear financial responsibility for the disaster. But how could this responsibility be prudently and fairly met? Complicating matters, some people believed that a dynamite blast had caused the dam to fail, thus making the city as much a victim as those who had drowned. And if the city was a victim of a terrorist attack (presumably by disgruntled ranchers from the Owens Valley), why should it be required to pay damages to other victims? No credible evidence of a dynamite attack ever surfaced, but the theory was widely aired and, at least in the short term, obscured the issue of legal responsibility.[10]

Los Angeles Mayor George Cryer understood the political dimensions of the disaster, fearing a public relations nightmare if the city was seen as indifferent to the flood-borne misery. In his view, the city needed to step up quickly and take responsibility. "Los Angeles cannot restore the lives lost," Cryer declared two days after the dam's collapse, "but the property damages should be paid. The dam was part of the city facilities for supplying Los Angeles with water. The responsibility is ours... [and] a way must be found to pay the money for the property damage done."[11] Los Angeles City Attorney Jess Stephens took a completely different tack, opposing the expenditure of any relief funds without a protracted investigation justifying the city's liability. Stephens haughtily lectured the press:

> It is not within the province of any officer or board of the city to determine liability in accordance with personal desires. That will be determined by the law and the facts...If after a complete and thorough investigation it appears that the city is liable for damages, that situation will be met in a fair and proper manner, having in mind both the interests of unfortunate people who have been directly visited with this calamity and the taxpayers of Los Angeles. This is all that can be said until we have before us all the facts which will result from the exhaustive investigation now underway.[12]

Nonetheless, Stephens' seemingly unequivocal position was undermined by city-sponsored work already underway in San Francisquito Canyon and at a "water bureau camp" near Santa Paula. Early recovery and restoration efforts led by Mulholland's deputy Harvey Van Norman were being subsidized by the city, whether Stephens liked it or not.[13]

Los Angeles' mixed messages on liability and reparations did not sit well with people in the Santa Clara Valley, especially in Santa Paula, where some 200 homes had been destroyed and many more damaged. The preeminent figure there was Charles C. Teague, the local patrician who led the Santa Clara River Protective Association (predecessor of the Santa Clara Water Conservation District), managed the Santa Paula–based Limoneira Ranch, and was president of the politically formidable California Fruit Growers' Exchange, popularly known as Sunkist. More than a provincial businessman from a citrus-growing hinterland, Teague was highly respected in Republican Party circles (in 1929, President Herbert Hoover would appoint him to the Federal Farm Board). This political prominence, combined with Teague's activism in promoting California's citrus and walnut crops, ensured that Los Angeles would not run roughshod over the business interests of the Santa Clara Valley.[14]

As head of Ventura County's hastily formed Rehabilitation Conference (or County Committee as it was often called), Teague took charge of helping local residents. As he later explained:

> The first problem presented by the disaster was naturally that of providing relief for the homeless and recovering the bodies of the dead. To meet the emergency a County Committee was set up, of which I was chair-

> man. Subcommittees were appointed to obtain temporary housing, clothing, food and medical assistance. All citizens...worked heroically day and night until the critical phase of the emergency was over.

With time short and the county's resources limited, Teague welcomed the Red Cross "to come in and take charge" of relief services.[15] The issue of how to fund restoration and reparations now came to the fore. Teague targeted Los Angeles, counseling members of the Rehabilitation Conference that "the great city of Los Angeles which caused the damage should be approached first of all for funds for rehabilitation." Initially pessimistic about getting cooperation from the urban leviathan to the south, he grumbled to the press two days after the disaster, "Los Angeles will try to minimize the damage and to prove that we are not entitled to anything... We probably will have to appeal to the courts."[16]

Teague was not alone in his distrust of Los Angeles. A. J. Ford of the Department of Water and Power visited Fillmore on March 14, later reporting that the local morgue "was in [the] charge of American Legion men, acting as a service body club. They were not communicative and left us with the impression that they were resenting the presence of Los Angeles people."[17] Former state Senator D. W. Mott of Santa Paula also shared Teague's skepticism, suspicious because of the city's earlier move to drain 5,000 acre-feet of water annually from the San Francisquito Creek watershed (see chapter 3).[18] While the collapse of the St. Francis Dam may have rendered this particular water rights battle legally moot, valley leaders were loath to forget. Mott forcefully made the point: "Responsibility for the catastrophe [rests with]...a selfish city that took the water belonging to us...We want Los Angeles to know that it has taken millions from us."[19] Ventura County District Attorney James Hollingsworth lined up behind Teague and Mott, angrily calling the St. Francis Dam "a creature of the city of Los Angeles," while the valley faced the brutal consequences of "a tragic fact and not a theory." Warning that Santa Clara Valley residents were "not going to rely upon promises or expressions of sympathy," he demanded that Los Angeles "be big enough to admit liability in this matter and do everything within its power to pay the loss."[20]

Before the flood, the valley's position on the San Francisquito Creek water dispute had created hardly a ripple in Los Angeles. But now, the hardline pronouncements by Teague, Mott, and Hollingsworth galvanized city leaders into action. Many, including John Richards and Peirson Hall of the City Council, and George Eastman, president of the

Los Angeles Chamber of Commerce, joined with Mayor Cryer in recognizing the city's indisputable responsibility for the flood damage.[21] A few days after the disaster, Richards and Hall drove to Santa Paula to meet with Teague and disavow the city attorney's stonewalling. In their view, "the right course was for the City of Los Angeles not only to assume restoration work [but also to]...assume complete and absolute liability for all damages that were caused." They believed that any other response risked "arousing animosity to the City of Los Angeles." They were driven not so much by altruism as by their fear that political backlash against the city could undermine the Boulder Dam bill, "an absolute necessity for the future water supply for the city of Los Angeles, [which] was pending before Congress."[22]

Teague was prepared for a contentious encounter, but Richards and Hall surprised him: "We assured Mr. Teague that we would...do everything in our power to see that the policy of accepting entire and absolute responsibility and liability was followed by the City of Los Angeles."[23] Upon returning to Los Angeles, Hall contacted a young attorney who disagreed with Stephens on the need for a time-consuming investigation into legal liability. In his view, there was already "legal authority which would permit the city to pay 'upon reasonable apprehension of liability.'"[24] As chairman of the City Council's Committee on Water and Power, Hall invited the attorney, William C. Mathes (later chief judge of the Los Angeles Superior Court), to meet with the council and the city attorney. Mathes proved persuasive and, in a dramatic about-face, Stephens soon publicly advised the council that

> [it is] a well established principle...that the right of the city to sue and be sued involves necessarily the right to compromise and settle claims. Under the city charter your honorable body has control of all litigation of the city and may direct the City Attorney therein. Under the decisions of the courts your honorable body may, pursuant to the above principles, compromise bona fide claims against the city in all cases in which there is a well founded apprehension of legal liability.[25]

Stephens' turnaround preempted a bitter struggle with Santa Clara Valley leaders. Nonetheless, his emphasis on "bona fide claims against the city" intimated that he still hoped to limit the financial impact of restitution.

The City Council's willingness to accept liability was well received in the Santa Clara Valley and, on March 19, Teague announced in the *Santa Paula Chronicle*: "Officials with whom I have conferred have given me every assurance within their power that the city will assume full responsibility for damage done to life and property and that they will make complete restoration and compensation."[26] Later that day, the Los Angeles City Council publicly endorsed the new policy: "It is just and proper that the city should immediately commence the adjustment and compromise of claims for damages resulting from the destruction of said dam." Concurrently, the council appropriated $1 million for settlement of claims. These funds, drawn initially from the city's Harbor Department reserves and later from appropriations supported by Bureau of Power and Light revenue and by general obligation bonds, were deposited in a St. Francis Dam Claim Fund under Stephens' supervision. In overseeing disbursements, he was to cooperate with both a committee of Los Angeles business and civic leaders and a Ventura County committee; final approval for all payments recommended by the joint committees was to come from the Los Angeles City Council.[27] On March 20, a week after the flood, the *Los Angeles Times* broadcast the news under the bold headline "City Ready to Pay Losses," and praised the decision as "the quickest large transaction ever made by a Los Angeles City Council."[28]

"Whatever Hysteria There Was, Is Now Over"

In committing to reparations, the city sought to complete the cleanup, rebuilding, and claim settlements without involving the civil courts. If successful, this strategy would cut out interference from plaintiffs' lawyers and eliminate their troublesome posturing and demands. On this issue, Teague proved to be in lockstep with his Los Angeles colleagues; their shared conviction that private attorneys were impediments to the reparations process soon forged an enduring bond between the business and political elites of Ventura County and Los Angeles. While some claimants resisted moves to limit their access to legal representation, they would find little solace among leaders in either the city or the Santa Clara Valley.

On March 21, six representatives from Los Angeles, led by Stephens and George Eastman, journeyed to Santa Paula to meet with Teague and other valley leaders at the National Bank building. Speaking to the press, Eastman emphasized his desire that there be "no delay in developing a practicable working plan so that the community can be placed back on a normal basis as soon as possible." In turn, Teague "was emphatic in his declaration that the Los Angeles Committee [was] to expedite restoration to the fullest extent of its ability," effusing that "there is not the slightest doubt in the minds of our people but that they will get a sympathetic and fair deal from Los Angeles and are not worrying. Whatever hysteria there was, is now over."[29]

At this initial meeting—which included a "goodwill luncheon" at the Glen Tavern in Santa Paula—two seven-member committees were formed. One represented the city and the other the Santa Clara Valley; collectively they were known as the Joint Restoration Committee. Eastman chaired the Los Angeles committee and Teague served as his counterpart for Ventura County. Each of the committees oversaw five subcommittees responsible for restoration of agricultural lands, reparations for land no longer usable, building reconstruction, personal property claims, and death and personal injury claims.[30] With a joint committee structure agreed upon, Stephens set September 12, 1928—exactly six months after the dam collapse—as the deadline for claims to be filed against the city.[31]

The civic leaders selected for the county committees represented a distinct social stratum that did not reflect the diversity of the flood victims. No elections or public meetings were held to determine committee membership. In Los Angeles, members were appointed by the foursome of Mayor Cryer, City Council President William G. Bonelli, banker Joseph F. Sartori, and President Eastman of the Chamber of Commerce. In Ventura County, members were appointed under the direction of the Board of County Supervisors. The Los Angeles committee included a bond broker, the president of the Los Angeles Harbor Commission, a contractor representing the Associated General Contractors, a vice president of the Union Oil Company, and a vice president of the Chamber of Commerce. Joining Teague on the Ventura County committee were the manager of the Newhall Land and Farming Company, the manager of the Sespe Ranch, the manager of Berylwood Investment Company in Hueneme, and three other ranchers from Santa Paula, Fillmore, and Piru. There were no agricultural workers, no small-scale landowners or homeowners, and—to use the ethnic signifier of the time—no Mexicans. In the Santa Clara Valley, the process of reparation was to be defined and controlled by men of means closely tied to the business of agriculture. It also appears that no relatives of any committee members had died in the flood.[32]

Once Los Angeles became officially engaged in reconstruction, other agencies and groups scaled back their labors. "With the action of the city of Los Angeles in making funds available to begin the immediate restoration of losses," announced Alexander Heron, chairman of the State Board of Control, "the time has arrived when State responsibility should be completely withdrawn from the stricken area with the exception of the control of health, sanitation and traffic."[33] Other groups, such as the Red Cross, the Salvation Army, and the American Legion, as well as hundreds of volunteers, also began winding down their operations (the Red Cross closed its Newhall office on April 1 and its office in Santa Paula on May 31). A tremendous amount of restoration work remained unfinished as April approached but, as Heron signaled, significant progress was underway.[34]

At the first meeting of the Joint Restoration Committee, Stephens' insistence that the city had a "right to compromise and settle claims" took center stage. Teague opposed such a policy as it applied to damaged land or damaged improvements, such as homes, barns, roads, orchards, and irrigation equipment. In his view, "We [are] dealing with the problem of complete reconstruction and restoration...and not with the matter of claim adjustments [which]...are usually made on the basis of the best bargain that can be obtained." Digging in his heels, Teague argued that this latter practice would constitute an injustice and declared that he "would have nothing...to do with the [restoration] work if it was to be undertaken on that basis." The Joint Committee agreed with Teague and endorsed a policy of restoring property "to its original shape as nearly as possible, for the benefit of the community as well as the individual."[35] In the face of broad opposition, Stephens backed down. Avowing an "ambition to settle every claim growing out of this disaster without a single lawsuit," he agreed that "our work having to do with property will be on the basis of replacement...rather than a cash settlement." But Stephens held fast to the view that death and injury claims constituted an entirely different matter: "Settlement for life losses and personal-injury losses necessarily must be on a cash basis."[36]

The Joint Restoration Committee faced a complex problem, chiefly because the damage varied enormously,

> from land being completely washed away, with no value left, to farming land and orchards heavily strewn with debris, in some cases piled ten or fifteen feet high; soil...[was] badly eroded in some places, in others heavy deposits of sand and silt have been left; orchards in some cases are completely destroyed, in others

> partly; trees completely washed out, [while in] other cases soil [was] washed away from roots.[37]

The problem was exacerbated, in Teague's view, by "the Los Angeles committee [which] consisted of city men, mostly unfamiliar with agriculture." The two groups finally concluded that they should identify an impartial agency "representing neither Los Angeles nor the Santa Clara Valley" to prepare appraisals. Each committee then promptly selected a different state agency to complete the work. The impasse ended only when the chief of the state Department of Agriculture, the agency chosen by Los Angeles, conceded: "We haven't the men qualified to do this type of job." Left with little room to maneuver, the Los Angeles committee accepted the Santa Clara Valley's choice of the California Agricultural Extension Service.[38]

Within a few days, about twenty "farm advisers" from the Extension Service "arrived, set up an office, and began their survey. Each piece of property affected by the flood was separately mapped and the damage described in detail."[39] The assessments prepared by the farm advisers were not intended to provide a dollar value to determine claim settlements. In Teague's words, "the Farm Extension Service will have nothing to do with appraising in dollars the amount of damage, but will simply find all of the facts." Nonetheless, the advisers provided a foundation for the work of the Joint Restoration Committee in settling hundreds of property damage claims.[40] Above all, Teague was proud that agricultural property damage reparations were handled beyond the purview of the court system. As he recounted in his memoir, "The joint committee had no authority from victims of the flood to settle their claims for damages. The committee was only seeking to determine what constituted fair compensation for individual losses. Thus it is all the more remarkable that these claims were settled without recourse to the courts."[41]

"Shyster Lawyers"

Teague's antipathy to the court system was not confined to compensation for agricultural and property damage and, in ways hard to imagine from a twenty-first-century vantage point, he severely restricted the ability of people in the Santa Clara Valley to seek redress for death and personal injury. Taking the offensive, he castigated "shyster lawyers who had appeared on the scene and attempted to get assignments of claims on a percentage basis from injured parties."[42] To suppress "parasites," Teague aggressively distributed a circular to newspapers and the public "request[ing] that no arrangements be made [by local citizens] with anyone to handle claims as such action will only result in complicating the situation and reducing the amount that those injured will eventually receive." The circular also demanded that "the names of any lawyers or claims agents...be immediately handed to me and I will see that their names are published and that they are branded as parasites seeking to take advantage for personal gain of a situation which should call for the unselfish cooperation of every honest citizen." And what might befall any lawyer who somehow found his way into the flood zone in search of clients? At a remove of fifteen years, Teague recalled that "when the first of these so-called 'parasites' appeared on the scene in the early days of the disaster, a committee, accompanied by the sheriff, took them to the county line and told them not to come back—in very emphatic terms." He acknowledged that such action "probably was not legal" and "perhaps savored a little of vigilante days," but made no apology because, he said, "in my opinion it was warranted under the circumstances."[43]

Teague's belligerence correlated perfectly with the interests of Los Angeles. As an editorial in the April 7 *Los Angeles Times* crowed:

> Conspicuously absent from the picture are the harpies and jackals who occasionally congregate at the scene of a disaster, the shyster lawyers who take advantage of distress for the benefit of their own pocketbooks, who induce the filing of extravagant suits...Such [lawyers] as have appeared have been driven out by the vigilance of the committees in charge.[44]

Teague worked to ensure that justice was served for victims in the Santa Clara Valley. But it was a form of justice in which he and other civic leaders exercised remarkable power in determining how claims would be heard and in impeding the access of citizens to the court system. While lawyers did find ways to connect with prospective claimants, their operations were significantly constrained by Teague and other elites in the community.

Cost Plus 6 Percent

Work undertaken at the expense of Los Angeles, including the removal of debris and restoring land, officially began on March 26. Of course, city-sponsored restoration had begun soon after the flood, undertaken by Department of Water and Power crews under the direction of Harvey Van Norman. But now the city could officially charge Van Norman with responsibility for construction, cleanup, and related field work. Expenses for debris removal incurred before March 26, paid directly by the department, came to about $200,000 and were never charged against the St. Francis Dam Claim Fund.[45] Much of the city's early reconstruction work focused around Power House No. 2, located just a mile and a half below the former reservoir. By the end of the year the city's Bureau of Power and Light had expended $1,090,305 to bring the plant back online at full capacity and to rebuild the workers' housing compound.[46]

Once officially appointed to his new post, Van Norman ordered a survey of the entire devastated area, down to the Pacific Coast. The findings convinced him that "the city's forces and equipment were entirely inadequate to cope with the situation." The task ahead, he concluded, "had to be completed with extraordinary rapidity, in order to save from complete loss properties which were damaged by being covered with debris and silt and were dying from suffocation or lack of irrigation." To provide the necessary manpower and to avoid the expense of leasing or purchasing heavy equipment (eventually some 500 tractors, drag lines, caterpillars, and similar machinery were used in the cleanup), Van Norman and his assistant J. E. Phillips negotiated a "cost plus 6 percent" contract with the Associated General Contractors. Nine AGC-affiliated contractors signed on to collect and burn debris, excavate and remove sand and silt, clear roads, level land, refurbish irrigation systems, erect work camps at Piru, Fillmore, Bardsdale, and Santa Paula, and to do anything else necessary to restore the valley to a modicum of normalcy. The AGC force reached 1,300 men within a few days and numbered 3,800 at peak employment.[47]

In City Attorney Stephens' words, the goal "was to free all lands of debris, to restore all pumps, pipe lines and irrigation systems...to restore all ditches, to reconstruct and replace all diversion dams in the Santa Clara River...[and] thus protect properties from additional loss."[48] The biggest task facing the work crews was to unearth and gather the hundreds of destroyed orchard trees and, after checking for bodies, set them afire. A seemingly straightforward job, the destruction of organic debris nonetheless required some innovative techniques:

> The debris piles did not burn readily, as they contained much wet earth and most of the wood was green. After

> proper piling, however, the piles could be sprayed with oil under high pressure thus securing deep oil penetration and a very intense heat could be developed which usually burned the debris entirely.[49]

Some land in the flood zone was not covered by piles of debris but was scarred by intense scouring:

> In another instance the lower part of an orange grove was eroded over an area of five acres to an average depth of five feet and as no silt deposits occurred in the immediate vicinity, a hill above the flood line was excavated into and the earth hauled about a mile and filled in the eroded area. 15,000 cubic yards of earth was necessary to fill the eroded area.[50]

As of October 31, 1930, the cost of agricultural restoration in the Santa Clara Valley came to $1,284,598. Along with the reconstruction of bridges and the repair of pumps, motors, wells, and farm implements, the work included:

> Removal and destruction of driftwood and debris from 7,564 acres;
>
> Excavation of 63,096 cubic yards of silt and sand;
>
> Leveling of 4,305 acres;
>
> Restoration of 57,458 linear feet of water supply ditches;
>
> Restoration of 85,475 linear feet of pipe and irrigation lines;
>
> Rebuilding of 581,711 feet of fencing; and
>
> Restoration (partially or entirely) of 38,966 trees.[51]

A major beneficiary, the Camulos Ranch Corporation, received $197,392 in November 1928 as restitution for its damaged fields and orchards east of Piru. The Newhall Ranch (formerly Rancho San Francisco), the largest single property in the path of the flood, did not receive its damage award of more than $737,000 until late 1930.[52]

The job of restoring houses or replacing those deemed uninhabitable was completed by mid-October 1928. Some 740 homes were repaired or replaced, many in Santa Paula, where Teague and other members of the Ventura County committee could readily keep watch on the work. The goal was to provide owners with homes "in as good condition as they were in prior to the flood." For houses "entirely destroyed and washed away, new structures were planned, which, as nearly as practicable, were a reproduction of the structure that had been destroyed."[53] The city hired local craftsmen and used locally purchased construction material, a policy favored by Santa Clara Valley residents. In addition, the Joint Architectural Committee adopted a flexible approach to the rebuilding process, reporting in November 1928:

> Some claimants suffered complete loss on two or three parcels of land and requested that restoration be made by combining their total losses into an equivalent building restoration in other locations...In some instances portions of buildings that were torn apart and badly damaged were used to restore the buildings of other claimants. This was found to be not only economical for the City of Los Angeles but profitable for the claimants as well. With very few exceptions, the claimants preferred to have restoration effected by actual building rather than by cash...The plans and specifications provided for building materials [were] both logical and economical for the particular type of building and the particular community in which it was to be constructed. Local labor and materials [were] used throughout the building operations, in so far as that was possible.[54]

In the end, the replacement and reconstruction of houses cost $615,088, a sum reflecting the relatively modest size of most homes despoiled by the flood.[55]

Along with agricultural losses and house reconstruction, the city also invited claims for property losses sustained by individuals, municipalities, utilities, and corporations. Some 1,855 such claims were filed, 1,168 of which involved clothing, furniture, household wares, yard equipment, and automobiles. Of the $1,133,959 sought by claimants for personal items, $560,813 was authorized for payment, with the average award a little under $500. The damage claims of a group of twenty utilities (including Southern California Edison, Southern California Gas Company, Pacific Telephone and Telegraph Company, and the Southern Pacific Railway) and the municipality of Santa Paula reached $1,426,667 and resulted in awards totaling $907,121. Settlements in this category were agreed upon with little public notice, and it appears that these large enterprises did not desire widespread airing of either the negotiations or the terms reached.[56]

Excessive or Bona Fide?

Restoration of land and rebuilding of homes proceeded relatively quickly and—once the city accepted financial responsibility—with little apparent acrimony. But this was not the case when it came to claims for deaths and personal injuries. Money could repair farmlands, orchards, and buildings, but no amount of treasure could restore life to the dead. In the city's view, compensation for grieving relatives of the deceased was acceptable, though not necessarily to the extent permitted for rehabilitating inanimate fields, buildings, and hydraulic infrastructure. The now-common legal phrases "pain and suffering" and "emotional distress" do not appear in the historical record. The city's bywords regarding death and injury claims reflect a very different perspective: there were to be no "excessive" or "unjust" awards.

"It is an almost universal human failing," observed George Travis of Santa Paula, "to consider an injury greater than it really is, and to consider payment as excessive of justice."[57] Everything depended upon whether you were the victim or the perpetrator. While there was no precise or official definition of what constituted an "excessive" or "unjust" award, City Attorney Stephens stated that he would recognize only "bona fide claims against the city." Not surprisingly, claimants were often disappointed by what the city proposed as a "bona fide" settlement.[58]

Workmen's Compensation

In many cases, the procedure for making a death claim was fixed by the state's Workmen's Compensation, Insurance, and Safety Act of 1913.[59] This law, which did not require a specific determination of liability or negligence, applied to all victims who died in California in the course of doing their jobs. Presumably the law was envisaged primarily as a means of compensating the dependents of workers killed in incidents such as factory explosions or construction accidents. But it also applied to those killed in other circumstances directly tied to their employment. Thus, the twenty-three employees of the Bureau of Power and Light who drowned when the flood engulfed Power House No. 2 and the nearby city-owned bungalows were deemed covered by the Workmen's Compensation law. The same held true for the eighty-four employees of Southern California Edison who died at the Kemp construction camp. Although asleep when the flood swamped their tents, they were present at the company-sponsored camp as a result of their work building Edison's transmission line down the Santa Clara Valley.

Death claims for both Edison and city employees were handled by the state Industrial Accident Commission. The first such claim was filed within a week after the flood by the mother of Earl Pike, who had worked with Tony Harnischfeger as an assistant dam keeper. Pike was reported to be twenty-nine years old and had drawn a monthly salary of $120 from the city (his annual pay, $1,440, is a useful gauge for evaluating the relative value of settlements for other flood victims).[60] The key issue for the Industrial Accident Commission was not liability but the extent of financial support. Simply being a victim's parent, sibling, or relative was not enough to justify an award; bona fide claimants were required to demonstrate that they had been economically dependent upon the victim. Claims were limited to a total of $5,000 per victim (plus $150 for burial costs) and, in general, were capped at three times the annual support that the victim provided to the claimant.[61]

For example, it was announced in May that "Mrs. Helen W. Bogue, widow of Beryl Bogue, camp clerk [at Kemp] was awarded $4760, three times the deceased's annual salary," and that an award of $300 was made to "the father and minor brothers of Orval R. Westbrook, a tower man. The evidence showed that Westbrook contributed about $100 per year to support his family."[62] That same month, the *Los Angeles Times* reported a cumulative award of $18,630 for claims involving four Edison employees. The dependents of three of the victims—tool maker R. B. Shields, camp cook Edward Crumley, and steward George Hawkins (who had come to Kemp from Long Beach the day before the flood). Their dependents received the maximum allowable, $5,000 (plus $150 for burial costs); but because laborer Edward Ritchie drew a smaller paycheck, his wife Florence received only $3,630.[63] Such payments proved substantial, and by September, Edison had disbursed over $105,000 to thirty claimants and estimated that claims involving fifty other employees would exceed $200,000.[64]

City employees who had worked at the dam or at Power House No. 2 received comparable consideration from the Industrial Accident Commission. The commission awarded the parents and minor brother of twenty-three-year-old Basil Bross, an operator at the power house, $1,500, because "the testimony showed [they] were receiving $500 from him toward their support." At the same time that the Bross family got word of their compensation, the commission awarded Lillian Curtis and her surviving son $5,000 for the loss of her husband, Lyman. She had also lost two young daughters, but their deaths lay beyond the purview of the Workmen's Compensation Act. Curtis chafed at the idea that what she suffered the night of the flood could be remedied by payment of a mere $5,000, and she soon began contemplating court action.[65]

Settlements and Investigations

Legally, there was no reason death claims should be limited to $5,000. But the Industrial Accident Commission's actions set a de facto standard that appeared to guide the work of the Joint Restoration Committee's Subcommittee on Death and Disability Claims. It also set a benchmark for City Attorney Stephens, who was averse to settlements that significantly exceeded the commission's limits. Stephens did not see his role as opening up the municipal checkbook to anyone who felt aggrieved or damaged by the flood. His job—which allowed little space for compassion or human emotion—was to protect the fiduciary interests of Los Angeles and its taxpayers.

To implement the city's settlement policy, Stephens worked with the Death and Disability Claims subcommittee to create an investigative department charged with ascertaining "all facts pertinent to the various claims." The process began at one of the department's three offices—in Santa Paula, Fillmore, and Los Angeles—where personnel reviewed the forms submitted for settlement. The files were then checked by appraisers who negotiated directly with claimants to reach what the city considered a fair settlement. If claimants believed they were being low-balled, their options were limited. They could implore the city to reconsider, or they could threaten to file a lawsuit in civil court. But lawsuits were no panacea. They represented a course of action that many felt they could not sustain financially, while settlement of a claim meant relatively fast payment. Litigation entailed delay, without any guarantee that an award would exceed the proffered settlement.[66]

Under Stephens' direction, the city's effort to fend off what appeared to be unjustified claims proved a major undertaking. It is documented in a multitude of records that survive in the Los Angeles Department of Water and Power Archives. An example is the file on a death claim involving Leona Johnson, one of the first people to fall victim to the flood.[67] She lived with dam keeper Tony Harnischfeger in the San Francisquito Canyon a short distance below the dam. Had they been married, any claims presented by family members would have been straightforward. But Leona Johnson's marital status was unclear, as she had been married to Henry F. Johnson, of Los Angeles. And soon after Leona moved into Harnischfeger's cabin in January 1928, the couple informed several people that they had been married during a recent trip to Oceanside. In the abstract, the private life of Leona and Tony was not of concern to the city of Los Angeles. However, when Henry Johnson filed a death claim for $17,100 premised on Leona's legal status as his wife, it became a matter of civic import. A memo in file #2538 explains the city's interest:

> Claim filed for $17,100, $15,000 for death, $1400 for P[roperty] D[amage] and $700 funeral. First reports were that Leona Johnson was the wife of Tony Harnischfeger. (See File #2285) It later developed that she was merely living in the valley, her body being found and identified by her husband, Henry F. Johnson, above claimant. We have sent to Santa Paula for a thorough investigation on case, upon receipt of which we will get in touch with Mr. Johnson and adjust claim. Of course we will have to pay something, but if we can get out of paying more than $1000 we feel we are lucky. There was a rumor of their obtaining a divorce, which rumor we are now running down.[68]

At least eleven people were interviewed as the city dug deep for information about Leona's personal habits, how she and Harnischfeger presented their relationship to neighbors, and whether she possessed any significant diamond jewelry that might have been lost in the flood.[69] This investigation and "a friendly chat" with the attorney representing Johnson uncovered evidence that, while an interlocutory judgment for a divorce had been issued on January 31, 1928, it had never become final. The city also determined that "on January 24, 1928 Anthony Harnischfeger and Leona B. Johnson filed a certificate of 'Intention to Marry' but no license was issued. Reports that they were married in

Oceanside or San Diego were investigated and no trace of such a marriage can be located."[70]

All of this was known to city officials by mid-November 1928, but they were in no rush to reach a decision as long as Johnson and his attorney held out for a windfall. When informed in December that "$500.00 was what we [the city] figured the pecuniary loss in this case was worth," Johnson's attorney bluffed an intention to sue while also stating that he would accept "$4,000.00 or nothing at all." In response, city officials opted "to let the case drag for a while before having the [City] Council act on the claim."[71] Finally, in April 1929, the two sides reached an agreement whereby $1,350 would be paid by the city "in full settlement and compromise of the claim of Henry F. Johnson for the death of Leona B. Johnson."[72] The city did not escape unscathed, but there would be no $17,100 payoff.

The fact that Henry Johnson was legally married to Leona at the time of her death was central to his claim. On this key point, his status differed from that of Gladys Harnischfeger, the former wife of Tony, who was also the mother of both his son, Coder (who died in the flood), and his daughter, also named Gladys (who was not in San Francisquito Canyon the night the dam collapsed). After the disaster Gladys rushed to the flood zone in hopes of recovering the remains of her six-year-old son (Coder's body was never found). She subsequently sought compensation from the city for her daughter, for whom Tony had been providing $30 per month. Gladys, as a former wife (who apparently received no alimony support), had no standing to make a claim for herself in regard to her ex-husband. But Tony's mother, Mary, had been receiving $15 per month in support from her son, and she could present a valid claim. And, regardless of her divorce from Tony, Gladys remained the mother of Coder and thus could make a claim on his account. So how were these claims in regard to Tony and his son resolved?[73]

Tony Harnischfeger was an on-duty employee of the city of Los Angeles when he died. Therefore, his death was covered under the Workmen's Compensation Act, with payments capped at $5,000 (plus $150 for burial costs). His mother was awarded $270.79 to be paid out over twenty-six weeks at $10.415 per week; daughter Gladys was to receive the same amount. The remaining $4,458 was to be disbursed from a fund deposited at the Union Bank and Trust Company in payments of $20.83 per week. (It is uncertain whether this fund was to be transferred to Gladys at the age of majority or could be drawn upon by her mother for child support.)[74]

As the surviving mother of Coder, Gladys filed a wrongful death claim. She received a total of $1,680 from the city as restitution for his death and as compensation for personal property in Tony's cabin.[75] The pecuniary cost of losing a child was impossible to calculate, but the city generally paid out a couple of thousand dollars in damages for each child victim. The emotional distress suffered by Gladys Harnischfeger because of her son's death was apparently given only modest weight in the city's settlement calculus. But if she was unwilling to actively pursue her claim in civil court, her ability to leverage a higher payment was essentially nil.

There were 172 death claim settlements endorsed by the Death and Disability Claims subcommittee as of July 15, 1929. These awards averaged $5,138, which was in line with payments authorized by the Industrial Accident Commission. On the high end, Chester and Velma Rogers received $24,000 for the deaths of four minors, and Aaron Coffer received $9,000 for one adult and one minor. But no one walked away rich. The city was determined to adjust claims considered "unreasonably high" and most claimants, often grudgingly, acquiesced. Nonetheless, some resisted and pursued justice through the civil court system.[76]

As the months passed, people were once again traveling and engaging in the social and business activities of daily life. It was consequently much harder for Teague and other civic leaders to impede access to legal representation than it had been in the tumultuous weeks following the flood. Inroads made by lawyers or "claim agents" by early summer are attested in a June 27 memo to Teague from his assistant, George Travis. Travis reported that Sisto Luna of Santa Paula's "Mexican Camp" wanted to talk with Teague because "a number of claim agents have been to see him and have promised $50,000 or $60,000 for the five children which he lost . . . Luna has told [others in the Mexican Camp] that you were looking after their interests and that they did not need claim agents." Luna was apparently concerned that the Death and Disability Claims subcommittee had only "offered him $5,000 ($1,000 each for the children) and Luna wants to talk to you about whether he should accept."[77] A meeting with Teague was held the next day and, although the details of the conversation are unknown, the result is clear: as of June 30 Luna had accepted a settlement of $9,000 for his five children.[78] By all appearances, Teague had counseled the subcommittee to increase its settlement offer to Luna by $4,000—an 80 percent bump—as a means of countering the efforts of plaintiff lawyers to recruit clients.

Teague's efforts to dissuade victims from filing civil lawsuits may have worked in the case of Sisto Luna, but he did not achieve universal success. A mid-July memo to W. B. Mathews, special counsel to the Board of Water and Power Commissioners, provides a "Notice of Liens Filed by Edward P. Garrett" showing that the Fillmore-based lawyer held "liens upon the claims of [thirteen people] on account of damages arising out of the flood occasioned by the breaking of the St. Francis Dam." The memo reveals that, to the city's chagrin, Garrett had convinced a sizable cadre of aggrieved persons in the Santa Clara Valley that they could benefit by pursuing civil court action; by September 1928 more than thirty-five death claims had been filed in civil court, and two of the clients signed by Garrett—Ray Rising and Lillian Curtis—would be the most tenacious litigants the city faced over the next two years.[79]

How well Ray Rising, an employee of the Bureau of Power and Light, and Lillian Curtis, whose husband also worked for the BPL at Power House No. 2, were acquainted before March 12 is unknown. But in the wake of the disaster they shared a conviction that the city owed them substantial compensation for the tragedy they endured. Rising's wife, Julia, and their three daughters were killed in the flood, as were Curtis' husband and their two daughters. Memories of how the flood had viciously ripped their families apart sustained the two litigants in a legal odyssey that dragged on for two and a half years.

Ray Rising

Early on, Rising adopted an aggressive posture and, as soon as the Death and Disability Claims subcommittee opened its doors in Los Angeles, he filed a claim for $175,000 ($100,000 for the death of his wife and $25,000 for each of his three children).[80] He immediately met disappointment, with the subcommittee countering that his petitions had to be settled "upon a compromise."[81] Rising soon determined that the process was not structured so that the two parties could engage on a level playing field. Deciding to seek justice through the courts, by early July he had engaged the legal services of Edward Garrett. As is common in tort actions, Garrett soon transferred Rising's case to a specialist in personal injury law, Lawrence Edwards of Stockton. But referrals of this sort provoked the wrath of both Los Angeles officials and the State Bar Association, who denounced lawyers such as Garrett and Edwards as "ambulance chasers...who hover over the scene of wrecks and solicit quick settlements for a nominal fee...The St. Francis matter had crystallized a situation recognized for a long time as a growing evil."[82]

Rising ignored such fulminations, and by August 1928 Edwards had brought his case to civil court. With attorney Len H. Honey of Los Angeles, Edwards filed suit on Rising's behalf in Bakersfield, the Kern County seat, located about 100 miles north of Los Angeles in the agricultural heartland of the San Joaquin Valley. Rising was not alone in making the legal trek to Bakersfield. In that same month in Kern County, Edwards and Honey filed thirty-six other suits against Los Angeles, with claims totaling $1,690,000. The *Los Angeles Times* explained that the suits were filed in Kern County Superior Court both "because of its neutrality" and because "Los Angeles Superior Court, under usual legal procedure, would be barred from deciding cases involving its principality [that is, the city of Los Angeles]."[83]

For the city, the actions of Rising and the other plaintiffs were decidedly unwelcome. Faced with the prospect of making arguments to jurors in a faraway rural enclave, city attorneys moved to have all the cases, including Rising's, transferred to Los Angeles. As reported in the *Times*, the city apparently adopted a practical argument as to why Bakersfield was unsuitable for St. Francis litigation:

> More than 1500 damage cases, each one entailing a jury trial, looms for Kern County if Judge Erwin G. Owen denies a motion made here today for a change of venue from Kern County to Los Angeles County in the hearing of thirty-seven suits against the city...If this possibility develops, it means that Kern County will be the scene of the most extensive damage litigation in recent years, if not the entire history of California.[84]

The threat to Bakersfield appeared clear: either embrace the city's entreaty for a change of venue, or be prepared for the Kern County legal apparatus to grind to a halt. The city's argument proved persuasive, and the lawsuits filed in Bakersfield were soon directed to Los Angeles. City attorneys then embarked upon a vigorous campaign to resist the impaneling of juries to hear the plaintiffs' cases. The strategy of delay worked well for the city, and through 1929 almost all who had filed cases in Bakersfield were forced to reconsider and accept out-of-court settlements.[85] But Ray Rising and Lillian Curtis held fast and continued to pursue their claims.

In March 1929 Rising's attorneys filed a new claim in Los Angeles Superior Court seeking damages of $100,000 for the death of his wife alone (and nothing for his daughters), perhaps in hopes of spurring a serious settlement offer from the city. This stratagem prompted no concessions and he soon modified his suit to include $25,000 for the death of each daughter, the total of $175,000 matching what he had sought at the outset.[86] In their suit, Rising's attorneys minced no words in laying blame for the disaster on the city: the "defendants so negligently constructed, maintained, operated, controlled and managed...St. Francis Dam [that it]...collapsed."[87]

As will be discussed in chapter 6, this accusation was in accord with—and perhaps even a bit milder than—the judgment reached more than a year earlier by the Los Angeles County Coroner's Jury. That jury found that "the St. Francis Dam was defective due to the very poor quality of the underlying rock structure upon which it was built," and moreover, determined that Mulholland and the city had failed to have "the design and foundation conditions passed upon by independent engineers and geologists" and had failed to implement "thorough and systematic methods of design, supervision and inspection."[88] Rising's case did not require any dramatic leap beyond what the Coroner's Jury found in April 1928. His lawyers simply took those findings and applied them in a civil court action.

The city's defense team, headed by Jess Stephens (now retired from his post as city attorney but retained as legal counsel to handle the Rising and Curtis suits), mounted a vigorous denial of the plaintiff's brief. In Stephens' telling, the dam "was located and constructed in accordance with plans and specifications prepared and drawn under the supervision and direction of experienced, qualified and competent engineers." Moreover, the plans "were based upon a sufficient consideration of all the surrounding conditions and...[were] adequate to provide for a dam capable of withstanding all stresses." The charges against the defendants were spurious, concluded Stephens, and the "plaintiff [should] take nothing by this action." This position was not in accord with the Coroner's Jury verdict and it is not surprising that the city fought long and hard to keep the case from going to trial.[89]

"An Act of God"

Legal maneuvering lasted for more than a year, until May 22, 1930, when the case finally came before a jury in Los Angeles Superior Court. Reporting on the start of the trial, the *Los Angeles Times* noted that "several similar suits brought against the city as the result of the flood...have been settled out of court."[90] But attorney Lawrence Edwards and his colleague Warren Atherton had persevered on Rising's behalf and were now prepared to force the city to answer for the quality and efficacy of William Mulholland's dam. In what appears to have been his last professional action as an engineer, Mulholland was called by Stephens to testify that the dam had been built properly and in accord with general engineering practice. Stephens tipped off the press, announcing to the *Times* his overarching strategy: "Efforts will be made by the city to show that the dam gave way because of an earth movement and not because of faulty construction; that the disaster was what is legally termed 'an act of God.'"[91]

When the city was pressed to the wall and needed to make arguments to a civil jury, the willingness to stand up and take responsibility for the dam disaster faltered. Assuming accountability for agricultural property damage and for "reasonable" out-of-court death settlements was one thing. But acknowledging in open court that the traumatic deaths of Julia Rising and her three daughters were the city's responsibility was something quite different. Generic liability as the city wished to define it was acceptable, but when Rising's attorneys argued that the city—and by extension Mulholland—was legally negligent, Stephens invoked "an act of God" as the proximate cause of all the flood deaths.

To rebut the notion that a vaguely defined "earth movement" undermined the dam, Edwards and Atherton called upon F. L. Ransome, a respected geologist at the California Institute of Technology. Ransome had served on the Governor's Commission that investigated the dam disaster (see chapter 6) and was a prominent consultant for the Bureau of Reclamation on the Boulder Canyon Project. When he "testified that there was no seismograph record of an earth movement or an earthquake at the time immediately preceding the breaking of the dam," it became much harder, almost impossible, for the city to sustain the notion that an unforeseen "act of God" caused the disaster. In Ransome's view, the dam collapsed because of defective foundations that Mulholland had failed to take into account when designing and building the structure.[92] This was in alignment with the findings of the Governor's Commission and could not have come as a surprise to Stephens or city officials.

In the trial's most dramatic moment, Ray Rising himself took the witness stand to describe the destruction of his family's wooden bungalow:

> We could hear a terrible roar. The sound grew louder and louder. Then we heard trees snapping. We went to the door to look out. Water was coming. We hurried

> back to get the children. When we got back to the door and tried to open it we could do nothing, as the force of the water held it shut...I suddenly felt myself thrown in the air with a force like from an explosion. I caught the edge of the roof of our house and climbed to the top of it. I ran back and forth calling to my wife and children.[93]

The next time Rising saw his wife and children they lay dead in the Newhall morgue. It was exactly such transfixing testimony that had energized Stephens to fight so tenaciously to keep the Rising lawsuit—and all death claims—from going before a jury. The city's position on death claims had always been predicated on a coolly rational approach, to the exclusion of emotional trauma. But Rising was relentless in insisting that the city be held accountable for the human dimension of the disaster.

The Rising trial lasted less than two weeks and, when handed over to the jury, a verdict came swiftly. After an hour of deliberation the jury found in favor of Rising, but the damage award came to only $30,000 ($20,000 for the death of wife Julia; $5,000 for daughter Dolores, age eight; $3,000 for Eleanor, age five; and $2,000 for Adeline, age two).[94] The relatively modest judgment (at least compared with his claim of $175,000) so angered Rising that he moved for a new trial. Stephens also filed for a new trial, reaffirming the city's contention that the plaintiff's attorneys had failed to prove negligence and, hence, that Rising was entitled to nothing.

While awaiting judgment by the court on the dueling motions, the city began a strategic reevaluation. "I do not think it would be necessary for attorneys on either side to argue the motion for a new trial," City Attorney Erwin Werner told his hired counsel Jess Stephens and Lucius Green, "as I think the Judge already has his mind fully made up that the trial was in all respects fair." Werner surmised correctly, and at the beginning of August, the court dismissed both appeals. Werner then directed Stephens and Green not to pursue further appeal to a higher court. "Even if we were successful in getting a new trial the hazards of suffering a greater judgment against us...would be too great," he reasoned. The goal now was to secure "a final settlement and payment of these judgments [to Rising], providing you can induce the other side to abandon their appeals."[95] Rising continued to resist, but by the end of December he relented under the pressure of an unsympathetic court and opposition from Stephens and Green. Finally, the city agreed to an additional $450 to cover loss of personal property and, close to three years after losing his family, Rising accepted a settlement of $30,450.[96]

Through all of the turmoil accompanying his battle with the city, Rising was not averse to maintaining a professional relationship with the Bureau of Power and Light (which was under the administrative domain of electrical engineer Ezra Scattergood and not led by either Mulholland or his protégé, Harvey Van Norman). After the flood the BPL kept him on as an employee, and in late 1928 he helped in the rebuilding of Power House No. 2. When that job was completed, Rising left the city's rolls for two years. After the trial and claim settlement in 1930, he returned, serving first as a station operator in the San Fernando Valley and then as an employee once more at Power House No. 2. Accompanied by his second wife, he lived there until his retirement in the 1960s, near to where he, Julia, and their three daughters had gone to bed on the night of March 12, 1928.[97]

Lillian Curtis

During the time Rising pursued his case, his attorney, Lawrence Edwards, was also legal counsel for Lillian Curtis. As described in the prologue, she lived with her husband, Lyman, and their three children in a bungalow near Power House No. 2. As the floodwaters engulfed the compound, Lyman pushed her and their son, Daniel, through a window in time for them to scramble to high ground. He then turned and grabbed their two young daughters, but too late. The three were trapped in the torrent and, a week later, they were buried together at Forest Lawn Memorial Park in Glendale.

Once recovered from minor injuries, Lillian Curtis petitioned for compensation. She sought $252,627 for the deaths of her husband and two daughters, for injuries to herself, and for property loss.[98] Like Rising, she was dissatisfied with the reception given her petition, and especially upset by the subcommittee's adversarial approach in insisting "upon a compromise of the claim for the alleged wrongful death[s]." Equally disappointing was the city's ruling that its liability for employees of the Department of Water and Power (including her husband) "was limited to the liability provided...under the Workmen's Compensation Act"—a total of $5,150, as noted above.[99]

By early July, Curtis had signed on with Edward Garrett, and soon her case was referred to attorney Edwards of Stockton. On August 18, 1928, Edwards filed suit on her behalf in Bakersfield, asking for damages of $260,000 ($100,000 for both her and her husband and $60,000 for

FIGURE 5-1. The reconstructed workers' compound at San Francisquito Power House No. 2, July 1931. The new houses for Bureau of Power and Light employees were built close to the site of homes destroyed by the flood. After settling his lawsuit against the city in late 1930, Ray Rising returned to work for the Bureau of Power and Light at Power House No. 2. He remarried and lived there with his family until retiring in the 1960s. [Los Angeles Department of Water and Power]

the two children).[100] Like the other litigants who had filed there, she was subsequently ordered to return to Los Angeles if she wished to pursue legal claims.[101] But unlike the majority of claimants sent packing from Bakersfield, she persisted in her quest for a civil trial. In March 1929 her attorneys filed suit in Los Angeles Superior Court and, as it did with Ray Rising, the city fought hard to delay the case and keep it from coming before a jury.

The Curtis case continued to languish for more than a year. After Ray Rising's suit came to trial in May 1930 and resulted in an award of only $30,000, Curtis began to temper her expectations. Forced to consider anew what would constitute a realistic settlement, she and the city agreed that "her group of claims [were] worth at least the gross verdicts in the Rising cases."[102] On October 21, 1930, she accepted an award of $31,662. Of that, $26,512 was to compensate her for the deaths of her two children, her injuries, and property loss; $5,150 (the amount dictated by Workmen's Compensation) was to cover the death and burial of her husband. By this time she seemed content to get anything at all. "I...thank you kindly for your effort spent in the matter," she wrote Lucius Green, one of the attorneys who had fought her from the outset. "With great pleasure" she acknowledged the "check I will receive."[103]

After a battle of two and a half years, Lillian Curtis was ready to move on—which she did, eventually remarrying and establishing a new life as wife and mother. But memories lingered, and there were occasions when the tragedy returned to center stage, as in 1978 when she and her son, Daniel, attended a gathering of survivors on the fiftieth anniversary of the flood. Decades had passed, but she could still vividly recall her reaction as the waters rushed toward her: "I grabbed my husband and screamed...the dam has broken!"[104]

Thanks to their tenacity, Lillian Curtis and Ray Rising did better than many other claimants. Though they did not get everything they wanted, they were modestly rewarded for their willingness to persevere in the face of the city's intransigence. Despite Charles Teague's protestations, their experiences lend credence to the view that if claimants had held out for civil judgments they would have received more than the out-of-court settlements favored by the city. But as Teague would have been quick to point out, it is also necessary to consider what the "shyster lawyers" who battled the city were paid. Exactly how much of the damages awarded to Rising and Curtis went to their attorneys is unknown, but at a standard contingency rate of 30 percent it would have come to a bit less than $10,000 for each of the two cases. Not a small sum, but it would have been for work carried out over a period of almost thirty months.[105]

"A Disaster That Killed Outright"

When the flood overwhelmed the Santa Clara Valley, many people died, but those lucky enough to survive typically suffered only minor injuries. A nurse at the Red Cross hospital in Santa Paula reported on March 14 that "she had [had] only two cases requiring medical care of any nature. Shock and exposure, necessity for clothing and food seemed to be the principal cause of anxiety for the refugees."[106] Or as Teague's assistant George Travis described it: "This was a disaster that killed outright. Most of those who were swamped by the water drowned; those who escaped were [generally] uninjured, physically...Naturally some who thought they had escaped were subsequently affected by the exposure and confined to bed."[107]

Not surprisingly, personal injury claims paled in comparison to those for death or property loss and, as of July 1929, the claims for injuries suffered by forty-three people totaled only $263,340. Nonetheless, the injured were subjected to a screening just as intense as that faced by those seeking compensation for dead relatives and, of the amount sought for personal injuries, nearly all—some $231,254—was denied by the Death and Disability Claims subcommittee. In total, settlements came to $32,087, with an average award of only $823. At the high end, Lena Frazer received $8,000; only seven other settlements exceeded $1,000, and most were in the range of $300 to $800.[108]

"Rural People" and a Great Metropolis

While some claimants were disappointed with their awards, many were satisfied with—or at least accepted—the city's rehabilitation and reparation effort. Within a week of the dam's collapse, Los Angeles had publicly accepted responsibility for an enormous cleanup, rebuilding, and claim-adjustment program. Soon a workforce approaching 4,000 was helping to repair or rebuild nearly 800 homes, businesses, and public buildings, while also restoring thousands of acres of orchards and farmland. And when, on the first anniversary of the disaster in March 1929, a memorial service was held in Santa Paula, Los Angeles officials involved in the valley's restoration were invited to participate. As the *Los Angeles Times* reported in a story headlined "Santa Paulans Bow in Memory," a plane was to fly the length of the flood zone from the empty reservoir to the Pacific, dropping flowers in memory of unrecovered victims. On the ground, children from Santa Paula's Isdell School were to be "escorted to the cemetery by members of the Santa Paula Lions Club . . . [to] place flowers on the graves of more than half a hundred unidentified victims of the flood." In addition, the *Times* reported:

> Memorial services for the dead in the St. Francis dam disaster, combined with a tribute of thanksgiving for reconstruction and a return to normalcy in a period of twelve months, will be observed...in Santa Paula with a program during the day and a dinner in the evening... [The local Southwest Improvement Club] will conduct a formal meeting in the Isdell schoolhouse, where just

> one year ago water and mud stood three feet deep. The club will have as their guests at the memorial dinner Los Angeles city officials who aided in the rehabilitation and reconstruction.[109]

Los Angeles Assistant City Attorney Lucius Green was on the docket to make a short address at the memorial dinner, as was local leader Teague. Of course, Green (along with Jess Stephens) was fighting Ray Rising's wrongful death suit, and Teague was a strong supporter of Green's efforts. In the weeks and months following the flood, Santa Clara Valley leaders had come to perceive Los Angeles as a partner, not an overlord, in the rehabilitation effort.[110] While Teague was at first skeptical of the city's intentions, his view of the metropolis evolved dramatically: "Never before in the history of the world, so far as I am able to learn," he proclaimed in his memoirs, "was [so] complete and equitable [a] restoration and rehabilitation made by a great metropolitan people to a rural people, where damage had been done and where large sums of money were involved, without recourse to court action."[111] Perhaps some of the "rural people" Teague boastfully championed would have benefited from legal counsel, but on this issue he would tolerate no dissent. While Los Angeles assumed financial responsibility for reparations, Teague and his leadership cohort used their social standing and political influence to guide the allocation and disbursement of these reparations.

As of October 31, 1930, the cost of rehabilitating the flood zone was recorded by the city as $9,392,487.57.[112] Did this constitute the actual cost incurred by the city as a result of the disaster? No. At a minimum, this figure did not include the $4 million or so later expended on the Bouquet Canyon Dam to replace the storage capacity lost when the St. Francis Dam collapsed. Nor did it take account of the monetary value of the 12 billion gallons of water lost when the reservoir burst (estimated in 1929 at over $1.25 million), nor the $352,940 spent to buy electricity from Southern California Edison during the time it took to rebuild San Francisquito Power House No. 2.[113] But $9.4 million is a reasonably accurate figure for what it cost the city to compensate victims for damage and to restore the Santa Clara Valley.

Given the scope of the disaster—stretching more than fifty miles from the dam to the sea and leaving some 400 people dead—it is remarkable that the city's payments for damages were ultimately so low. For this, the city owed a considerable debt to Jess Stephens, who as city attorney relished the role of parsimonious curmudgeon, ever watchful over the interests of municipal taxpayers. And even after Stephens retired as city attorney, he led the fight to make it as difficult as possible for litigants such Ray Rising to get their day in civil court. Rising may have won a damage award larger than Stephens and co-counsel Lucius Green believed justified, but Stephens won the larger war, as Rising proved to be the only plaintiff to present a claim to a jury.[114]

In critiquing the city's response to the St. Francis disaster at a distance of eighty years, it is easy to perceive shortcomings and hypocrisy in the way damage claims were handled. But considered in the context of the 1920s, the city's willingness to bear financial responsibility was not insubstantial, especially in comparison to what New Orleans did for rural communities in southeastern Louisiana after a levee was purposefully breached during the Mississippi River flood of 1927. As vividly described in John Barry's book *Rising Tide: The Great Mississippi Flood of 1927*, city leaders in New Orleans made effusive promises to the people of St. Bernard and Plaquemines parishes that they would not suffer financial hardship if the metropolis was allowed to reduce river levels (and thus protect the city) by dynamiting the levee at Caernarvon. But once the levee was breached and waters were dispersed over low-lying bayous, the threat to New Orleans was averted, and city leaders initiated a mean-spirited and legalistic campaign designed to deprive downstream rural communities of any compensation. This was far different from the way Los Angeles treated the Santa Clara Valley a mere year later. No doubt, Los Angeles' actions were driven by a less-than-altruistic desire to avoid criticism that might threaten passage of the Boulder Canyon Project Act. And the city could have been more generous in compensating victims and in allowing plaintiffs access to civil trials. But Los Angeles' willingness to pay reparations for the St. Francis disaster bore no resemblance to the cynical scheming of New Orleans' leadership in the aftermath of the 1927 Mississippi flood. In this, the City of Angels and its boosters could take a modicum of pride.[115]

Long before restoration work in the Santa Clara Valley came to an end and long before the multitude of damage claims were resolved, people in Los Angeles and California as a whole were forced to confront other challenging questions posed by the disaster. City leaders had recognized the need to take financial responsibility for the flood but, beyond bromides and generalities, what actually caused the St. Francis Dam to collapse? How could people be sure that other dams would not fail in a similar manner? And what effect might the St. Francis debacle have on future dam-building initiatives, especially the proposed Boulder

Dam? The disaster's impact resonated across the West, a region where dams and water supply lay at the heart of a burgeoning economy. Questions about dam safety now came to the fore in ways unimaginable before the night of March 12. Much depended on how political and engineering leaders addressed these questions and explained why William Mulholland's big dam in San Francisquito Canyon had failed so catastrophically.

CHAPTER SIX

The Politics of Safety: Inquest and Investigations

> We overlooked something here. This inquiry is a very painful thing for me to have to attend but it is the occasion of it that is painful. The only ones I envy about this thing are the ones who are dead.
>
> William Mulholland at Coroner's Inquest, March 21, 1928[1]

> The failure of St. Francis Dam need cause no apprehension whatever regarding the safety of the proposed Boulder Canyon dam...On the contrary the action of the middle section of the St. Francis Dam which remained standing even under such adverse conditions is most convincing evidence of the stability of such structures when built upon such firm and durable bedrock as is present in Boulder Canyon.
>
> Governor C. C. Young to Congressman Phil Swing, March 27, 1928[2]

Reparations for damaged property—and in most instances even for the death of loved ones—could be handled with relative equanimity once Los Angeles found common ground with the civic leadership of Ventura County. Investigating the cause of the dam's collapse was related to the issue of reparations, but was in many ways quite distinct. Within hours after the flood, and as scores of bodies were being hauled to makeshift morgues, survivors were gripped by a simple question: how did this happen?

Among Santa Clara Valley residents, speculation about why the dam collapsed of course arose quickly, and not surprisingly, they often blamed William Mulholland. A hand-painted sign with the message "Kill Mulholland" reportedly appeared in front of at least one home near

Santa Paula.[3] Some people looked to state government as a perpetrator of the tragedy, and newspaper accounts at times avowed (mistakenly) that during construction the dam had been inspected by State Engineer Edward Hyatt.[4] And rumors abounded that a dynamite attack had destroyed the dam, presumably carried out by renegades from the Owens Valley.[5] Amid the growing confusion, a need to establish the cause of—and assign responsibility for—the dam's collapse assumed great urgency, not just in Southern California but also in the office of Governor Clement "C. C." Young.

Governor Young was in San Diego on March 13 when he got word of the disaster. That afternoon, he rushed north to Santa Paula to witness the destruction firsthand,[6] and swiftly acted to direct state officials to assist rescue and recovery efforts throughout the Santa Clara Valley. But the crisis Young faced was not confined to cleanup and restoration. He also perceived a more politically complicated problem, one provoked by widespread fear among the public. As newspapers across the state printed shocking stories of death and devastation, people normally unconcerned about the technology of water supply began to wonder if they too might become victims of a disastrous flood. After all, if a major dam built by a major city (and by a supposedly world-class engineer) could fail so horrifically, why should anyone have confidence in the hundreds of other dams spread across California? It was not simply dams already in operation that raised concern. What about major water projects on the drawing board, deemed vital to the state's economic growth? A titanic effort was underway to win authorization for what would become the Central Valley Project (including Shasta and Friant dams).[7] And what about the biggest project of them all, the one focused on building a huge storage dam across the Colorado River that would allow Southern California to draw water and hydroelectric power from the largest waterway in the Southwest?

Mulholland and Boulder Dam

The story of Hoover Dam (prior to September 1930 it was called Boulder Dam) and its relationship to the political economy of Southern California is one of the most important chapters in the hydraulic history of the American West.[8] As Congressional authorization for Boulder Dam hung in the balance in the spring of 1928, a question pervaded the political landscape: how could anyone be certain that the fate of the St. Francis Dam would not befall the huge concrete dam proposed to hold back the mighty Colorado? The question took on special urgency because, through the 1920s, William Mulholland had been a prominent advocate for the Boulder Canyon Project Act (also called the Swing-Johnson Act in recognition of its sponsors, Congressman Phil Swing and Senator Hiram Johnson).

As chief engineer of the Bureau of Water Works and Supply, Mulholland was responsible for developing new sources of water for the city. Anticipating that the Los Angeles Aqueduct would eventually tap the full capacity of the Owens River (as well as the proposed Mono Basin extension), he and other city officials saw the Colorado River as the best means of sustaining the region's economic growth. In the early 1920s he began planning for a new aqueduct to transport Colorado River water more than 200 miles across the Mojave Desert to coastal Southern California.[9]

Mulholland was particularly outspoken in promoting the Boulder Canyon Project because, without the water storage and the hydropower provided by the proposed dam, operation of the Colorado River Aqueduct along the most advantageous right-of-way would prove impossible.[10] As early as 1921 he journeyed with electrical engineer Ezra Scattergood to explore power dam sites along the Colorado, and the next year he spoke at the city's Sunset Club on the importance of the Boulder Canyon Project.[11] In November 1923 he took a well-publicized five-day journey down the river, proclaiming that a dam should be built at Boulder Canyon (or at an alternative site in nearby Black Canyon) and that conveying Colorado River water to Los Angeles was eminently feasible. This excursion was followed by a trip to Washington, D.C., in early 1924, where he testified before Congress in support of the Swing-Johnson legislation.[12]

That year the city of Los Angeles formally filed a water rights claim to 1,500 cubic feet per second of Colorado River flow (almost four times what the city was drawing from the Owens River).[13] Soon Mulholland began lobbying California legislators in support of the Metropolitan Water District of Southern California—the regional agency that would manage the Colorado Aqueduct and convey this river flow to greater Los Angeles. At the end of the year, he again testified at a hearing related to the Boulder Canyon Project and made a second trip to the proposed dam site.[14] In January 1928, when it appeared that the Swing-Johnson bill was close to approval in Congress, Mulholland returned to Washington to lobby for enactment. As it turned out, the proposed hearings were delayed and Mulholland left for Los Angeles without testifying, but he gave Congress a written report arguing that Los Angeles' economic future depended upon Boulder Dam.[15] He arrived back on the Pacific Coast only a few weeks before the collapse of the St. Francis Dam.

FIGURE 6-1. This 1925 publicity photo of a visit to the Colorado River just above the proposed dam site highlights Mulholland's association with the Boulder Canyon Project Act. Although he played no role in designing Boulder (later Hoover) Dam, his plans for the Metropolitan Water District of Southern California's Colorado River Aqueduct relied upon hydroelectric power from the dam for its pumping stations. [DC Jackson]

To avoid misunderstanding, it should be emphasized that Mulholland was never directly involved in the design of Boulder/Hoover Dam. But his support for the Boulder Canyon Project and the Colorado River Aqueduct imbued the St. Francis disaster with political meaning that transcended the effects of the flood on both the Santa Clara Valley and the political hierarchy of Los Angeles. In the public's mind, Mulholland was one of the Boulder Dam project's most prominent engineering advocates.

Governor Young: "Champion of Boulder Dam"

Governor Young's interest in Boulder Dam did not suddenly arise once the St. Francis Dam collapsed, but dated back at least two years. He had been

FIGURE 6-2. Boulder (later Hoover) Dam, shown here soon after completion in 1935, was designed by the Bureau of Reclamation. Boulder is a concrete curved gravity dam, but, in detail, the design bears little resemblance to St. Francis. Opponents of the Boulder Canyon Project Act nevertheless linked the two structures and raised questions about using a design type for Boulder that had failed so catastrophically at St. Francis. [DC Jackson]

elected California's chief executive in 1926 on a platform that strongly endorsed the Boulder Canyon Project Act (his campaign literature labeled him "Champion of Boulder Dam"), and he reiterated this support in his inaugural address in January 1927.[16] On March 12, the day the St. Francis Dam collapsed, he had addressed the young men's division of the Long Beach Chamber of Commerce and told the audience of more than 700 people that "diminishing supplies of water in Southern California make the Boulder Dam a positive necessity for this section."[17] The long-sought legislation was entering a crucial stage in March 1928, and the timing of the St. Francis disaster could not have been worse for Governor Young and other project supporters. The front page of the March 15 *Sacramento Bee* juxtaposed photos of flood damage in Santa Paula with a story headlined "Boulder Dam Favorably Reported to Congress."[18] Six years after it was first proposed to Congress, the Swing-Johnson Act appeared poised for passage. And then the St. Francis tragedy upended the sponsors' well-laid plans.

Today Hoover Dam is often seen as a great national landmark, a symbol of America's wondrous ability to harness the gifts of nature through

FIGURE 6-3. A political leaflet distributed during C. C. Young's gubernatorial campaign in 1926, trumpeting the candidate as "Champion of Boulder Dam." [DC Jackson]

technological prowess. But in the 1920s it was perceived by investor-owned utilities as a baneful incursion by the federal government into the business affairs of the nation. One prominent engineer with close ties to the private power industry went so far as to brand the Swing-Johnson Act "the most vicious piece of legislation that has appeared in Congress since the close of the Civil War...The outstanding feature of this bill is to definitely commit the United States Government to the policy of going into the power business...[It] is wholly socialistic [and] I must oppose the Swing-Johnson bill as a cancerous growth upon the nation of which I am a citizen."[19] If nothing else, this inflammatory rhetoric shows why the supporters of Boulder Dam feared that their opponents might manipulate the St. Francis disaster for their own purposes.

Though Young and State Engineer Hyatt had played no role in the St. Francis Dam's construction, the governor considered it his responsibility to investigate the cause of the failure and, if possible, assuage public

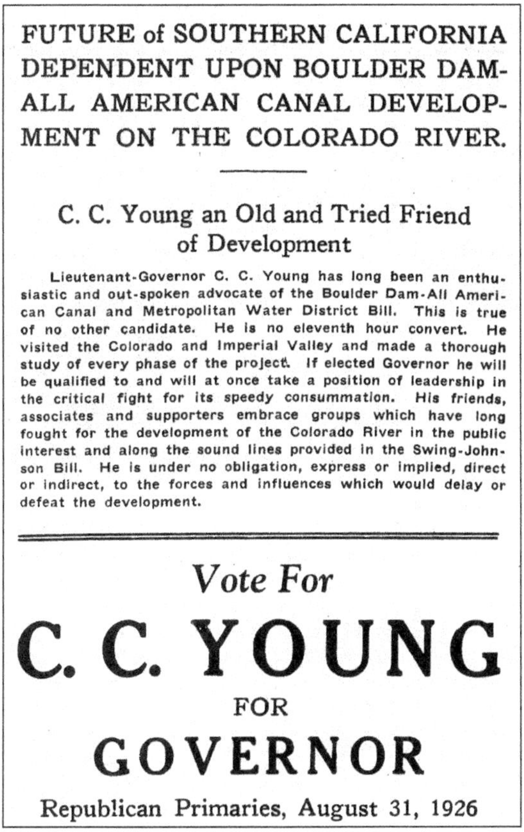

FUTURE of SOUTHERN CALIFORNIA DEPENDENT UPON BOULDER DAM-ALL AMERICAN CANAL DEVELOPMENT ON THE COLORADO RIVER.

C. C. Young an Old and Tried Friend of Development

Lieutenant-Governor C. C. Young has long been an enthusiastic and out-spoken advocate of the Boulder Dam-All American Canal and Metropolitan Water District Bill. This is true of no other candidate. He is no eleventh hour convert. He visited the Colorado and Imperial Valley and made a thorough study of every phase of the project. If elected Governor he will be qualified to and will at once take a position of leadership in the critical fight for its speedy consummation. His friends, associates and supporters embrace groups which have long fought for the development of the Colorado River in the public interest and along the sound lines provided in the Swing-Johnson Bill. He is under no obligation, express or implied, direct or indirect, to the forces and influences which would delay or defeat the development.

Vote For

C. C. YOUNG

FOR

GOVERNOR

Republican Primaries, August 31, 1926

FIGURE 6-4. A prominent advocate of the Boulder Canyon Project Act (often referred to as the Swing/Johnson Act), C. C. Young understood that the St. Francis Dam disaster might be exploited by opponents of the proposed legislation. [DC Jackson]

anxiety about the proposed Boulder Dam. Other investigations sponsored by organizations such as the Los Angeles City Council and the Santa Clara Water Conservation District also released reports in the weeks after the flood. But the report from the commission authorized by the governor would become the best known and most widely publicized. Unfortunately, it would prove less than insightful in identifying the cause of the failure.

Rumors of Dynamite

Even to experienced engineers, the evidence at the dam site—with both sides washed out flanking a surviving center section—was not amenable to easy analysis. The concrete fragments and their location within the debris field, as well as other evidence, had to be carefully studied before a convincing explanation of the failure could be deduced. In the immediate aftermath of the flood, however, theories came quickly, often fanned

FIGURE 6-5. Owens Valley vigilantes attacked the No Name Siphon in 1927, using dynamite to obstruct operation of the Los Angeles Aqueduct. The siphon was back in service within a few weeks, but the attack heightened fears among city officials that other sites would be targeted, including the St. Francis Dam. [*Los Angeles Times* Photographic Collection, Department of Special Collections, Charles E. Young Research Library, UCLA]

by newspaper editors willing to offer bold pronouncements based on a few presumed facts. In the heat of the moment, the disaster was at times ascribed to an earthquake or earth movement, or, as noted earlier, to a dynamite attack. Vigilantes in the Owens Valley had carried out attacks on the Los Angeles Aqueduct since 1924, and in May and June 1927 dynamite explosions at No Name Siphon and at the aqueduct near the Cottonwood power plant had seriously disrupted the city's water supply.[20] In this light, the notion of an attack targeting the St. Francis Dam was not so far-fetched.

Immediately after the flood, the *Los Angeles Examiner* mentioned a "slight earth tremor" as the possible cause, with "another report, unconfirmed...that it might have been dynamited."[21] The *Los Angeles Daily*

FIGURE 6-6. Mulholland visited the site of the collapsed dam twice in the week after the disaster; this view shows him in the company of Assistant Chief Engineer Harvey Van Norman. In his first day of testimony to the Coroner's Jury, Mulholland alluded to "human aggression" as a cause of the failure. [Peirson Hall Papers, Huntington Library]

News also fueled such speculation, noting that "the dynamite theory was given support by the report to deputy sheriffs by a motorist who stated that he was on the ridge road below the dam when he saw a flash that illuminated the heavens." The *Los Angeles Herald* was more circumspect, acknowledging that "rumors of dynamiting also were started but were disproved by the statement of the water bureau officials."[22]

Mulholland appeared willing to embrace the dynamite theory and in a March 14 newspaper story headlined "Mulholland Lays Break to Ground Shift," he was quoted: "From the short study we were able to make, it appeared that there had been a major ground movement in the hills forming the western buttress of the dam."[23] A "major ground movement" could be caused by an earthquake tremor or by a dynamite blast, and Mulholland left open the possibility that either event could have precipitated the collapse. When his colleague Arthur P. Davis met with him privately soon after the flood, Mulholland "indignantly denied" that the dam was poorly built and, in Davis' words, "appeared to believe that something other than natural causes had led to the failure."[24] An earthquake would fall into the category of "natural causes," leaving dynamite as the presumed cause—at least in Mulholland's mind.

By the end of the first week after the disaster, the dynamite theory was given some specificity, but the story told was—at best—barely credible. As recounted in newspapers under headlines such as "Dynamiting Theory as Map of Dam, Rope Found," a segment of rope was reportedly found on the ridge above the dam. Hardly momentous proof—except that this rope was alleged to be similar to one found near the site of the attack against the No Name Siphon the year before; alternatively, the rope was linked to a hand-sketched map found on a street in Hollywood (some forty miles from San Francisquito Canyon) that supposedly illustrated the assault on the dam. This limited evidence—along with reports of dead fish in the drained reservoir that were supposedly killed by an explosive blast—constituted the basis of the theory that dynamite caused the collapse.[25]

Skepticism arose almost immediately. Assistant District Attorney E. J. Dennison stated directly that "the St. Francis Dam was not dynamited," and in the same news story Deputy Sheriff Harry Wright unequivocally proclaimed: "Nothing has been presented to this office in the way of evidence to substantiate the rumors of dynamiting...There is no foundation for the wild rumors being 'hopped up.'"[26] The *Los Angeles Record* decried the supposed dynamite attack as a "red herring" intended to divert attention from the city's responsibility for the disaster.[27] Nonetheless, once the dynamite hypothesis was brought into the public arena, like so many conspiracy theories that litter the historical landscape it proved impossible to eradicate. It reappeared near the end of the Coroner's Inquest and lingered in the realm of public discourse for years after.[28] But no credible evidence of a dynamite attack has ever surfaced.

"We Must Have Reservoirs"

Although State Engineer Edward Hyatt and his predecessor Wilbur F. McClure were never formally involved in reviewing or approving the St. Francis Dam design, reporters and the public looked to state officials for information about the disaster. While it appears that Hyatt was judicious in his initial comments, the ever-eager press credited him with statements that he probably never made. On March 13 the *Los Angeles Herald* announced that Hyatt was to "probe [the] dam collapse" and also asserted that he had "inspected the dam...during construction."[29] Two days later the paper reported on Hyatt's visit to the site, claiming that he had "blamed" the collapse on "faulty construction," declaring that "this improper construction would never have been possible under state supervision."[30]

Exactly what Hyatt did say to the press is uncertain, but he clearly believed that reporters had taken liberties. In a letter to R. F. del Valle, president of the Board of Water and Power Commissioners in Los Angeles, he explained that "[my] statements regarding the St. Francis Dam [have been] misinterpreted and misquoted...As you are aware it is not uncommon for newspaper statements to become garbled in retelling."[31] He also dispatched a letter to Harvey Van Norman professing his unhappiness "over the widespread and loose use of my name in connection with the failure of the St. Francis Dam, and particularly over the misquotation to the effect that the dam itself was faulty, and that the responsibility was to be placed upon the city. Such statements were not made... I cannot help but feel that I have been victimized by a rabid press." For clarification, Hyatt restated his views: "first, that the dam was designed and constructed by the Los Angeles City Water Department without state supervision, and second, that in my opinion failure resulted from trouble arising in the foundation."[32]

By March 15 Hyatt, his superior B. B. Meek, director of the Department of Public Works, and Governor Young had decided that a special commission, authorized by Young, would investigate the dam's collapse.[33] With the press jumping at every opportunity to publish accounts of the disaster and its cause, the ensuing journalistic chaos threatened to upend public trust in the state's water supply systems. In addition, it appears that the governor deemed it unacceptable that all formal investigations into the dam failure be tied to or sponsored by parties in Southern California. Although no specific state law could be drawn upon to justify such a commission, Young believed that the interests of citizens throughout California were at stake, and thus the state itself had reason to initiate an investigation. Personnel accepting assignment to the Governor's Commission would do so gratis, thus limiting expenses while also lending the panel an air of impartial public service. On March 15, Young telegrammed A. J. Wiley of Boise, Idaho, requesting that he serve as commission chair; Wiley accepted, making plans to come to California immediately.[34]

Wiley was a highly respected consulting engineer who, after coming west from Delaware in the 1880s, worked on numerous dam and water projects.[35] He had served as a consultant for several Reclamation Service (later Bureau of Reclamation) projects, most notably the proposed Boulder Dam, and was well known to the state engineer's office through his design of two major concrete gravity dams (Exchequer Dam for the Merced Irrigation District and Don Pedro Dam for the Turlock and Modesto Irrigation Districts). In accordance with the state's 1917 dam safety law, these two projects had been built under the supervisory authority of that office. Wiley, now sixty-five, unquestionably had the experience and gravitas to stand up to Mulholland and the city of Los Angeles. The others on the commission—engineers F. E. Bonner of the U.S. Forest Service, H. T. Cory of Los Angeles, and F. H. Fowler of San Francisco, and geologists F. L. Ransome of the California Institute of Technology and George Louderback of the University of California, Berkeley—were also seasoned professionals, but Wiley stood out as the clear leader.

In charging Wiley and the other commission members with "learning just what caused the failure of the St. Francis Dam," Governor Young stressed that "the prosperity of California is largely tied up with the storage of its flood waters. We must have reservoirs in which to store these waters if the state is to grow. [And] we cannot have reservoirs without dams."[36] As a result, the commission's task proved to be twofold: find the cause of the failure, and discern "the lesson that it teaches us [that] must be incorporated into the construction of future dams."[37] In other words, faith in dam construction needed to be affirmed so that the state, with the support of its people, could move forward with major water projects.

Records documenting the work of the Governor's Commission are essentially nonexistent, as it appears that all drafts and other records relating to its deliberations were destroyed upon completion of the final report. However, a surviving three-page memorandum provides a bare-bones account of the committee's activities, beginning, "Mr. A. J. Wiley, chairman, arrived at Sacramento the morning of March 18th, [and] spent that day in consultation with Governor Young, Director of the Department of Public Works B. B. Meek, and State Engineer Edward Hyatt."[38] Unfortunately, the memo provides no specific account of what was discussed at these or any other meetings, but it makes clear that Wiley and the governor had extensive interaction prior to the start of the investigation. That evening Wiley headed south to Los Angeles, where the committee as a whole was to start its investigation the next day. On Monday morning, March 19, he issued a brief statement to the local press avowing that "we seek the facts," and he and the committee offered no further comment. Even after the report became public a week later, Wiley and his colleagues held no press conferences and granted no known newspaper interviews.[39]

"We Overlooked Something Here"

Because of the many deaths caused by the dam's collapse, it was always understood that the Los Angeles County Coroner's Office would convene an inquest to investigate the disaster and weigh the possibility of criminal charges. Coroner's inquests are not uncommon. What distinguished the St. Francis disaster, however, was the scale of destruction. Sixty-nine bodies had been found in the county within a week of the collapse and many more presumed victims remained lost. (Of course, bodies recovered in Ventura County were the subject of inquests in Fillmore, Santa Paula, Moorpark, Oxnard, and Ventura; see chapter 4.) District Attorney Asa Keyes also had a legal obligation to determine if any crimes had been committed in the design, construction, or operation of the St. Francis Dam, and his responsibility overlapped with the work of the coroner's office. At first it appeared that Keyes would carry out a separate inquiry, and to that end he appointed a board of engineers and geologists.[40] But not long after he began participating in the Coroner's Inquest, he opted to drop his own investigation, with the understanding that his board would also present its findings to the inquest jury.[41]

The Coroner's Inquest started at 10:00 a.m. on Wednesday, March 21, at the Hall of Justice in downtown Los Angeles.[42] With Coroner Frank A. Nance presiding, a jury of nine citizens, accompanied by District Attorney Keyes, Assistant District Attorney E. J. Dennison, Assistant City Attorneys Herman Mohr, S. B. Robinson, and K. K. Scott (the latter two representing the Board of Water and Power Commissioners), and Ventura County District Attorney James C. Hollingsworth, solicited testimony from more than sixty witnesses, most notably William Mulholland.[43] The court's transcript of the inquest logs in at more than 800 pages and offers a unique window into the disaster and into the ways that Mulholland and his underlings were publicly queried about the dam. Some of the testimony is enlightening and extremely useful from a historical perspective. In contrast, long stretches are bogged down by meandering answers to less than cogent questions. But despite limitations, the transcript documents incontrovertibly what Mulholland said at the inquest regarding the dam, its condition on the morning of March 12, and its collapse. For this alone it stands as one of the essential records of the disaster.

The inquest started with a short series of questions posed to Ray Rising, one of the three people who survived the deluge that engulfed Power House No. 2. In accordance with legal protocol, Rising was asked by the coroner if he had identified his wife's body in the Newhall morgue on March 13, and he responded, "Yes sir."[44] Julia Rising served as a proxy for all the flood victims and as the legal basis for the inquest, but her body was not present there, as one writer has claimed (she and her three daughters had already been buried at the Oakwood Cemetery in Chatsworth).[45] After Ray Rising's testimony, which lasted no more than a minute or two, Nance read into the record the names of all sixty-nine victims recovered in the county (three were listed only as "unidentified Jap," "unidentified boy, white, 4," and "Jane Doe–75"). Then Dr. Frank R. Webb, the assistant autopsy surgeon for Los Angeles County, testified on his examination of Julia Rising's body on March 15 in a San Fernando mortuary. His finding: "death was due to drowning." That concluded the preliminaries, and Mulholland was called to the stand.[46]

Mulholland's testimony lasted for over an hour—filling thirty-one pages in the inquest transcript—with Coroner Nance and Assistant District Attorney Dennison leading the inquiry. Their initial questions addressed background on the selection of the St. Francis site and on the dam's design and construction.[47] Soon the coroner homed in on dam keeper Tony Harnischfeger's March 12 phone call and Mulholland's subsequent trip to the dam. Here, the issue of water seeping under or through the west abutment came to the forefront for the first (but certainly not the last) time, and the questions proceeded from the assumption that failure began on the west side of the canyon. Mulholland remained adamant, almost defiant, insisting that the leakage he observed during his March 12 visit was clear water, not muddy, and that he had been unfazed by its possible effect on the dam's stability. As Nance and Dennison probed further, asking about the west side leakage, the general character of the dam site, and the quality of construction, three memorable exchanges occurred. In modern parlance, these constitute the "sound bites" of Mulholland's first day of testimony:

Q. [Coroner]: What was the object of the construction of the St. Francis Dam?

A: [Mulholland]: The conservation of water. Sometimes our water is derived as the Lord makes it. Sometimes it is in buckets full in the mountains and as fast as we can use it and there are periods when there is a dearth of water and then we have to use the reservoirs. The usual use of a reservoir is to conserve water. Some cities do not have to have them because they are fronting on great rivers or lakes or great sources

FIGURE 6-7. Illustration depicting the first day of testimony at the Los Angeles County Coroner's Inquest. Coroner Frank Nance presided over the hearing. The first to take the stand was Ray Rising, who testified that he had identified the body of his wife, Julia, in the Newhall morgue. But the key witness was William Mulholland, who answered questions about the dam, its construction, and possible causes of the disaster. In his testimony, he alluded to a "hoodoo" that may have haunted the ill-fated dam site. [California Scrapbook No. 8, Huntington Library]

of inexhaustible supplies. The City of Los Angeles is one of the unfortunate cities in regard to a water supply. We live under a hazard here. The people of the city take this as a matter of course, that there is going to be water in the faucet the next morning. There have been various times in my experience here—I have been in charge of these works for nearly fifty years—where we have been pretty close to the bottom. That is the purpose of reservoirs, and we could not exist without reservoirs at all.

Q: In the construction of the dam state what precautions and prudence were exercised, as an engineer who has constructed dams, in the construction of this dam?

A: The conventional observations that are used in building dams. I have had rather more experience than most engineers in building dams. I have built nineteen of them and took the usual precautions, judging the [geological] formation as best I could, the hazards which the dams must be exposed to and all the things that relate to the continuous safety of the dam. We overlooked something here. This inquiry is a very painful thing for me to have to attend but it is the occasion of it that is painful. The only ones I envy about this thing are the ones who are dead.[48]

What is usually remembered about this exchange is Mulholland's "envy" for the dead. But in context this sentiment does not seem particularly contrite, as he also boasts of nineteen dams that he built with "the usual precautions." As discussed in chapters 2 and 3, he appears oblivious to the fact that the St. Francis design fell woefully short of what contemporary gravity dam designers would consider best practice.

Questioning eventually returned to the west side leakage that prompted Harnischfeger's phone call:

Q [Dennison]: You say that the water was pouring through there or cutting through there [on the west side abutment], that there would be a natural erosion in the conglomerate?

A [Mulholland]: We at the time thought there was not because the water is very clear.

Q: Of course, I don't know anything about dams and you know something about them.

A: I am here to give you all that I know, and I swear to God that my oath is as binding on me as—

Q: Have you an explanation as to the cause of the failure of this dam?

A: I have no explanation that could be called an explanation, but I have a suspicion, and I don't want to divulge it. It is a very serious thing to make a charge—to me it is a sacred thing to make a charge, even of the remotest implication.

Q: Colonel Mulholland, this is a very important matter to everybody.

A: Yes sir. It is most important to me. Several human beings are dead.

Q: Of course, if it is only a suspicion it does not amount to anything, but can you—that is all you have to offer, that you have a bare suspicion?

A: Yes sir, I don't want to offer it even.[49]

From this vague invocation of "suspicion," it is commonly inferred that he believed the dam had been the target of a dynamite attack, akin to assaults on the aqueduct in the Owens Valley. A few minutes later, when the issue of whether he would build another dam on the same site was raised, his allusion to "human aggression" became explicit. But it was complicated by reference to a supernatural force that Mulholland, apparently in the grip of a superstition, termed a "hoodoo."

Q: [Dennison]: You said that you would not build this dam again in the manner in which it was constructed. Will you tell us why not?

A: [Mulholland]: In the manner in which it was constructed?

Q: Yes.

A: I build all dams in that manner.

Q: I understood you to say if you had to do it all over again you would not build this dam in the same way it was erected.

A: Not in the same place.

Q: Why not?

A: Well, it fell this time and there is a hoodoo on it, that would be enough for me.

Q: A hoodoo?

A: Yes, it is vulnerable to human aggression and I would not build it there.

Q: You don't mean that because it went out on the morning of the 13th?

A: Perhaps that, but I did not think of that before, but that is an additional hazard. I had not thought of that.[50]

Had the coroner, the jury, or any of the lawyers at the inquest wished to publicly excoriate Mulholland for believing that a "hoodoo" haunted the St. Francis dam site, they had every opportunity to do so. But in a telling sign that the inquest would not be used to destroy Mulholland personally, the comment passed without rebuke. Deputy City Attorney K. K. Scott quickly changed the subject to the way the reservoir impacted power generation, and the word "hoodoo" never came up again.

Near the end of Mulholland's time on the witness stand, Coroner Nance asked him, "Where do you believe the dam broke first?" Mulholland responded, "I am inclined to think it broke on the west side." And to the follow-up question, "How do you account for the wing on the east side being broken out from the center, almost identically like it did on the west side?" he replied, "The reason is that the west side let go a tremendous water and over against the east side, and there being there below the dam was rocks [*sic*] in the formation—great caves came down from above—that made the least little bit of a easement there, and if you take and ease up on an arch, the arching effect is gone."[51] It is noteworthy that Mulholland had accepted, and was espousing, the "west-side-first" failure mode. But in his less than lucid response he refused to give an opinion on exactly what had precipitated the collapse.

While he expressed envy for the dead, in his first day of testimony Mulholland did not take any responsibility for the dam's collapse, implying instead that "human aggression" or perhaps a "hoodoo" was the root cause of the disaster. Mulholland would not return to the witness stand for almost a week, and only then, after the release of the Governor's Commission *Report,* would he acknowledge responsibility for the disaster.

The local press gave the first day of the inquest major coverage and Mulholland's testimony was the big story. The *Los Angeles Times* treated his time on the stand with dignity, labeling a photo with the headline "Builder of Nineteen Dams Defends One That Failed" and leading the main story with "Mulholland...Defends Work and Hints at Dynamiting."

The same story paraphrased his declaration as "I envy the dead" and described him as a white-haired engineer nearly eighty years old (he was actually only seventy-two) who "walked to the stand with feeble steps."[52] No mention is made in the *Times* of a "hoodoo," which may have been an editorial decision designed to shield him from ridicule. In addition, a short piece printed near the main inquest story was headlined, "Praise Accorded to Mulholland," which came from a group identified as "California Progressives"—further evidence that the *Times* had little interest in discrediting The Chief.[53] Other papers were not so deferential. In particular, the *Los Angeles Herald* made pointed reference to the "hoodoo" comment in a story headlined "Mulholland Sobs Story of Dam Disaster Inquest." The *Herald* also observed that "Mulholland looked ten years older today than he did a week ago." But, overall, the person most responsible for the St. Francis Dam survived the inquest's first day without major opprobrium.[54]

After Mulholland left the stand, the jury viewed a film made at the dam site a few days earlier. Although many still photos were examined in the course of the inquest, the coroner deemed it important that the jury also see a motion picture of the disaster scene.[55] Two days later the jury traveled as a group to San Francisquito Canyon and witnessed post-disaster conditions firsthand.

Lengthy questioning of construction superintendent Stanley Dunham and James Phillips, the engineer in charge of field surveying, consumed the rest of the first day. The questions elicited basic facts about the construction and also drew out information from Dunham about the east side tunnel (discussed in chapter 3). But the east side abutment was of limited interest to the jury and the reference to the tunnel passed without attracting much attention. Seemingly of more interest were descriptions of the concrete and whether or not any dirt or clay had been mixed in with the cement and aggregate.[56] On this topic, the jury and the various attorneys did not exhibit a particularly sophisticated understanding of gravity dam technology.

Starting the inquest's second day, Assistant Chief Engineer Harvey Van Norman took the stand, offering a limited amount of information about the dam and reporting on his March 12 inspection trip with Mulholland.[57] Van Norman had been on leave from the Bureau of Water Works and Supply from May 1923 through August 1925, when he was in charge of building the city's north outfall sewer line and then served as city engineer.[58] After an absence of twenty-seven months, he returned to the water bureau as assistant chief engineer. As a result, Van Norman possessed what a lawyer might call plausible deniability about issues involving the St. Francis Dam: he simply wasn't present during the design phase and most of the construction. By the time he returned to service as Mulholland's top assistant, the dam was approaching completion and he had little to do with it. Van Norman's testimony is not particularly informative except in relation to the March 12 trip to the dam, where he offered nothing that contradicted Mulholland's account. Excused by the coroner before lunch, he was never recalled for further questioning.[59]

Following Van Norman, other members of the BWWS were called to testify, including surveyor Harold B. Hemborg, assistant civil engineer and surveyor R. R. Proctor, concrete chute rigger Edward V. Hendrick, concrete mixing foreman David S. Menzies, and concrete mixer Richard Bennett. From a historical perspective, Bennett's testimony is perhaps most notable among these because it corroborated Dunham's description of the east side tunnel. Overall, the testimony of these workers provided a reasonably straightforward description of the basic regimen followed during the dam's construction.

"Dirty, Reddish-Colored Water"

Before adjournment on the inquest's second day, a different sort of testimony began, focused not so much on how the dam was constructed as on what conditions were like at the dam and the canyon below in the days preceding the disaster. Some of the witnesses were employees of the city and others were ranchers who lived near the dam or traveled past it. They were asked to describe the site prior to the flood, and perhaps present views that contrasted with or contradicted Mulholland's assertions. Of course, after the dam collapsed it was easy to offer recollections of how unsafe it might have appeared. Aspects of these recollections may well have been exaggerated or embellished, but they cannot be simply dismissed.[60]

The most famous such testimony was that of David C. Mathews, the last witness to appear on Thursday, March 22. The forty-seven-year-old Mathews lived in Newhall but worked as a general laborer at Power House No. 2, and he was in San Francisquito Canyon the afternoon before the dam broke. Historian Charles Outland gives prominent attention to Mathews in his book *Man-Made Disaster* as the one witness to claim that Mulholland knew the dam was in perilous condition on March 12 but did nothing to warn or protect people living downstream.[61]

Mathews was part of a crew that, on the afternoon of March 12, had been directed to block off the adit that allowed water flowing down

San Francisquito Creek to enter the aqueduct at Power House No. 2. This blockage was intended to keep brush and debris carried by storm runoff from flowing into the aqueduct.[62] It also made sense at the time because more than enough water was descending through the penstocks to serve downstream demand in Los Angeles. Mathews recounted that his co-worker Harley Berry went up to see dam keeper Tony Harnischfeger to retrieve equipment needed to help close the adit. When Berry returned, Mathews testified that he was told by Berry that, according to Harnischfeger, Mulholland and Van Norman had said the dam was unsafe and could fail at any time. But Berry was not to tell anyone because this would only cause trouble at Power House No. 2. When Mathews heard the warning from Berry, he too chose to tell no one this seemingly momentous news. At least, he did not tell anyone about it until after the flood.

In Outland's analysis, Mathews' story is less than persuasive because, if Harnischfeger and Berry were persuaded that Mulholland thought the dam was on the verge of failure, why did both of them stay in the canyon, with their loved ones, below the dam that night? Because both men perished, they could neither impeach nor confirm Mathews' account, and there the matter stood. However, what Outland was unaware of was the statement made by Archie Eley in October 1928, describing a phone call made by Leona Johnson, Harnischfeger's companion, at 6:30 p.m. on March 12 (see chapter 3). Leona reportedly said to Eley that Tony "had orders to stay on the job because the dam was leaking so bad and he had orders to report three times a day."[63] Eley's statement—made six months after the disaster—does not stipulate that Mulholland said he knew the dam was going to fail. Nonetheless, it lends some credibility to Mathews' account of what Berry told him regarding Harnishfeger's alleged warning. However, Mathews was clearly distraught over the deaths of his family members, and this may have colored his post-disaster recollection of the conversation.[64]

Mathews' testimony carried over to the next session of the inquest, but before he returned to the stand, the jury had seen the dam site firsthand.[65] Reporters were allowed to tag along during the jury's daylong trip on March 23 and, although interviews were not allowed, they could observe what attracted the jurors' interest. From these observations came newspaper headlines such as "Jury Views Dam Site: Inspectors Test Canyon Rock" in the *Los Angeles Times*; "Jury Analyzes Samples of Dam Rock" in the *Los Angeles Herald*; and "Take Red Samples" and "Experts May Blame Site for the Disaster" in the *Examiner*. In particular, the *Examiner* article flatly stated: "The coroner's jury is more interested in the

FIGURE 6-8. Members of the Coroner's Jury examined the dam site on Friday, March 23. They were shielded from the press during their visit, and officially they were to refrain from publicly discussing what they observed. But their interest in the geology of the west side conglomerate was soon reported in local newspapers. [DC Jackson]

site upon which the dam was built than they are in the structure itself."[66] The inspection trip affirmed a shift in the inquest's focus, away from the design and construction of the dam and toward the nature of the foundation rock, especially the red conglomerate forming the west abutment.

The importance of the jury's site visit and the rock samples they gathered became apparent on Monday, March 26, not long after Mathews returned to the witness stand. The inquest transcript provides a limited portrayal of that morning's drama, as it reflects only spoken questions and responses. But when given context by a newspaper account, the importance of what occurred becomes clear.

After questioning Mathews about how he had observed "soaked" and "muddy" conditions along the west side abutment, Assistant District Attorney Dennison directed his attention to a rock sample:

Q. [Dennison]: I am going to show you a piece of earth or rock or whatever you call it, and ask you to look at it and tell the jury whether or not that is the kind of formation of that hill where you saw the water, when you speak of the red.

A. (Witness examines a piece of rock and tests it with his knife) Yes sir, that resembles that clay very much.[67]

Then the questioning returns to leaks and their location, and the "piece of earth or rock" presented to Mathews is seemingly forgotten. But two pages later in the transcript (representing a few minutes in real time), Dennison suddenly refers to a muddy tumbler of water:

Q. [Dennison]: Is that the color of the water that was running down there (indicating a tumbler of dirty, reddish colored water)?

A. [Mathews]: Yes, it resembles it. It had a reddish color to it.[68]

This is the only time Mathews refers to the "tumbler of dirty, reddish colored water" in his testimony, but it came to represent one of the defining episodes of the inquest, at least in popular retellings. To better understand what took place that morning, it is useful to turn to a more dramatic newspaper account, while recognizing that reporters were not immune to a bit of embellishing. This description is from the *Los Angeles Record*:

> [D. C. Mathews] was on the stand when Dennison worked the experiment to prove that the rock at the west end of the dam was not solid and impervious as the water board engineers [that is, Mulholland] testified...Then Dennison placed a tumbler of water before Coroner Frank A. Nance and dropped the fragment of rock about the size of a walnut into the water which had taken on a reddish, muddy hue and it was seen the serpentine had disintegrated. When the water cleared,

> a layer of reddish sediment was deposited in the tumbler. Mathews was asked whether the water he saw issuing from the west hill of the dam was similar in color to that in the tumbler. He said it was. Dennison made no comment, but left the glass of water in full view of the jurors.[69]

The dissolving conglomerate specimen seemed to explain ever-so-graphically what had caused the disaster. Mathews himself made this argument as his testimony concluded:

Q. [Dennison]: Of course it is only an opinion of your own, but have you an opinion as to how the [west side] hill became saturated?

A. Yes sir.

Q. Tell the jury what you feel about it and the reason upon which you base your opinion, if you can?

A. In my experience in handling water, the condition of the hill and that soil—it was very plain to me that the soil was of such a nature that it would sub there, and after a certain length of subbing, the whole thing would get soft.

Q. You mean by that, that this great structure which was placed on the hillside there, had pressed into it that it opened the fissures of the hill, is that what you mean by subbing?

A. Subbing is the natural soaking of the water underneath the hill. In handling the water anywhere and putting in small dykes or dams of any kind, my experience is when a dam becomes so soaked and saturated it will naturally become softer and softer until it gives way.

Q. [By the coroner] In other words, the formation under the dam would become full of water and the water would seep out side?

A. Yes sir.[70]

Although a layperson, Mathews had described a process leading to the dam's collapse that seemed to make perfect sense. The red conglomerate on the west side softened because of leaking water. Then the softened muddy rock disintegrated, like the specimen in the tumbler, and the dam collapsed. It was a seemingly logical analysis, but if true, Mulholland's testimony was almost certainly false, or at least wildly inaccurate in claiming that the leakage was clear. The stage was now set for Mulholland to make a dramatic return to the witness stand. But before this occurred, the Governor's Commission issued its report.

Defective Foundations

After A. J. Wiley met with Governor Young in Sacramento on Sunday, March 18, he set out by train for Los Angeles. The next morning he and the commission joined B. B. Meek, director of the Department of Public Works, and State Engineer Edward Hyatt to make "plans for a procedure." The group then "adjourned to meet Mr. Mulholland by appointment at his office." There is no record of what they discussed, but it may have been largely a bureaucratic formality, with the committee questioning Mulholland about the basic chronology of the dam's design and construction and also requesting access to all of the water department's files related to the dam. In this context, it should be understood that the Governor's Commission had no legal right to demand access to any BWWS records; they were voluntarily made available by city leaders. However, had Mulholland or other officials declined to facilitate the committee's work, they would almost certainly have been excoriated by the press and by civic leaders in the Santa Clara Valley.

On Tuesday, March 20, the commission spent a day inspecting the St. Francis site, and the next day it reconvened in Los Angeles with geologist F. L. Ransome, who had just arrived from Nevada. The full committee returned to San Francisquito Canyon on Thursday in hopes of better identifying "the various fragments of the dam that had been washed downstream."[71] The following day members witnessed "testing of the samples of concrete and conglomerate" and then started work on their report. Clearly the investigation was proceeding at a whirlwind pace, with Wiley seeking to get the committee's conclusions into the hands of Governor Young as soon as possible. Without crediting any individual as a source, on Saturday, March 24, the *Los Angeles Express* leaked the news:

> Declaring they were in thorough agreement on the essential facts leading to the collapse of the St. Francis Dam, members of the engineering and geological commission appointed by Governor Young to investigate the catastrophe will lay the full report before the governor Monday.[72]

The committee worked steadily through the weekend and, at 5:00 p.m. on Sunday, signed the "completed report agreed upon by all members [although] photographs, maps, etc. were not yet complete." By the next

FIGURE 6-9. This post-disaster photograph shows the west side/conglomerate formation in the foreground. The integrity of the conglomerate when saturated became an issue of great concern to the Coroner's Jury when small samples readily dissolved in a glass of water. Whether this reflected the condition of the foundation as a whole remained a contentious issue. [Richard Courtney Collection, Huntington Library]

day all details were cleared up and "three copies were taken north" to Governor Young in Sacramento.[73] Late Monday, the report became public and the next day its conclusions were broadcast in newspapers across the state. Within a month, the State of California formally published the report in a pamphlet that included fifty-one pages of illustrations and photographs. Widely distributed, it became the de facto standard account of why the dam failed.[74]

Although written under an extremely tight schedule, the Governor's Commission *Report* was a very polished and, at least on the surface, very complete document. In straightforward prose it takes the reader through an explanation of the site geology and of how the dam was constructed; it provides a timeline of reservoir levels from March 1926 through March 1928; it details the flood's passage down the valley from the inundation of Power House No. 2 (five minutes after Edison's Lancaster-Saugus power line went down) through the destruction of the Edison camp at Kemp, the inundation of lower Santa Paula, and the destruction of the Montalvo Bridge near the Pacific Ocean shortly before daybreak. It also discusses the propensity of the red conglomerate to disintegrate when submerged in water and notes that the flood gauge attached to the surviving center of the dam appeared to record a gradual drop in the reservoir level in the hours preceding the collapse.[75] Significantly, the report rejects the possibility that an earthquake or major earth movement shook the dam prior to the disaster. And, giving no credence to any alleged dynamite attack, it implicitly dismisses theories predicated upon, in Mulholland's phrasing, "human aggression."

Within the report's descriptive matrix lies the commission's assessment of the reasons underlying the dam's collapse. Anyone who had been reading newspaper accounts of possible failure modes in the week after the flood would not have found the commission's focus on a west-side-first foundation failure surprising. However, because the report was sponsored by the highest level of state government, the commission's conclusions took on special import and authority.

While asserting that "the foundation under the entire dam left very much to be desired," the commission tagged the west abutment as the culprit in the calamitous collapse: "The west end was founded upon a reddish conglomerate which, even when dry, was of decidedly inferior strength and which, when wet became so soft that most of it lost almost all rock characteristics."[76] The report averred that the softening of this conglomerate undermined the west side and precipitated the first breach and the first massive release of water:

> Many of the available data indicate that the initial foundation failure occurred near or at the old fault or contact between the conglomerate and the schist under the west end, and was due to the percolation of the water into and through this section of the foundation, with the resulting softening of the conglomerate under the dam. Either a blowout under, or a settling of the concrete at this place, or both, occurred, quickly followed by the collapse of large sections of the dam.[77]

Report of the Commission

Appointed by

GOVERNOR C. C. YOUNG

to Investigate the

Causes Leading to the Failure of the St. Francis Dam

NEAR SAUGUS, CALIFORNIA

A. J. WILEY, Chairman, Boise, Idaho
Consulting Engineer

GEO. D. LOUDERBACK, Berkeley, California
Professor of Geology, University of California

F. L. RANSOME, Pasadena, California
Professor of Economic Geology, California Institute of Technology

F. E. BONNER, San Francisco, California
District Engineer, U. S. Forest Service
California Representative of Federal Power Commission

H. T. CORY, Los Angeles, California
Consulting Engineer

F. H. FOWLER, San Francisco, California
Consulting Engineer

FIGURE 6-10. Title page of what is known as the Governor's Commission *Report*. The text was completed on Sunday, March 25, and it was formally submitted to Governor Young the next day. Although the report, which strongly favored the west-side-first hypothesis, was released to the public, no press conference was held at which commission members might have faced questions about their conclusions, and none of them were called to testify at the Coroner's Inquest. [Governor's Commission *Report*]

How then to explain the huge landslide that swept away the east side abutment? In the words of the commission, "It is probable that the rush of water released by failure of the west end caused a heavy scour against the easterly canyon wall...and caused the failure of that part of the structure." There "quickly followed...the collapse of large sections of the dam."[78]

The accessible, coherent prose of the report imbues the commission's forensic analysis with considerable authority, and the average citizen would have likely found it a convincing document. But a more careful reading reveals some major caveats. Most significantly, the report acknowledged that "as yet the manner and chronological order in which the failure of the various sections of the structure occurred are not yet certain."[79] Although the commission had returned to the dam site a second time, on March 22, to further study "the various fragments of the dam that had been washed downstream," they were unable to explain how the pieces of the dam had broken away and how they ended up at their final positions in the debris field.[80] In addition, the commission had difficulty explaining exactly how water being released from a breach on the west side would travel perpendicularly across the downstream face of the dam and, on a massive scale, erode the east abutment. In general terms, the report proposed that flow originating from the west abutment had somehow crossed over to the other side of the canyon and "caused a heavy scour against the eastern wall." But as to how such scour could erode 500,000 cubic yards of schist, the Governor's Commission could only say that it was "not yet certain."[81]

Despite the haste with which the Governor's Commission completed its study, and its admission that "the manner and chronological order" of the collapse remained uncertain, the report was nonetheless confident in its findings:

> With such a formation [the red conglomerate] the ultimate failure of this dam was inevitable, unless water could have been kept from reaching the foundation. Inspection galleries, pressure grouting, drainage wells and deep cut-off walls are commonly used to prevent or remove percolation, but it is improbable that any or all of these devices would have been adequately effective, though they would have ameliorated the conditions and postponed the final failure.[82]

As far as the Governor's Commission was concerned, the poor quality of the foundation material on the west side of the canyon rendered all other design issues—including uplift and all the ways it could be mitigated—essentially irrelevant, thus requiring no further consideration.[83]

The final page of the report offers three summary conclusions. The first reads in its entirety: "The failure of St. Francis Dam was due to defective foundations," without any distinction between the geology of the east and west abutments. In using the generic phrase "defective foundations," the commission apparently sought to deflect attention away from its inability to determine the "manner and chronological order" of the failure. The third conclusion admonishes that "while the benefits accrue to the builders of [dam] projects, the failures bring disaster to others who have no control over the design, construction, and maintenance of the works. The police power of the state certainly ought to be extended to cover all structures impounding any considerable quantities of water." The days of the municipal exemption allowed by the state's 1917 dam safety law were clearly numbered and, in making this recommendation, the commission was giving voice to a common and widespread belief.

But the second, and most remarkable, of these conclusions reassured the public that "there is nothing in the failure of the St. Francis Dam to indicate that the accepted theory of gravity dam design is in error...when [structures are] built upon even ordinary sound bedrock." The commission even went so far as to claim that "on the contrary the action of the middle section which remains standing even under such adverse conditions [offers] most convincing evidence of the stability of such structures when built upon firm and durable bedrock."[84] In other words, the failure of the St. Francis Dam was not be interpreted as an indictment of concrete gravity dams, but rather as a vindication—"most convincing evidence"—of the technology's inherent strength.

The Politics of Safety

In championing gravity dam technology, the Governor's Commission intended to offer more than a dispassionate analysis of why the dam failed. The worst dam disaster in twentieth-century American history involved the failure of a poorly designed concrete gravity dam, and the most prominent engineering investigation took special care to reassure the public that the technology itself was not to blame for the tragedy. Interestingly, the attribution of great strength to the surviving center section was not unique to the Governor's Commission, but can also be seen in a letter from Walter L. Huber, a San Francisco–based hydraulic engineer who later became president of the American Society of Civil Engineers, to the grand patriarch of massive gravity dams, John R. Freeman. Writing on March 21—five days *before* the commission's report became public—Huber waxed eloquent in describing the center section as "the one great witness of the stability of a gravity section founded on a solid foundation."[85] The "one great witness"—a memorable phrase that reveals a strong proclivity to protect gravity dam technology from opponents and naysayers.

But why would gravity dam technology need defenders? Here, the issue of Boulder Dam and its fortunes in the halls of Congress comes to the fore. Once the Governor's Commission had completed its work, Governor Young and State Engineer Hyatt sought to quickly communicate with Congressman Phil Swing to aid him in his fight for Boulder Dam. The imagery of "convincing evidence" ascribed to the surviving center section was a key part of the proffered defense. And while the Governor's Commission *Report* made no direct comparison between St. Francis and the proposed Boulder Dam, this was not the case in private telegrams to Swing.

Early Monday, March 26, Hyatt wired Swing in Washington with good news:

> Report of the Investigating Committee St. Francis Dam just completed but not yet in hands of Governor Young Stop Statement to you to the effect that there is absolutely no relation between the failure of the St. Francis Dam and the safety of the proposed Boulder Canyon Dam can be sent best advantage tomorrow morning after conference between Governor Young and A J Wiley Chairman of the investigating commission Stop Please wire advice if this is satisfactory or if statement absolutely necessary today.[86]

By the next day Young did indeed meet with Wiley, and soon thereafter he sent his own telegram to Swing. He too vowed that what occurred at St. Francis bore no relation to anything that could be ascribed to the Hoover/Boulder site:

> I have positive assurance from A. J. Wiley, Chairman of Commission and of Dr. F. L. Ransome Professor of Economic Geology at California Technical Institute, who is also a member of the St. Francis Dam investigating commission, both of whom have examined the Boulder and Black Canyon dam sites that the bedrock there is so sound, hard and durable and so different from the very soft foundation of the St. Francis Dam, that the failure of St. Francis Dam need cause no apprehension whatever regarding the safety of the proposed Boulder Canyon Dam.[87]

Young further counseled Swing that the failure was not related to gravity dam design:

> The report of the investigating committee also states that there is nothing in the accepted theory of gravity dam design that is in error or that there is any question

Form 1211A

CLASS OF SERVICE DESIRED
TELEGRAM
DAY LETTER XX
NIGHT MESSAGE
NIGHT LETTER
Patrons should mark an X opposite the class of service desired; OTHERWISE THE MESSAGE WILL BE TRANSMITTED AS A FULL-RATE TELEGRAM

WESTERN UNION TELEGRAM

NEWCOMB CARLTON, PRESIDENT GEORGE W. E. ATKINS, FIRST VICE-PRESIDENT

NO. CASH OR CHG
CHECK
TIME FILED

Send the following message, subject to the terms on back hereof, which are hereby agreed to

1111 SUN FINANCE BUILDING
LOS ANGELES CALIFORNIA
MARCH 26TH 1928

HONORABLE PHIL D SWING
CONGRESSMAN ELEVENTH CALIFORNIA DISTRICT
WASHINGTON D C

REPORT OF INVESTIGATING COMMITTEE ST FRANCIS DAM JUST COMPLETED BUT NOT YET IN HANDS OF GOVERNOR YOUNG STOP STATEMENT TO YOU TO THE EFFECT THAT THERE IS ABSOLUTELY NO RELATION BETWEEN THE FAILURE OF THE ST FRANCIS DAM AND THE SAFETY OF THE PROPOSED BOULDER CANYON DAM CAN BE SENT TO BEST ADVANTAGE TOMORROW MORNING AFTER CONFERENCE BETWEEN GOVERNOR YOUNG AND A J WILEY CHAIRMAN OF INVESTIGATING COMMISSION STOP PLEASE WIRE ADVICE IF THIS SATISFACTORY OR IF STATEMENT ABSOLUTELY NECESSARY YET TODAY

EDWARD HYATT
STATE ENGINEER
1111 SUN FINANCE BUILDING
LOS ANGELES CALIFORNIA

FIGURE 6-11. The political character of the Governor's Commission *Report* is evident in this telegram sent to Congressman Phil Swing by State Engineer Edward Hyatt on March 26, 1928. Before the report's public release (and even before it was presented to Governor Young), Hyatt notified Swing that "there is absolutely no relation between the failure of the St. Francis Dam and the safety of the proposed Boulder Canyon Dam." [Division of Safety of Dams, St. Francis Dam File]

about the safety of concrete dams designed in accordance with that theory when built upon ordinarily sound bed rock but that on the contrary the action of the middle section of the St. Francis Dam that remained standing even under such adverse conditions is most convincing evidence of the stability of such structures when built upon such firm and durable bedrock as is present in Boulder Canyon.[88]

An appreciative Swing thanked the governor for the assurance that the St. Francis disaster could not (or at least should not) jeopardize authorization of Boulder Dam in Congress: "I think the report of the Engineering Commission will give me just the information I desire."[89]

"This Thing Has Got Away With Me"

Once the Governor's Commission announced its belief that the faulty red conglomerate was the essential cause of the collapse, the west-side-first failure mode assumed great authority. Other prominent investigations (including the one sponsored by Los Angeles District Attorney Asa Keyes and the Los Angeles City Council's investigation led by Bureau of Reclamation Commissioner Elwood Mead) fell into lockstep and embraced the hypothesis. In particular, the Mead committee appears to have been guided by the conclusions of the Governor's Commission, as its report (completed on March 31) proclaimed that "the dam failed as a result of defective foundations."[90] On April 19 *Engineering News-Record* confidently published a story under the headline "All St. Francis Dam Reports Agree as to Reason for Failure," targeting defective west side foundations as the key culprit.[91] Eventually the west-side-first hypothesis came under fire, and some objections as to its plausibility were raised by witnesses in the final week of the inquest. But for the time being, the Governor's Commission *Report* and its analysis dominated understanding of the collapse.

Upon release of the report, the tenor and focus of the Coroner's Inquest changed. Now the issue was not so much how the dam had failed—it was generally accepted that the Governor's Commission had resolved that—but how Mulholland had been blind to what the commission called "defective foundations." In the immediate aftermath of the report's release, and with dramatic demonstrations of conglomerate samples dissolving in water, the cause of the collapse appeared clear. Mulholland was now obliged to return to the witness stand to answer for his failings.

If looked at closely, the "dissolving red conglomerate" hypothesis posed a conundrum: if this rock was so prone to disintegration, why was there no evidence that the west side abutment had been subject to extensive dissolution during the almost two years that water had been stored behind the St. Francis Dam? In addition, Mulholland had been adamant in his first round of testimony on March 21 that the west side leaks he examined the morning before the collapse were not muddy. How could this possibly be? While Mulholland was in no position to

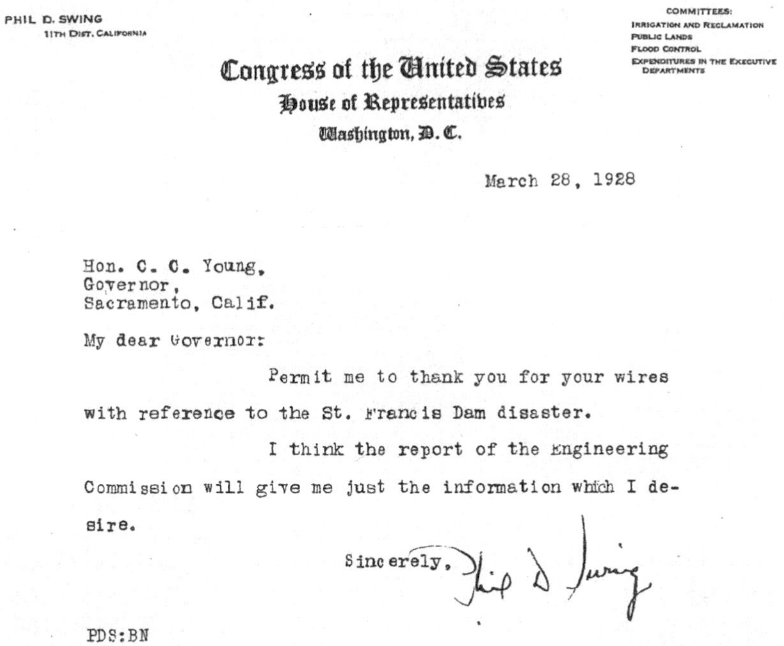

PHIL D. SWING
11TH DIST. CALIFORNIA

COMMITTEES:
IRRIGATION AND RECLAMATION
PUBLIC LANDS
FLOOD CONTROL
EXPENDITURES IN THE EXECUTIVE DEPARTMENTS

Congress of the United States
House of Representatives
Washington, D.C.

March 28, 1928

Hon. C. C. Young,
Governor,
Sacramento, Calif.

My dear Governor:

Permit me to thank you for your wires with reference to the St. Francis Dam disaster.

I think the report of the Engineering Commission will give me just the information which I desire.

Sincerely, Phil D. Swing

PDS:BN

FIGURE 6-12. In response to both Hyatt's telegram of March 26 and one sent to him by Governor Young on March 27, Congressman Swing advised the governor that "the report of the Engineering Commission will give me just the information...I desire." What Swing needed was ammunition for the fight against opponents of the Boulder Canyon Project Act. [Division of Safety of Dams, St. Francis Dam File]

directly criticize the Governor's Commission, he was flummoxed that its conclusions appeared to directly contradict his previous testimony. Viewed in hindsight (and with an understanding that the west-side-first hypothesis is untenable), there is every reason to believe that Mulholland told the truth when he described the March 12 leakage as clear. Nonetheless, it would at this point be difficult for him to explain how his tests provided verification that the red conglomerate would remain firm when subjected to sustained saturation.

There were no preliminary niceties when Mulholland testified for the second time, on March 27. The coroner and the jury immediately questioned him about the west side conglomerate, and Mulholland explained how he had taken measures during the start of construction to test whether the red conglomerate could safely withstand sustained inundation:

Q. [By the coroner]: You were at the dam during its construction frequently?

A. Very frequently.

Q. And you observed all the formation there, all the work that was intended before it was even started?

A. Everything.

Q. Will you answer some questions of the Jury, give them the information they especially seek?

Q. [A juror]: I wonder if Mr. Mulholland examined the holes [excavated into the red conglomerate], after you put water in to see the effect of the water on the conglomerate?

A. I ordered the holes in, was very interested in their behavior, put a big hole at the top [of the west abutment], filled it full of water, and it was there about two weeks, and the conglomerate was no softer than it was afterwards; had to bail the hole out, the conglomerate was very tight. We dug holes, had holes dug, I saw the holes, holes dug into that red conglomerate all over, the north flank of the hill...They talk about that conglomerate part of the soil laying on top of it. That has been utilized as vegetation. It is pliable and reached to a depth of an inch, two or three inches. It is hard, been wet for two and a half years, perfectly firm to walk on. You have all been up there, perfectly firm, doesn't run to mud, anything else. You will have geologists who perhaps are better posted than I am. I have examined it under a microscope and everything else. I know that it is vitreous, not a clay at all.

Q. In your examination of the result of your holes, you felt from your experience that it would be good practice to erect the dam at that site?

A. Yes sir, and the behavior of the dam justifies that. That part [the center section] is there yet. I have been over that hill, not as much as I would like to—have only had time to go out twice since the disaster—all of you have been over it. It is not muddy, it is firm stuff, doesn't make mud. If you go on the top soil where the soil is not stripped away, a great many places there will get mud on your feet.[92]

Questions about the conglomerate and its supposed softness continued for a few more minutes—Mulholland was adamant in defense of his testing of the west side foundation—and then the issue of drainage was raised. This exchange led to Mulholland's first apparent acknowledgment of responsibility for the disaster, although his testimony wandered off into references to other geologists retained by the Board of Water and Power Commissioners. This phase of the inquest also aroused intense emotion, with the court reporter noting at one point, "witness weeping."

Q. [By a juror]: Mr. Mulholland, why didn't you underdrain?

A. There was no water in the formation. That formation is dry as a bone. We drove tunnels from miles above. It is anhydrous formation. There is another thing is a very convincing fact, there is no leakage down in those riffling edges here at all. This is anhydrous and there is no flowing there, the dryness, the aridity of this base here—this is common where the stimulation of heat will make a spot green that has a lot of moisture in it, and there was no moisture out there at the time, only simply the moisture of the rain, neither one of those flanks.

Q. We have the fact that the dam fell.

A. I have a deep appreciation of the job before you, gentlemen. Don't imagine for a minute that I would throw you off the scent. I am willing to take my medicine like a man. If there is anything I can say to help you in your disclosures, I will be the first to point it out if I see it first. I have nothing to conceal.

Q. Would it have been possible if there were a soft streak in this bank, a streak where the water softens it, and the water stood—in softening up the formation, wouldn't it—suppose that some vitrified clay and water was up against this clay, there was a streak through the dam. It would then take a long time to penetrate through clay?

A. Yes sir. I have a very strong opinion myself as to what was the approximate cause of that failure.

Q. We would like you to tell us.

A. We have three geologists, one of them the best in the country, two of them the best in the country, the third I don't know much about. I am talking now. I know some geology myself, and I am curious to see if they have the same knowledge as I have.

Q. [By the coroner]: Who are these geologists?

A. Three of them, one of them the very best in the union, Mr. Hill.

Q. Do you intend to have them impart information to this jury?

A. I intend whatever they find they will publish to the world, or, at least, the board—I think they will consent to that. I am quite sure we have a public spirit in the matter. I know I have (witness weeping).[93]

Exactly what Mulholland thought was the cause of the collapse and how the three geologists were to corroborate his belief are left hanging in this exchange—presumably he is making another oblique reference to a dynamite attack or earth movement. A juror quickly interceded and, at least for a few minutes, brought the discussion back to geological conditions. But it was not long before Mulholland offered his famous acceptance of blame for the disaster. It came after a long, defensive account of how hard he had worked during his career:

Q. [By a juror]: I would like to ask you—it has been intimated to me that for the last two, three, or four years, you have been letting up, and delegated more to your subordinates.

A. I don't think anybody that has been associated with me—I haven't been letting up—I believe I have been working harder than I ever did in my life. I haven't had a day off—the only vacation in my lifetime, about three months ago, the only vacation I ever took, the only time I have taken a vacation in fifty years, went through the Panama Canal to New York. I am the first up in the morning, the last to go to bed, with my men. There are a very few beat me in the office in the morning. As far as letting up is concerned, I wish I could. I believe I will have to very shortly, this thing has got away with me.

Q. I have asked that question because I have come in contact with you in connection with the Water and Power Committee of the Chamber of Commerce a great many times, and you have always been in close touch with every question I have asked. I wanted to bring that question out, because it has been intimated to me that you have not been active.

A. Perhaps I welcome the news. If there is any inactivity creeping over me, I will enjoy myself. I am the first at the office in the morning. I go wherever the work is. Some things I don't observe, the actual pipe laying [down city streets], anything like that. I don't like this anyway. Whether it is good or bad don't blame anyone else, you just fasten it on me. If there is any error in human judgment, I was the human. I won't try to fasten it on anybody else.[94]

Thus Mulholland's acceptance of blame came in the context of questioning about the way responsibility for the dam may have been diffused through his engineering staff. He would have none of it. He would not allow any of his underlings to be blamed for the tragedy, and in this he exhibited a noble sense of loyalty. But Mulholland's desire to protect his staff did not preclude assignment of responsibility to State Engineer Wilbur F. McClure.

A Half-Day Stumbling Around

Immediately after Mulholland's acceptance of blame, Coroner Nance followed up with a line of questions that led to McClure. Taking the offensive, Mulholland asserted that he had secured an outside inspection of the St. Francis Dam project comparable to the state supervision mandated by the 1917 dam safety law.

Q. [Coroner]: This morning we had Mr. B[ayley] on the stand, to try and determine—Mr. B[ayley] and Mr. Hurl[but] [of the BWWS staff]—to determine who was the designer of the dam that was erected for the St. Francis Dam. We learned it was an adaption of the dam originally designed for Hollywood, and, with some changes, the general scheme was applied for the St. Francis Dam.

A. [Mulholland]: Yes sir, very little difference.

Q. Was that done at your direction?

A. I think we all concluded it was the case. I can't say the idea originated with me, we collaborated with the thing, as I do with most of my work.

Q. It was altogether wholly and exclusively a departmental project, and handled in your office?

A. Yes sir, we don't let contracts.

Q. You had no inspection of the site by any state authority?

A. Yes sir, the State Engineer examined the site, examined it carefully.

Q. Who was the engineer?

A. McClure.

Q. He examined the site before any work was done?

A. Yes sir, he was there. I think there was some little excavation, and my men went around there, stumbled around there over the country, and never a word to say about it.

Q. You are not required to have state inspection?

A. No sir, not with us, we are not required to.

Q. Why did you call for state inspection when you didn't require it?

A. I am not a strict caviler about the law, I like to comply as far as I can and go over the mark in conformity to the law, recognize there ought to be state inspection of such things, whether it is a municipality or not.

Q. [By a juror]: How much time did Mr. McClure spend?

A. Didn't spend but half a day, and he saw all there was to see in half a day, because there wasn't much to see.[95]

A week later, Mulholland returned to the witness stand and provided the inquest jury with further details on McClure's visit to the dam site.

Q. [By a juror]: You also testified that the State Engineer had passed upon this?

A. [Mulholland]: Yes sir, went on the ground, looked this thing over.

Q. [Coroner]: That was Mr. McClure?

A. Yes sir.

Q. He didn't make any geological test?

A. Don't know what you call it, looked as I did, exposed rock, excavations that were made, I don't really know if he is a geologist or not. He examined the exposure there, and it was his official business.

Q. [By a juror]: Was he acting in an official capacity?

A. Yes sir.

Q. Didn't you testify that you were not subjected to any inspection by the State Engineer?

A. Some question brought out the fact. I said I was willing to be subjected to any inspection, I believe if you remember.

Q. He actually did spend a half day at the dam, but it wasn't his duty to do so?

A. No sir.

Q. [Coroner]: Did he come at your request?

A. Yes sir, came there by my request.

Q. [By a juror]: With the specific object of examining the dam?

A. Precisely, I don't like to be stubborn about things, I wouldn't think of telling him it was none of his business, I did insist it was his business.

Q. [Coroner]: Although it was not required by law to be inspected by the State Engineer.

A. No sir.

Q. Did Mr. McClure see the finished work?

A. I think he has, pretty sure he has been down here several times while they were working on it.[96]

In this last exchange Mulholland and his questioners speak as if McClure is alive, but in fact he had been dead for almost two years, since June 1926.[97] Mulholland's remarks do not specify that he ever talked with McClure in any substantive way about the latter's site inspection, and there is no evidence that McClure sent a written report or written commentary to Mulholland referring to the visit. Most important, Mulholland's testimony offers scant support for any assertion that he had called for a "state inspection," going "over the mark in conformity to the law." Particularly untenable is the notion that having the state engineer "stumble around" the site for half a day might constitute a substantive review comparable to those done under formal authority of the 1917 dam safety law.

To appreciate the absurdity of Mulholland's contention regarding the state engineer's alleged review, one need only consider the history

FIGURE 6-13. A design for the multiple-arch Littlerock Dam was presented to the state engineer in the fall of 1918, but construction did not begin until November 1922, following a review by the state engineer that took four full years. The half-day visit to the St. Francis site by State Engineer Wilbur McClure, which Mulholland alluded to in his testimony, was in no way comparable to the review given other projects, this one in particular. This photo of the 170-foot-high Littlerock Dam was taken in the late 1970s, after it had been in operation for more than fifty years. [DC Jackson]

of the Littlerock Dam. Located about thirty-five miles east of the St. Francis site in the Antelope Valley of northern Los Angeles County, this reinforced concrete multiple-arch dam was built by the Littlerock and Palmdale Irrigation Districts in 1922–24. In stark contrast to the privilege afforded the city of Los Angeles, the farmers in these districts could not build their dam until they obtained explicit approval from the state engineer. The approval process stretched over four years, starting in 1918. During that time McClure engaged three outside engineers to review plans with his staff. He also solicited the advice of J. B. Lippincott, a consulting engineer for the bond house that was to finance construction.[98]

After lengthy deliberation, McClure at last approved plans for the Littlerock Dam in May 1922. The following August his engineering representative visited the site and noted that the contractor was making slight adjustments in the design to make it conform better to the topography. He quickly informed McClure that the contractor had been told to "suspend operations...until the changed plans were submitted to the state Department of Engineering and Irrigation for approval and action thereon." In addition, the contractor was told "that the foundation would have to be cleared, viewed, and passed as satisfactory by a representative of the State Engineer" before construction of the dam could commence.[99] Two more site visits took place before McClure issued final design approval on November 4, 1922. Thereafter, a representative of the state engineer visited the site regularly and reported on the construction until work was completed in June 1924.[100]

Mulholland had expressed willingness to take the blame for the disaster in order to shield his staff. But he felt no such compunction about trying to tie McClure to the disaster, an instance where his motives appear neither magnanimous nor noble. Mulholland's assertion that a half-day site visit by the state engineer might constitute compliance with the 1917 dam safety law is specious. The pear and almond farmers of Littlerock and Palmdale, who weathered years of delay and scrutiny in order to build the Littlerock Dam, could vouch for that.

After he described McClure's brief visit to the St. Francis site, Mulholland's March 27 appearance on the witness stand was extended, and he was confronted with a sample of red conglomerate. Much like the scene the day before during D. C. Mathews' testimony, he was forced to watch what happened when the specimen was immersed in a glass of water. The *Los Angeles Examiner* described the event: "A juror had brought back a chunk of reddish rock taken from a seam between two formations. A fragment was placed in glass of water and disintegrated. 'Yes, I see,' said Mulholland. 'But the whole foundation wasn't made of that. That sort of thing is uncommon.'"[101] The court reporter's transcript offers no evidence of this alleged statement by Mulholland, but it does record District Attorney Keyes saying, "You saw that piece I put in the glass. What do you call that?" with Mulholland tersely responding, "A piece of dirt."[102]

Here we can begin to appreciate the difficulties posed by simplistically assuming that each and every "sample" of red conglomerate taken from San Francisquito Canyon necessarily represented the structural character of the west side abutment. Although Mulholland did not have his BWWS workforce excavate a deep cutoff trench up the west side canyon, the loose soil on the surface was removed prior to the casting of concrete, and such surface soil—which was most readily available for visitors to pocket and take home for ad hoc immersion tests—did not reflect the structural character of the rock comprising the cleared foundations. While it was easy to extrapolate from the dissolution of small samples of the red conglomerate to a disintegration of the west side abutment—which then supposedly fostered major leakage directly leading to the collapse—in fact no major leakage dissolving the west side canyon wall ever occurred. Later attempts during the inquest to document this supposed leakage would prove extremely problematic—evidence that the quality of the west side foundation was nowhere near as faulty as small tumblers filled with reddish sediment might lead us to believe.

After some further exchanges involving the red conglomerate and west side leakage (which the chief engineer again described as "clear water, absolutely clear"), Mulholland was excused and the inquest moved on to other witnesses.[103] Newspaper coverage of his second day on the stand was generally favorable, praising him for his resolute willingness to take blame for the disaster. A *Los Angeles Herald* headline proclaimed, "Ready to Face Music—Mulholland," and the *Los Angeles Times* described him as "unflinching" and "calm" in answering questions (although it noted that "the veteran engineer [shows] signs of breaking under the strain of the failure of the dam and the resultant inquiry.")[104] And the *Los Angeles Examiner* gushed: "Gallantly lifting his gray head, his voice ringing out strongly, William Mulholland, chief engineer, yesterday accepted the responsibility for whatever mistakes of man may have had to do with the breaking of the St. Francis Dam."[105] The big story of the day was not so much the egregious failings of the dam design and its foundation, nor the chief engineer's attempt to tie State Engineer McClure to the disaster. Instead, the focus was on Mulholland's desire to protect his underlings from blame and, in his words, "take my medicine like a man."[106]

March 27 set the high watermark of Mulholland's dominance over post-disaster inquiries, and his stature soon began to fade. Anything he might say about the cause of the collapse had now been eclipsed by the authority of the Governor's Commission *Report*. In this light, inquest questions seeking to portray him as distant and removed from the work of his department foreshadowed his future prospects. Mulholland nominally held the position of chief engineer for another six months, but he would no longer be The Chief in fact and deed. He would remain a presence at the inquest until the release of the jury's verdict. And he would, at his request, again take the stand to further explain his use of—or failure to use—consulting engineers. But release of the Governor's Commission *Report* marked a professional watershed. After fifty years, the era of Mulholland as a dynamic leader had passed.

Scattergood, Brome, Del Valle

The great majority of city employees who testified at the inquest worked for the Bureau of Water Works and Supply, and thus answered to Mulholland. But three witnesses with significant ties to the Department of Water and Power were not subordinate to him. Ezra Scattergood, who served as chief electrical engineer for the Bureau of Power and Light, was Mulholland's presumed bureaucratic equal. However, Scattergood's influence over the siting of the St. Francis Dam had been trumped by Mulholland, and his testimony on this issue proved to be among the most contentious at the inquest. James P. Brome was secretary of the Board of Water and Power Commissioners, a role that placed him in the position of presenting evidence that, however briefly, constituted the only time the change in the dam's height was mentioned. R. F. del Valle was president of the Board of Water and Power Commissioners and, at least nominally, Mulholland's supervisor. But Del Valle's testimony makes clear that, with respect to engineering decisions involving the city's water supply, the tail wagged the dog.

When Mulholland first proposed building the St. Francis Dam, Scattergood had objected on the grounds that deployment of the new reservoir would significantly reduce electric power generation at San Francisquito Power House No 2. In essence, any water diverted into the St. Francis reservoir could never pass through and spin the power house turbines and thus no electricity could be generated and sold to municipal customers. A loyal city employee, Scattergood had expressed his objections privately to the Board of Water and Power Commissioners and had not leaked them to the press.

In the aftermath of the disaster, evidence of opposition to the St. Francis Dam within the Department of Water and Power in 1922–23 was not something that the city wanted to discuss. Scattergood's complaints about use of the site had had nothing to do with safety—they focused on efficient power generation—but if his earlier opposition became widely known, it would be easy for people to assume that there were other reasons for it, including safety. The city and Scattergood wanted to avoid this if at all possible, which apparently accounts for his combative testimony:

Q. [Coroner]: The reason I called you here today is because it has been reported to me that you had examined the site and did not approve of it, and I wanted to get it into the record, the actual facts.

A. [Scattergood]: No sir, I have not examined the site before the dam was built or since with any such view in mind, or at all, and I never expressed any opinion that it was not a proper dam site or any opinion regarding it, as proper or improper, as a dam site.

Q. [Assistant District Attorney Dennison]: You are the engineer in charge of what part of the power and light[?] There is a department over there and it has two heads?

A. The Department of Water and Power has two branches; one the Bureau of Water Works and Supply, of which Mr. Mulholland is the head, and one the Bureau of Power and Light of which I am head, under direction of the Board.

Q. Does this Bureau of which you are the head, erect dams?

A. No sir.

Q. You don't have anything to do with that?

A. No. I made one little regulating reservoir at the head of the pipeline of Power Plant Two. You might call it a reservoir. It has less than one hundred acre feet or thereabouts, and is more of a regulating puddle than a reservoir.

Q. And you had nothing to do with the selection of this [St. Francis] site?

A. No, I have not been consulted in any capacity in regard to it.

Q. And your advice was not taken on it?

A. No sir.

Q. You did not see it before it was constructed?

A. Yes, I have been up and down that canyon many times, but have never stopped and examined the site for any purpose.

Q. As an engineer do you think you would be capable of selecting a site?

(No response)

Mr. Scott [assistant city attorney]: I think the witness has stated he had nothing to do with the dam.

Mr. Dennison: Assuming now, that you were called upon to build a reservoir up there. You have seen the place and you would make some exploration preliminary to doing it?

(No response)

Scott: What do you want to bring out? We want the light here, but this character of examination is uncalled for, I think.

The Coroner: Mr. Scattergood has said that he has never been engaged in building dams or selecting sites for dams, and that this is out of his line.

Mr. Dennison: I will withdraw the question if he don't [*sic*] want to answer it. Do you think you are capable of answering it?

A. I did answer the question. I said I would make a thorough investigation, the same as I would in selecting a power house site, or any site.

Q. Would you mind telling me how you would make that investigation?

A. Not being an expert or having experience in the construction of dams I would, of course, consult those who have, the same as any manager would employ experts, even though he was an engineer and engaged along a line in which he had no experience.

Q. Do you mean by that that you would call in geologists to tell you what kind of soil and formation it was?

Mr. Scott: I think we are taking too much time here. What someone else would have done—we are trying to find out Your Honor, why this dam went out, and what someone else has done would have nothing to do with that investigation. I don't know what counsel's purpose is, whether to get publicity or get the light. I feel he is not exactly adhering to Your Honor's ruling.

The Coroner: I think you are correct.[107]

Within a few minutes, Scattergood left the witness stand, never to return. For his part, he had kept his objection to the reservoir on the grounds of power system efficiency out of the public record. Of equal historical interest, the questioning he faced also reveals that there were tensions among the inquest lawyers and that they appreciated how they might "get publicity" as a result of their performance.[108]

James Brome was also a city employee, but his position in the civic hierarchy was quite different from Scattergood's, and the significance of his testimony is also quite different. As secretary of the Board of Water and Power Commissioners (a position he had also held with its predecessor, the Board of Public Service Commissioners), Brome was called upon to testify about actions the commissioners had taken vis-à-vis the St. Francis Dam. For the most part, Brome's contribution to the inquest was quite perfunctory, as essentially all board actions were already in the public record as part of its annual reports. In his testimony he simply read aloud from the reports. But Brome's testimony is intriguing because, as noted above, it was the only time during the inquest that the issue of the dam's raised height was at least briefly acknowledged—and that it was raised after construction began.

Near the beginning of Brome's time on the witness stand, Coroner Nance asked him: "What record do you have of the adoption of the site for St. Francis Dam?" and he replied, "The only other record [that] appears, outside of the purchase of land, are the approval of the annual reports of the Board to the Council, and I have it compiled complete."[109]

Much of his subsequent testimony consisted of a reading of the annual reports, so they could be incorporated in the inquest transcript. He first read from the report for the period ending June 30, 1923, and then started in on the report for the next year. This exercise apparently taxed the jury's patience and Nance suggested an alternative: "Gentlemen…do you want to hear any more of these read? If not, I think we will file these and let the Jury read them at their leisure, however, we will copy them in the record." With no objections made, the annual reports for the years ending June 30, 1924, 1925, and 1926 were then simply entered into the transcript without a public reading.[110] This decision by Nance and the jury is noteworthy because it came just before Brome would have read evidence that the dam's spillway level was raised from 1,825 feet above sea level to 1,835 feet and that the reservoir capacity had been enlarged from 32,000 to 38,000 acre-feet. Presumably the jury could have dug the reports out of the record and, as Nance phrased it, "read them at their leisure," but apparently this was never done. The increased height of the dam was acknowledged in an exchange with Keyes near the end of Brome's testimony:

Q. [By District Attorney Keyes]: Who employs the subordinates or office help of the Engineering Department?

A. [Brome] It is changed somewhat.

Q. At the time of the erection of this dam, who employed them?

A. It would be recommended by Mr. Mulholland and the Board would approve it.

Q. Mr. Mulholland recommended all the employees there?

A. All the principal ones.

Q. I want to find out if Mr. Mulholland was the Board, or was the Board the Board.

A. It is quite obvious the Board was the Board.

Q. Did Mr. Mulholland change the plan of this dam?

A. I am sure I can't tell you, I have no record in the minutes.

Q. In reading your record, you said it would be a certain height?

A. I have in there [a] description showing as it progressed it increased a little bit. It is in the record there, but I didn't read it.[111]

It appeared that the heightening of the dam in mid-construction would at last be publicly discussed. But then City Attorney K. K. Scott abruptly changed the subject by interjecting, "They are under Civil

Service?" to which Brome replied, "Yes, the employees, with the exception of Mr. Mulholland."[112] A moment later Brome was excused as a witness and never again—even briefly or obliquely—would the issue of the dam's increased height be broached at the inquest.

The most politically intriguing exchange prompted by Brome's testimony occurred when Keyes asked if "Mr. Mulholland was the Board, or was the Board the Board." Brome side-stepped the provocative query—"it is quite obvious that the Board was the Board"—but the question lingered when Del Valle took the stand the next morning. Before long, Nance sought to get on the record exactly what type of oversight the Board of Water and Power Commissioners gave to Mulholland and his staff.

Q. [Coroner]: Have any members of your board since the inception of the St. Francis Dam project been engineers, men who had practical scientific knowledge in the construction of dams?

A. [Del Valle]: None at all.

Q. Did your board at any time instruct the Chief Engineer as to any assistance he should employ, expert assistance, geologically or in any engineering way from the outside, outside your department?

A. Mr. Mulholland has had charge of the department ever since its inception, built the aqueduct prior to that, for the last forty-five years has been identified with the management of the Water Department of this city. During that time he conceived the construction of the aqueduct, built it, has built nineteen dams for the department, and during that whole time, the board has found that he has used the proper judgment, has been competent, efficient in every manner, and therefore the matter of detail as to whom he should consult—he consulted with the department occasionally...or what he should do in detail, has been left entirely to his judgment, because the board has had the utmost confidence, and has now, in his ability as an engineer.[113]

The district attorney soon returned to the issue of Mulholland's failure to engage consultants:

Q. [District Attorney Keyes]: Do I understand the Water Board had ample means to employ geological experts to make surveys of these dams.

A. [Del Valle]: Yes sir.

Q. But no such were employed.

A. If anything of that kind was necessary—we left that entirely to Mr. Mulholland.

Q. Nor were any hydrostatic [*sic*] engineers employed?

A. Not by the Board.

Q. And it was never suggested to Mr. Mulholland that in the construction and building of these stupendous structures [it] might require the assistance of other men—you left it—he had the sole responsibility?

A. Yes sir. I will also remark, up to this time the State of California didn't even consider the examination of reservoirs for that purpose.[114]

Here, Del Valle deftly deflected criticism by blaming the state for including the municipal exemption in the 1917 dam safety law. But some members of the jury seemed unimpressed with such a self-serving strategy. After Del Valle mentioned that Arthur P. Davis had engaged Mulholland as an adviser for Oakland's East Bay Municipal Utility District, a juror pressed the point about consultants and reminded Del Valle of Mulholland's limited experience with masonry and concrete dams:

Q. [A juror]: You made reference to Arthur P. Davis. Do you know his experience in the matter of building large works, dams?

A. Arthur P. Davis has been head of the Reclamation Service of the United States, is the man who first reported on the question of the Colorado River works.

Q. Is it a fact, the East Bay Utility District, that Davis as an engineer and general manager, considered it good policy to employ as consulting engineers Mr. Mulholland and General Goethals?

A. I think that is correct.

Q. Then the city of Los Angeles did not consider it necessary to follow a similar policy and employ men of equal standing to advise with it in regard to projects of major importance?

A. For two reasons, first, as I said, a man who had already built eighteen dams successfully, and who had the power to consult anyone he desired in regard to this matter, we left it entirely to him, and he had built eighteen dams successfully and had participated, to my certain knowledge, in the building of other dams of the different water departments of the state to which I refer.

Q. You know how many of these dams were masonry dams?

A. I couldn't tell you.

Q. Did you know up to the building of the Mulholland and St. Francis Dams there had been no masonry dams [built by Mulholland]?

A. Perhaps not. We had full information on these facts and, as I say, Mr. Mulholland always surrounded himself—even I have seen Professor Davis, Doctor [Elwood]

Mead, men of that standard with Mr. Mulholland, in connection with the future Colorado River Dam, and men who have charge of that matter, from Mr. Mead down, have been in consultation often.

Q. As a matter of fact, Mr. Mulholland is subservient to the water commissioners.

A. Absolutely so in every particular. He also expresses whatever is improper. He is a servant of that commission. I have never seen anybody connected with the institution that more willingly took orders from his superiors.

Q. [Coroner]: That is all, you may be excused.[115]

This was a strange and duplicitous performance. On the one hand Del Valle sought to reassure the jury—and by extension the public—that Mulholland was "absolutely" subservient to the Board of Water and Power Commissioners "in every particular." But on the other, under his leadership the board acquiesced to any and every decision Mulholland made because they believed they had no expertise, and no reason, to ever question his technological competence. The board was nominally in charge, but to what end? For the moment, Mulholland's bravado in choosing not to engage any outside engineers or geologists to review the St. Francis project, and the board's unwillingness to push him on such an issue, passed from the inquest's limelight. But the board's responsibility for the disaster would again arise when the jury delivered its verdict.

Where Did the Water Go?

Once the Governor's Commission Report had seemingly resolved questions about the cause of the collapse, and once Mulholland had taken the stand and, however conditionally, taken blame for the disaster, it appeared that the inquest would conclude rather quickly. But it was still necessary to hear testimony from the engineers and geologists engaged by Keyes to investigate the collapse. The jury was led to believe that the district attorney's investigation, headed by the engineer Edward Mayberry (and often called the Mayberry Committee), would be ready to present its findings at the end of the inquest's second week. But a series of delays kept pushing the date back so that it did not occur until the third week of the inquest.[116] When members of the committee began their testimony it quickly became clear that they endorsed the west side failure hypothesis espoused by the Governor's Commission. In this, Mayberry and his colleagues broke no new ground. But they were forced to confront questions that commission chairman Wiley and his cohorts had never faced in a public forum.

The testimony of Mayberry Committee members makes clear that there were major evidentiary problems with the conjecture that disintegrating red conglomerate fostered dramatic leakage through the west side abutment.[117] Assuming that the water gauge accurately recorded a significant drop in the reservoir level in the hours preceding the collapse, and that this drop reflected the leakage that precipitated disaster, then there should have been a significant flow of water down San Francisquito Creek during the evening of March 12. But witness after witness who had traveled up the canyon that evening and passed Power House No. 2 in the half-hour before the collapse testified that they had observed nothing unusual in the flow coming down the creek or the concrete conduit (see chapter 3). And a mere ten minutes before the collapse, the watchman at Power House No. 2 made a routine call to his counterpart at Power House No. 1 and did not mention any unusual stream conditions.[118]

This was no mere glitch bedeviling the west side first hypothesis. As hydraulic engineer Charles Lee subsequently testified, there should have been an impossible-to-ignore flow down the canal and lower streambed if heavy leakage had been seeping through the west abutment.[119] So if it wasn't in the streambed, where did the water go? Edward Mayberry and his committee members L. Z. Johnson, Allan Sedgwick, W. G. Clark, and Charles Leeds struggled mightily with this question, with Johnson proposing at one point that there was a giant "subterranean cavity" (or underground cavern) somewhere below the dam that had absorbed much of the leakage.[120] It is unnecessary to quote extensively from the testimony by members of the Mayberry Committee about the missing water, but when Charles Leeds took the stand the questioning took an interesting turn:

Q. [By the coroner]: Where did the water go to?

A. It went out of the dam. From that point I question if it is pertinent.

Q. We have the testimony that no stream of water flowed in the stream down that canyon. Where did it go?

A. If that amount went out—

Q. [By Assistant District Attorney Dennison]: I understand that during the twenty-three hours immediately preceding the failure of the dam, there was a certain amount of water which left the dam as indicated by that [water gauge] diagram. Is that correct? When the water left it is an utter impossibility to tell, except that it flowed out. During the last twenty-three hours there was apparently a greater flow

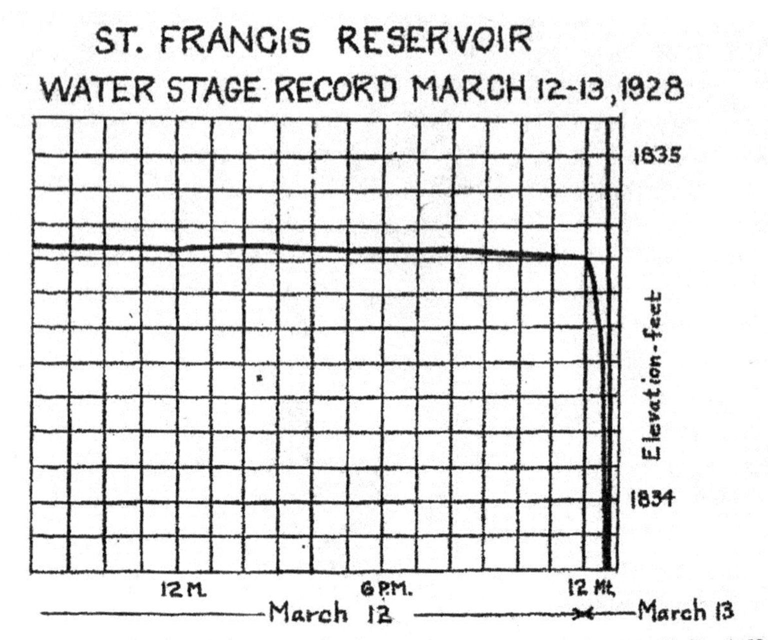

Fig. 1. St. Francis Reservoir Water Stage Record March 12-13, 1928

FIGURE 6-14. A graph of water gauge levels, illustrating what appeared to early investigators to be a gradual yet significant drop in the reservoir level before the collapse. What the gauge actually recorded was a gradual lifting of the dam in the hours prior to failure. [Governor's Commission *Report*]

than during the previous time. You don't know how much previously and how much during the twenty-three hours?

A. I think that is a significant fact, if over a relatively long period of time there was a less rapid flow and over a short period of time there was an increasing flow, showing that it was increasing in a mathematical ratio, and something greater than that.

Q. Would the flow be an unknown quantity?

A. Not an undeterminable quantity because the surface of the reservoir at the end certainly was not level, there was a much greater drop at the dam.

Q. And this seventy-four second feet [of flow] disappeared in the bowels of the earth in the last three and one-half hours?

A. No sir, I did not say that seventy-four second feet went out.

Q. The witness before you [W. G. Clark, a member of the Mayberry Committee] said that seventy-four second feet—

A. Seventy-four second feet is not a measure of volume, but a measure of flow [that is, seventy-four cubic feet per second (cfs)], and the average rate is indicated by the gauge, was seventy-four second feet.

FIGURE 6-15. Detail view of the dam's surviving center section, showing the top of the water gauge. During the hours before the collapse, this gauge purportedly measured a drop in the reservoir level. [DC Jackson]

Q. We are mystified as to where the water went to?

A. I think we should not take that seventy-four second feet as any too accurate.

A few minutes later, Leeds was again presented with the mystery of where all the supposed leakage from the disintegrating west abutment went and why, in the half-hour before the dam's collapse, this heavy flow was not noticed at Power House No. 2.

Q. [Assistant City Attorney Scott]: If there was a change of shift in Power House No. 2 at 11:30 that night and men had to pass the channel in which there was seventy-four second feet [of flow], and if there is a report by telephone from Power House No. 1 ten minutes before the dam went out and nothing was said about seventy-four feet passing there, we might assume that there was not seventy-four feet of water passing in this canal to Power House No. 2?

A. [Leeds]: No I don't think I might assume any such thing, because the men in the power house might not have been out to observe that water at that time.

Q. I said, assuming that he goes over and went on shift at 11:30—he crosses this channel, the only place where that water can come down, and there was seventy-four feet flowing, and if there is a report by telephone from Power House No. 2 to Power House No. 1, ten minutes before the dam went out, and nothing was said about it I think we might assume that there was not seventy-four second feet flowing, might we not?

Mr. Dennison: That is not an engineering question. That is to determine if these people were telling the truth.

The coroner: If you feel competent to answer that question you may answer.

A. [Leeds]: Personally I think that it is pertinent but I want to refer to the [Mayberry Committee] report. I still think it would be perfectly possible, taking into account the inaccuracy of this gauge, and the various records and the fact that it was 11:30 P.M. and very dark and over a road that a man was accustomed to walk every night, so that he could probably walk it with his eyes shut, I think there is very little significance to it myself.

To make sense of the west-side-first hypothesis, Leeds was left to postulate that workers could have walked to their shift at Power House No. 2 with "eyes shut." This was not a very convincing explanation, and Leeds himself apparently realized that it strained the bounds of credulity. When pushed by a member of the jury on what the water gauge readings actually meant, all he could do was look skyward for a helping hand: "After I get to Heaven I will ask God and will tell you."[121]

Earth Movement, Dynamite, and the East Side

As the Mayberry Committee struggled to make sense of the water gauge data, the inquest lawyers representing the city and the Water Department tried to broadly discredit the west side failure mode espoused by the Governor's Commission. They did not directly contradict the Commission's report, but they did use the inquest to raise questions that might make Mulholland—and by extension the city—appear in a better light.[122] Mulholland had been adamant that the west side seepage he observed on March 12 was not muddy, and witnesses who cast doubt on the presumed seventy-four cubic feet per second flow were in essence providing testimony that supported him on this point. The most impressive of those who questioned the existence of any excessive leakage was Charles Lee, a consulting hydraulic engineer from San Francisco hired by the Bureau of Water Works and Supply to analyze pre-collapse stream flow in San Francisquito Canyon. Lee presented the jury with a clear explanation of why any substantial flow of water would have been visible in the canyon below the dam the evening before the flood; his testimony boiled down to this brief exchange:

Q. [by a juror]: Have you given any thought to the matter in which fifty to one hundred second feet [cfs] could escape from the dam, without notice over a period of several hours?

A. [Lee]: I have given it considerable thought.

FIGURE 6-16. View of the concrete channel built to carry water released from the St. Francis reservoir downstream to the Los Angeles Aqueduct at Power House No. 2. In the minutes before the dam failed, motorcyclist Ace Hopewell rode parallel to this channel and noticed no unusual water flow. Had there been significant leakage through the west side foundation (as proponents of the "west-side-first" hypothesis believed) this channel would have carried a significant, and noticeable, stream of water. [Richard Courtney Collection, Huntington Library]

Q. What ideas have you formed as to the possibility of it having occurred?

A. My view is that it could not have occurred.[123]

Lee declined to offer the jury any opinion as to the cause of the collapse—that was not the issue that the city had hired him to study and report on at the inquest—but he understood that his testimony could not be easily reconciled with either the Governor's Commission *Report* or the Mayberry Committee report. Later that spring, he would publish an article in *Western Construction News* that would explain exactly why he believed the west-side-first hypothesis to be untenable.[124]

Through the efforts of city lawyers S. B. Robinson and K. K. Scott, two alternative theories that did not require excessive flow through a dissolving west side abutment were posited by the city to explain the dam's collapse. One involved a supposed "earth movement" of the west abutment, and the other reintroduced the notion of a dynamite attack, but this time specifically focused on the east side abutment.[125] The earth movement theory was consistent with Mulholland's early pronouncement, as quoted in the March 14 *Los Angeles Examiner*: "From the short study we were able to make, it appeared that there had been a major ground movement in the hills forming the western buttress of the dam."[126] And the resuscitated dynamite theory also accorded with Mulholland's earlier allusions to "human aggression."

Evidence of an earth movement on the west abutment was derived from work carried out by BWWS surveyor H. B. Hemborg, who compared a pre-dam survey of the canyon completed in 1923 with post-collapse measurements.[127] Although at least one key data point had been washed out in the flood, Hemborg's work appeared to indicate that points on the far edge of the west abutment had moved about an inch toward the east and that the west abutment may also have lifted a few inches. Mayberry was unconvinced by Hemborg's surveys, his skepticism about their accuracy deepened by the impossibility of replicating or rechecking the 1923 survey.[128] In addition, no seismic sensors at the California Institute of Technology or anywhere else recorded any tremors on the night of March 12–13, making Hemborg's results yet more puzzling. And even assuming that the findings of the surveys were correct, it was uncertain how a one-inch eastward movement of the west abutment would have precipitated the massive mica schist landslides on the east abutment. Despite these questions, Hemborg's testimony and the possibility of a slight shifting of the western abutment became part of the evidentiary mix.[129]

Like Charles Lee, Frank Reiber was a San Francisco–based engineer who had been hired as a consultant by the BWWS to study the collapse. But in contrast to Lee, Reiber had developed a theory on the cause of the collapse and was willing to testify about it. Starting late on Thursday, April 5, and continuing the next morning, Reiber took the stand to explain his reasoning. Through analysis of the way concrete fragments of the dam were dispersed in San Francisquito Canyon, he argued that it was much more plausible that the collapse had begun on the east side of the canyon than on the west. He also called attention to the access road running up the west side abutment, which had not been washed out in the flood, noting that this road could not have survived if the collapse had actually started on the west side when the reservoir was full. And he pointed out that the surviving section of the water gauge was bent toward the east, something best explained if the outrushing water had first flowed through the east side of the canyon.

Reiber was the first to articulate why an east-side-first hypothesis made much more sense than west-side-first theories. Using photographs and a model of the dam to illustrate his points, he outlined a theory to the jury that inherently discredited both the Mayberry Committee and the Governor's Commission. But the insights underlying his overall analysis were overwhelmed by the attention given to his claim that a dynamite blast had initiated the east side collapse:

> [Frank Reiber]: Having gone through what presumably happened I think there is some reason—we are forced to conclude that the failure took place up here [indicating on a photograph] at or near the east end. I would say that it was in this distance here (indicating). That to me would explain more clearly what happened than if the failure happened anywhere else.
>
> Q. [By the coroner]: What do you think happened?
>
> A. [Reiber]: Only two things could have happened. One is, by some earth movement, tremendous slide of schist—there have been slides of schist in other places—my reason for rather feeling this isn't the best explanation is this, the bedding planes of that schist are too parallel to the end of the dam, and if we had that end of the dam in place up here (indicating) the water pressure acting on the arch would have been holding that pretty firmly against here (indicating), so that wherever the schist slid, I wouldn't expect it to slide where it was being held tight against the wall. The rapidity with which that came out rather leads me to feel that the schist might have been very intensely shattered. If you had a slide first, the water would have taken out the slide material there. In any event, I think one of the best explanations, if it were not for the almost incredibly human side of it, I think if we had a charge of dynamite and placed in here (indicating) and shattered that schist, it would shatter it for some distance around here (indicating). Such a charge, if of the proper kind of explosive, wouldn't throw out the dam bodily.[130]

Once Reiber had proposed that a dynamite charge was responsible for the disaster, this became the highlight of his testimony.[131] No one thought to suggest that the east side tunnel excavated during the earliest stages of construction might have contributed to collapse. And the effect of uplift—exacerbated by the lack of grouting and the absence of a cutoff trench and foundation drainage up the east side abutment—was not considered by either the jury or Reiber in relation to the east side failure. Dynamite was the big news that Reiber brought to the inquest, and the story in the next day's *Los Angeles Times* was headlined, "Dynamite Dam Theory Backed—Explosive Expert Holds Evidence Plain—Sudden Exterior Violence Blamed in Break."[132]

Other testimony supporting an east-side-first hypothesis came from Halbert P. Gillette, president and editor of the Chicago-based journal *Water Works*, who spent his winters in South Pasadena and had traveled up through San Francisquito Canyon while the dam was under construction. Gillette was not connected to the BWWS as a consultant; he had developed theories on the collapse sequence as part of a series of articles for *Water Works*. His testimony to the jury was largely based on what he had drafted for publication in his journal and from his analysis of the surviving dam fragments.[133]

Gillette was decidedly unimpressed with reasoning that blamed the dam's failure on softened red conglomerate and excessive west side leakage.[134] But the clarity of his testimony on an east-side-first sequence was diminished by his desire to embrace the "earth movement" espoused by Hemborg. In addition, he had no convincing explanation for how a one-inch movement of the west abutment caused the massive landslide on the east abutment. While he did not advocate Reiber's dynamite theory, the jury was unsure of his position. At one point near the end of his testimony a puzzled juror asked him, "Do you think that somebody blew up the dam and there was an earthquake at the same time?" and Gillette responded, "No sir, I think it was a major earth movement that did the trick." But apparently not "major" enough to register on any seismic sensors in Southern California.[135]

Hemborg, Reiber, and Gillette had certainly added complexity to the inquest, proposing the possibility of a west abutment earth movement, an east-side-first collapse sequence, and—last but not least—a dynamite blast. In the final days of the inquest, the water board lawyers grabbed on to the dynamite theory as a way to protect the city's interests. Mulholland may have accepted "blame" for the disaster, but hope lingered that the city might somehow shunt responsibility off to some other party.

Decayed Dead Fish

To this end, they called to the stand Edwin Starks, a Stanford zoology professor who specialized in "fish life" and who had been hired to examine the fish stranded in the drained reservoir.[136] The point of his investigation was to determine if the carp and black bass trapped in the drained reservoir had died as the result of a dynamite-induced concussive force. But Starks did not visit the dam site until March 23, a week and a half after the collapse. By then, the stranded fish had been dead for many days. This fact alone rendered his fieldwork largely useless:

Q. [By the coroner]: Will you please state what inspection and investigation you made and the results of it, stating it just as briefly as is consistent with accuracy?

A. I found a good many fish. It is so long since the catastrophe that the fish were in very poor condition. The only thing which might show as to an explosion would be shattered bones. I saw nothing of that sort. The fish were simply reduced to empty skins or skins filled with maggots.[137]

And a few minutes later:

Q. [Coroner]: Did you make any observation of significance as to any fish which were found stranded, either in the reservoir site or in the pool—either one of the pools?

A. I opened a great many of them to see if I could find any shattered bones. I found the vertebral column apart, but that could be explained by decay, just as well as by dynamite.

Q. Is that a condition which might be caused by an explosion?

A. It certainly is.

Q. But, on account of the decomposed condition, you were unable to determine what caused it?

A. It might have been either.[138]

Just to be certain, Coroner Nance pressed the point one more time:

Q. You went up in the reservoir itself and found some dead fish?

A. Yes.

Q. And they were decayed?

A. Yes.

Q. And you could not get anything from them?

A. No.[139]

Starks contributed little to the inquest (except in a negative sense), but his testimony is notable because it convincingly eliminated "dead fish" as viable evidence of a possible dynamite blast.

The dynamite theory had one last moment in the sun when Zattu Cushing, a supposed explosives expert from El Paso, Texas (who years before had worked on the Los Angeles Aqueduct), testified that a submerged dynamite blast at the east side of the dam could have initiated failure. Cushing had traveled to Los Angeles at his own expense to appear at the inquest, but after his appearance on Friday, April 6, the BWWS hired him to investigate the dam site for additional evidence of a dynamite explosion. He returned four days later to update the inquest on his findings, but his weekend labors proved less than illuminating:

Q. [By the coroner]: In your testimony Friday, you stated you were having some excavation work done under a portion of the dam?

A. [Cushing]: Yes sir.

Q. Have you carried out that work?

A. Been carried on but not completed.

Q. You haven't reached any conclusion?

A. No sir.

Q. Haven't developed any fact that you care to disclose at this time, that would be of interest to the jury?

A. Disclosed there was an abrasion found.

Q. You haven't come to any conclusion as to what caused this abrasion?

A. No sir, because that would involve technical study of chemists.

Q. How long a time?

A. I don't know.

Q. You have had considerable experience in this type of thing?

A. Not in examination of concrete, not the technical end of it.

Q. Have you already submitted samples to a chemist?

A. Samples are taken, but I don't know whether they have gone to the chemist or not.[140]

Hardly an auspicious start, and City Attorney S. B. Robinson felt compelled to interject, "I might say [this] investigation is being very vigorously prosecuted, but our [the city's] feeling is this is a matter that should not be gone into piecemeal, there ought to be a complete investigation, lay it all before the jury, not lay a few fragments. It isn't that we want to keep anything back."[141] The inquest was on the verge of entering its fourth week, and the coroner was anxious that Cushing make a full report as quickly as possible, but it proved impossible to pin him down on any conclusive statement:

Q. [By the coroner]: You realize, of course, as everyone does, that both sides of the dam are out. Is it your contention that both ends must have been dynamited?

A. My contention is it isn't dynamited at all yet.

Q. You are not making a positive statement from your investigation, that it is dynamited?

A. No sir.

Q. [By a juror]: Is any effort being made to determine whether there was any such shock on the west side?

A. Can only do one thing at a time.

Q. Are men searching the valley floor there under your direction?

A. Not now, no sir, I do my own searching.

Q. [By the coroner]: Is it your opinion that if this jury went up there to see what you have just described, it would be of any value to them?

A. It might, I don't know. They might see something that is interesting. I will be glad to take them and describe it to them.

The coroner: That is all, you may be excused.[142]

The curtain on the inquest was about to fall, but one last witness was called for evidence of a possible dynamite blast. William J. Bright, chief of the homicide department of the Sheriff's Office, was asked to describe a recent visit to San Francisquito Canyon, where he had examined a hole in one of the concrete remnants that Cushing said might have been shaped by a dynamite blast. Bright expressed no enthusiasm for his quixotic assignment. When the coroner asked him, "If you hadn't heard of the possibility that [the hole was formed by dynamite], seeing that, would you have thought of explosives being used, or it might have happened in the destruction?" he laconically replied: "I wouldn't know." With that, testimony at the inquest sputtered to a close.[143]

"Monstrous to Place a Man Upon Trial"

On the afternoon of April 10, twenty days after Ray Rising had testified about identifying his wife's body, Coroner Frank Nance finally presented the St. Francis Dam case to the jury. By this time, it was clear that the likelihood of developing convincing evidence of a dynamite attack was nil. Over lunch, Nance, the jury, and the lawyers participating in the inquest had apparently reached a consensus that nothing could be gained by dragging out the hearing any further. The time for a verdict had come.

Shortly after 2:00 p.m., Nance announced to the jury that "it seems wise to submit the case to you now [because] one of your number has a very important engagement in San Francisco within the next day or two...and it is highly desirable that the full jury, which has heard this testimony should act on it." He also counseled, "I will give you some brief instructions and some advice as to what your verdict should consist of, and how it should be reached." Directing them that their task was to determine "how, when, and where" Julia Rising and the other sixty-eight victims found in Los Angeles County died, they were then to ascertain whether "these deaths were due to homicidal or accidental means, or natural causes, [and if] of homicidal means, who is responsible therefor, [and if due] to accidental means, whether the accident was due to negligence on the part of any person or persons." Reading from California's Penal Code Section 192, he described the elements of voluntary and involuntary manslaughter, with the latter defined as "the commission of an unlawful act, not amounting to a felony; or in the commission of a lawful act which might produce death, in an unlawful manner, or without due caution and circumspection."[144]

Nance then introduced Assistant District Attorney Dennison to provide "some elaboration as to the law governing negligence." This so-called elaboration provides remarkable evidence that District Attorney Keyes had no interest in pursuing criminal charges. Viewed in a cynical light, it might appear that the fix was in. Dennison addressed the jury:

> I think you have covered it, Mr. Coroner. Of course, here is one thing: If, in the erection and in the inspection and [care] of this dam, the men who were charged with it exercised ordinary care and prudence and honestly endeavored to do that which they were charged with doing, there could be no criminal action resulting from the misfortune or accident of the dam failing. In other words, if, in the selection of this dam site, and

> the building of the dam itself, Mr. Mulholland, who frankly stated he was responsible for the matter, made an error of judgment, an honest mistake, and a catastrophe resulted, of course, there could be no criminal responsibility attached to it. It would be monstrous to place a man upon trial for the crime of manslaughter or murder, who just merely made an error of judgment.[145]

With this entreaty, the district attorney's office made it essentially impossible for the jury to recommend criminal charges against Mulholland or anyone else.

Two days later, on Thursday, April 12, the jury issued its verdict, giving the district attorney no reason to be surprised or disappointed.[146] "After carefully weighing all the evidence," concluded the jurors, the dam failed for two fundamental reasons: "an error in engineering judgment" and "an error in regard to fundamental policy relating to public safety." The first error consisted of building the dam on defective "rock formations." Compounding these foundation problems was a dam design "not suited to [the] inferior foundation conditions"—a design that, among other flaws, did not carry the dam "far enough into the bedrock" and that lacked precautions against uplift, such as "cut-off walls," "pressure grouting of the bedrock," and "inspection tunnels with drainage pipes." In the jurors' view, responsibility for these lapses in engineering judgment rested "upon the Bureau of Water Works and Supply, and the Chief Engineer thereof."[147] Nonetheless, the jury found:

> no evidence of criminal act or intent on the part of the Board of Water Works and Supply of the City of Los Angeles, or any engineer or employee in the construction or operation of the St. Francis Dam, and we recommend that there be no criminal prosecution of any of the above by the District Attorney.[148]

As for the error in public policy, the jurors laid that at the feet of "those to whom the Chief Engineer is subservient," that is, "the Department of Water and Power Commissioners, the legislative bodies of city and state, and to the public at large." If these groups had insisted on "proper safeguards...making it impossible for excessive responsibility to be delegated to or assumed by any one individual in matters involving great menaces to public safety, it is unlikely that the engineering error would have escaped detection and produced a great disaster."[149]

So there it stood, with responsibility so widely attenuated as to be essentially meaningless. Mulholland was a mere pawn, and society as a whole was to blame for allowing him to act so recklessly. The corollary to this judgment on civic responsibility seemed clear: create an omnipotent state regulatory regime tasked with protecting the interests of California's citizens and unencumbered by an exemption for municipal dams. On this point the jurors were explicit, recommending "that steps be taken to the end that all existing dams be thoroughly examined... [and] that steps be taken to change state law so as to place the building of municipal and county as well as privately owned dams under the jurisdiction of state authorities."[150]

Finally, how did the jury address the possibility of a dynamite attack and how did they reconcile the west side and east side failure hypotheses? Acknowledging the "previous attacks upon the City's water supply system by use of explosives" the jurors conceded:

> While it is undoubtedly possible that the destruction of the dam could have been caused by an explosion, no conclusive evidence that such was the cause has been brought before us. [But] even if the failure had been precipitated by this cause, it would not change the situation so far as concerns the defects that have been described and which were the more probable cause of the disaster.

Although members of the Mayberry Committee were far from persuasive in defending the theory that disintegrating conglomerate initiated a west side collapse, the jury nonetheless accepted it. "The preponderance of expert opinion," they concluded, "favors the conclusion that the initial failure was on the west side." But they understood that this conclusion was far from incontrovertible, offering the caveat:

> Whether the [east side] landslide was the first event or only a later development is a subject concerning which there may be a difference of opinion, or at least, an uncertainty...The exact sequence of these events is

> of great engineering interest but has little bearing on the basic cause and responsibility. A susceptibility to landslide was one of the defects that should have been foreseen.

After three weeks of often labored testimony and legalistic Sturm und Drang, the Coroner's Inquest ended with an anticlimactic whimper. The Governor's Commission had taken the headlines and left the inquest to lurch to a finish, avoiding answers to hard questions and offering conclusions that in all significant ways conformed to the Governor's Commission *Report*. Newspapers covered the inquest's verdict, but a full month had passed since the flood, and the disaster had lost its hold on the public's attention. The *Los Angeles Times* dutifully reprinted the entire verdict, accompanied by an editorial that characterized the jury's findings as a "foregone conclusion," and instructed readers that "the lesson of the St. Francis disaster" should focus on "a faulty system which loads upon any man's shoulders, no matter how broad, responsibilities that are too heavy for any man to bear."[151] The *Examiner* soft-pedaled the verdict with the headline "Honest Error Made in Site, Verdict Says," quoting Assistant District Attorney Dennison's praise for the jury's "admirable, thorough, and fearless" work.[152] The *Los Angeles Record* took a more contentious stance, blaring the headline "Resign Now!" at Mulholland and the Board of Water and Power Commissioners and caustically asserting that "the St. Francis Dam failed—and 400 people died—because of the ignorance of the engineer who built it, and the stupid inefficiency of the board by which he is employed."[153] Strong words, but they had little impact. With the board's assent, Mulholland officially maintained his position as chief engineer for another seven months. Coming in the wake of the Governor's Commission *Report*, the inquest jury's verdict proved to be almost irrelevant, prompting no great public outcry—in fact, no response of lasting consequence.

The East Side Did *Fail First*

Upon its release in late March, the Governor's Commission *Report* had taken center stage, advancing a theory focused on the disintegration of the west side conglomerate. And, thanks to widespread dissemination of the report by the state printing office, it remains the most readily available historical account of the failure. However, the most insightful investigative reports detailing the mechanics of the collapse came later that spring from civil engineers Carl E. Grunsky and his son, E. L. Grunsky, Stanford University geologist Bailey Willis, and San Francisco–based consulting engineer Charles Lee. Their work, later affirmed and expanded upon, showed that the dam failed as a result of water infiltrating and saturating the fractured mica schist of the east abutment. This saturation created "uplift" pressure that destabilized the structure and fostered the massive, and destructive, east side landslide.[154] The analysis of the collapse presented in chapter 3 of *Heavy Ground* is based upon the insights that were first formulated and published by the Grunskys, Willis, and Lee in 1928.

Carl Grunsky was a major figure in the engineering world, serving as the first San Francisco city engineer, as a member of the Panama Canal Commission, as a consulting engineer for the U.S. Reclamation Service, and in 1922 as president of the American Society of Civil Engineers. He had also studied water supply issues on behalf of farmers along the Santa Clara River starting in the mid-1920s. His son, E. L. Grunsky, worked with him as a consulting engineer. Bailey Willis, with degrees in mining and civil engineering as well as "geological studies...directed primarily to the mechanical problems of rock structures," had a half-century of engineering and geological experience in the United States and South America, including service with the U.S. Geological Survey and, most recently, as a professor of geology at Stanford University. Such qualifications, bolstered by the Santa Clara Water Conservancy District's existing professional relationship with Carl Grunsky, prompted the district to hire the Grunskys and Willis to investigate the dam collapse.[155]

Their work, carried out with little fanfare, culminated in two reports completed in April 1928.[156] Willis' "conclusions and our own," observed Carl Grunsky, "were reached independently...[but nonetheless] are in substantial agreement." Both reports were soon published in *Western Construction News*, the Grunskys' in May 1928 and Willis' a month later.[157] Demonstrating a technical knowledge of the dam site, the three men identified four major factors that, in combination, precipitated the disaster:

1) *Unsuitability of the Foundation*: Foundations on both sides of the dam were deemed deficient, "but the critical situation developed more rapidly in the east abutment," where "the schist [was] traversed by innumerable minute fissures, into which water would intrude under pressure and by capillary action."

2) *Old Landslide*: The "east abutment was located on...the end of an old landslide."

3) *Uplift*: "When [the old landslide] had become soaked by the water standing in the reservoir against its lower portion, it became active and moved." That movement resulted from "a great hydrostatic force under [the dam's] foundation surface from end to end," which triggered the collapse of the east abutment.

4) *Inadequate Design*: "The old slide against which the dam rested at the east...offered only insecure support to the dam, and this was rendered more precarious by the [dam builders'] adoption of a design which did not include adequate foundation drainage."[158]

Willis, as the geologist on the investigative team, most likely discovered the "old slide" (his report discusses it at greater length and the Grunskys drew liberally on that discussion). Conversely, the Grunskys took the lead in describing the effect of uplift, and censured the design because "no measures [were taken]...which would have reduced percolation into the hillside material under the dam." They also emphasized precautions that could have been implemented to combat uplift, such as "thorough hillside and foundation drainage...fortified with deep cut-off walls along or near the up-stream face." Because of these design deficiencies, "at a full reservoir there was a great hydrostatic force under [the dam's] foundation surface from end to end, relieved but slightly by a few weep-holes [in the center of the canyon]. This hydrostatic pressure, the uplifting force of the swelling red sandstone at the west, and the horizontal and up-lifting pressure of the slide at the east, lifted the dam [and] broke it from its foundation."[159]

To help illustrate how the dam had been lifted, the Grunskys included a photo showing the base of the dam's surviving center section. A horizontal crack had formed as the section tipped downstream, on the brink of collapse. A wooden ladder had become lodged within the crack and then was crushed when the structure rocked back into place. It was powerful visual evidence of just how close the center section had come to failure (see figure 3-27). And it was evidence that had been ignored by the Governor's Commission and passed unmentioned in their report.[160] But in presenting this evidence, and in making a persuasive case for the east side failure theory, the Grunskys and Willis did not attack the logic underlying earlier studies—most notably, the Governor's Commission *Report*—that focused on the west side. They had been hired by the C. C. Teague–led Santa Clara Water Conservation District, and they were not tied to Los Angeles or larger state interests. Yet the civic leadership of Ventura County had reached an accommodation with Los

Angeles on the issue of reparations (see chapter 5), and there was good reason why they would have been reluctant to launch a broad assault on studies sponsored by the city or the governor. The Grunskys and Willis were not prevented from publishing their investigations in *Western Construction News*, but apparently they were not encouraged by Ventura County leaders to embark on a high-profile campaign to discredit the west-side-first hypothesis.

A month after the Grunskys published their findings, Charles Lee affirmed their conclusions in his own report. Lee had testified at the Coroner's Inquest where, as a consultant for the BWWS, he rejected claims that significant leakage had percolated through the west side abutment in the hours before failure but declined to postulate the cause of the collapse. After the inquest, Lee pursued a fuller investigation of the disaster and, by late spring, began to circulate his findings in public lectures and, most significantly, in an article in the June 1928 *Western Construction News*.[161] The most plausible collapse theory, Lee concluded, was that "the immediate cause of failure [derived from] a slide at the east abutment," and that there was probably "an upward thrust under the dam great enough to crack off a substantial portion of the east end or possibly raise the whole structure from its foundation."[162] He also addressed water gauge readings presumably documenting significant leakage from the reservoir before the collapse. Lee reiterated that no witnesses reported any increased flow in San Francisquito Canyon, and that the reservoir was not dropping in the hours before the collapse; instead, the gauge record was likely "due to a slow rising of the dam."

Lee was mindful that his work at the dam site had been supported by the BWWS (and hence the city of Los Angeles), and he couched his report in tentative, qualified language ("may have been," "it is possible").[163] But like the Grunskys, he gave no credence to a dynamite attack. And despite his gentle tone, he was forthright in critiquing the west-side-first hypothesis: "This theory does not satisfactorily explain many of the known facts...and is at variance with the fact that there was no appreciable leak from the dam prior to failure, or at least up to within a few minutes of failure." Instead, he detailed "a logical sequence...predicated upon a failure commencing at the east end of the dam," and he clearly explained how the dynamics of the collapse created the debris field left by the flood.[164] To bolster his argument, Lee explained how so-called "minor circumstances" reinforced his conclusion:

FIGURE 6-17. As Charles Lee noted in *Western Construction News* in June 1928, the top of the water gauge pipe attached to the surviving center section was bent toward the east side of the dam site (to the left in this photograph of the upstream face), strong evidence that when the disaster began, the heavy flow of water was toward the east side. [St. Francis Dam Disaster Papers, Huntington Library]

> There are a number of minor circumstances which also point to the east end having failed first. The remaining upper portions of the broken stilling-well pipe, which was bracketed to the upstream face of the dam, are both bent to the east as if by a current swiftly moving in that direction, while the water level was still high on the dam. The steam shovel fill on the hairpin turn in the construction road 50 ft. downstream from the base of the dam near the west end and about 60 feet below [the] spillway crest, was just awash at the highest stage of the water flowing from the gap at the west end. It was not eroded or undermined, thus indicating a reservoir stage considerably less than maximum at the time of highest level flow from the west side. Finally, the far greater depth of bedrock erosion on the

> east side, as compared with the west, indicates that a greater volume of water flowing for a longer period and under greater head, was pouring through this gap than through that on the west."[165]

Hemborg's survey of the dam site made much of a one-inch eastward movement of the west abutment. But Lee was far more impressed with how mica schist had been scoured from the east canyon wall to a depth of more than forty feet. He and the Grunskys were not constrained by preconceptions tied to the fact that, on the morning before the flood, dam keeper Tony Harnischfeger had called Mulholland to report leakage on the west abutment. Lee, Willis, and the Grunskys took account of the conditions at the post-collapse site and devised theories that made the best sense of the evidence at hand. The failure did start on the east side, and Lee, Willis and the Grunskys offered the most convincing explanation of the disaster.

The Best Christmas Present Ever

To those paying attention, the analyses presented by the Grunskys, Willis, and Lee were extremely compelling. But few seemed to be—or at least not to the extent that anyone invested in the west-side-first hypothesis wanted to revisit or revise their earlier pronouncement publicly. Silence reigned among those who early on had argued that the west abutment constituted the site of initial failure. Engineers like A. J. Wiley and Elwood Mead (who both had strong ties to the proposed Boulder Dam) and other engineers and geologists who served on their committees offered no immediate reaction either accepting or disputing the logic underlying the east-side-first failure sequence. The only way to explain their lack of interest in responding is to consider once again how the St. Francis collapse intertwined with the hoped-for passage of the Boulder Canyon Project Act.

It had taken six years to reach the point at which enactment of the Swing-Johnson bill seemed possible. Then the wave of uncertainty uncoiled by the St. Francis disaster threatened to disrupt and permanently doom plans to build Boulder Dam. With Arizona adamantly opposed to the dam (on grounds that it served the interests of Southern California at the expense of Arizona) and other states in the Colorado River basin ever fearful that California might take advantage of them, the political coalition cobbled together by Swing and Johnson was under assault in the spring of 1928.[166]

The privately controlled "Power Trust" that dominated America's electric power industry had long fought against the Boulder Dam, and opponents of public power were happy to exploit the St. Francis disaster. For example, the San Francisco–based *Argonaut* railed that "the catastrophe is an awful indictment of a system of municipal ownership that sets itself above the law."[167] In the House of Representatives, Utah Congressman Elmer Leatherwood took the lead in tying the disaster to the proposed Boulder Dam, seeking ways to delay—and preferably kill—the legislation. As the *Los Angeles Times* reported a week after the collapse:

> The fight to obtain a review of the Boulder dam project by a board of eminent engineers appointed by the President will be resumed today when the House Irrigation Committee meets...Representative Leatherwood of Utah announced today he will present an amendment... providing for such a review on the ground the safety and feasibility of the project have not been fully and satisfactorily established. He will point to the St. Francis disaster as fresh evidence of the need for caution. The amendment will be opposed vigorously by Representative Swing of California. [If necessary] Leatherwood said he later will carry the fight to the floor of the House.[168]

Swing and Johnson were helped by the speedy completion of the Governor's Commission investigation and the political cover it offered, but significant momentum had been lost and the bill languished in Congress through early spring and into May.[169]

Opponents of the legislation continued to insist that the St. Francis disaster foretold what would befall the lower Colorado basin if the plan to build Boulder Dam went forward. The arrogance of Los Angeles was often highlighted, and Arizona Congressman Lewis Douglas reminded people that Mulholland had failed to enlist any outside consultants for his project in San Francisquito Canyon: "Referring to the St. Francis Dam Disaster, the Arizona Congressman charged that responsibility went back to 1917 when the city of Los Angeles was successful in preventing the California Legislature from passing an act which would subject all dams in California to rigid inspection by State engineers."[170]

In the U.S. Senate, Reed Smoot of Utah proved especially antagonistic to the Swing-Johnson bill, marshaling arguments that coupled the

St. Francis disaster to Mulholland's advocacy of the Boulder Canyon Project Act. His blistering attack became national news and, under the headline "Smoot Warns of Boulder Dam, Death Trap," the *Chicago Tribune* reported:

> Senator Smoot made comparisons with the St. Francis Dam of the city of Los Angeles, [the] breaking of which cost the lives of between 400 and 500 persons. Incidentally, Senator Smoot pointed out that William Mulholland, the engineer who selected the site of St. Francis Dam and approved its foundation as safe and adequate, was one of the engineer witnesses before congressional committees who vouched for the feasibility of the Boulder Canyon project.[171]

In the House of Representatives, Swing was able to fend off the dogged objections of Douglas and Leatherwood and, on May 25, win approval of an amended bill.[172] But in the Senate project proponents faced an uphill fight. With congressional adjournment looming in late May, Hiram Johnson made an impassioned plea on the Senate floor, directly addressing Smoot's scathing criticism:

> It has been a rather sad thing that the prophecy I made in opening this case should have been justified in the remarks made by the senior Senator from Utah. I said then that certain individuals would seize upon the St. Francis Dam disaster in the San Francisquito Canyon in California in order to read a horrible lesson into dam construction in this country. He seized upon it with avidity and an enthusiasm that was worthy of Josiah Newcomb, of the Power Trust, and...dwelt upon the possibilities if the dam that should be erected at Boulder Canyon should thereafter be destroyed.[173]

Johnson invoked testimonial support from A. J. Wiley and F. L. Ransome (both had served on the Governor's Commission) and stressed their assurance that the Colorado River dam would pose no danger to the public.[174]

But despite all his politicking, Johnson could not turn the trick in the Senate and, at the end of the month, the *New York Times* reported the controversy in a story headlined, "Congress Ends Session with Senate in Uproar over Boulder Dam Bill."[175] While passage of the bill eluded him, Johnson was able to win agreement that consideration of the Boulder Canyon Project Act would be the first order of business when the Senate reconvened in December. In the interim, a "Colorado River Board" of engineers and geologists chaired by Major General William Sibert was authorized by a joint House-Senate resolution to independently evaluate the proposed Boulder Canyon Project (formation of such a board had been included in the House bill passed on May 25). Sibert's committee was to submit a report to Congress within six months, making it available for consideration when the Senate again took up the Swing-Johnson legislation at the start of the December legislative session.[176]

If the St. Francis tragedy had not occurred, it is unlikely that such an engineering board—as proposed by Congressman Leatherwood after the collapse—would have been created at such a late date in the legislative process. But, with some 400 people dead in the Santa Clara Valley, it was politically difficult for Swing and Johnson to oppose such a move. After all, why rush authorization of what would be the largest dam in the world? Why not take every precaution to ensure that Boulder Dam would not suffer a fate similar to the St. Francis tragedy? Swing and Johnson also likely anticipated (or hoped) that, by December, memories of St. Francis would have faded from public consciousness. It was therefore essential that nothing be widely aired or discussed that appeared to challenge the conclusions of the Governor's Commission and its affirmation of concrete gravity dams as a safe and viable technology.[177]

By November, Sibert and his Colorado River board members had completed their work and endorsed the basic design for Boulder Dam. However, seeking to be extra careful, they recommended that the maximum allowable compressive stresses in the structure be reduced from forty tons per square foot to thirty.[178] To the layperson this might appear as little more than a simple way to increase the strength and stability of the design, but strict adherence to a limit of thirty tons per square foot would significantly add to the already massive bulk of the dam and dramatically increase construction costs. The Bureau of Reclamation did not overtly resist the Sibert board's directive but made no meaningful changes to the design profile, eventually claiming that more sophisticated mathematical analysis affirmed that their proposed design afforded a maximum stress of about thirty-four tons per square foot, which was deemed sufficient to meet the thirty-ton criterion. In Commissioner Elwood Mead's words: "It is not believed that the maximum stress as finally calculated will appreciably exceed the 30-ton limit. It is believed that the general plan of the dam can be agreed upon without serious difficulties."[179] Thus, while the Sibert board made significant recommendations related to funding, in terms of basic design it had little impact on the structure that became Hoover Dam.[180] But most important, the board served the purpose that Johnson and Swing had hoped: it endorsed the Boulder Canyon Project Act and successfully neutralized arguments about safety posed by dam opponents.

When Congress reconvened on December 5, the political landscape had changed dramatically since the previous spring. On the local level in Los Angeles, Mulholland had resigned as chief engineer of the Bureau of Water Works and Supply (see chapter 8 for more on his resignation). Nationally, Federal Trade Commission hearings had brought to light nefarious lobbying by foes of public power, fomenting outrage against the "Power Trust" among the American electorate. The public had wearied of proselytizing by private power advocates and, by the end of the year, denunciations that blasted the Boulder Canyon Project as an unwarranted expansion of government into the hydroelectric power business carried little weight.[181] Senator Johnson's acceptance of a six-month delay before seeking Senate approval of the Boulder Canyon Project Act had proved prescient; Capitol Hill power brokers and politicians were now far more willing to embrace a massive federally financed hydropower dam on the Colorado River.[182] In the words of historian Donald Pisani: "By December, many senators who had opposed the legislation earlier in the year favored it out of fear that a negative vote would brand them as stooges of the vilified power trust."[183]

Using the Sibert board's endorsement as a foundation, in little more than a week Johnson won passage—by an overwhelming margin of sixty-four to eleven—of a Senate bill that was reasonably close to what the House had approved the prior spring.[184] Rather than seek a joint House–Senate committee to resolve differences between the two chambers, Swing adroitly presented the Senate legislation to the full House and, on December 18, it passed without amendment.[185] Three days later President Calvin Coolidge signed the Boulder Canyon Project Act into law. The *Los Angeles Express* memorably described this presidential stroke of the pen as "the best Christmas present a community ever received."[186]

At no time during the legislative whirlwind that swept through Congress in December 1928 does it appear that opponents raised the specter of the St. Francis disaster.[187] Less than a year after floodwaters teemed in the Santa Clara Valley, the collapse of Mulholland's dam had lost force as a means of politically opposing Boulder Dam and large storage dams in general.

Let Sleeping Dogs Lie

In early June 1928, the nationally distributed *Engineering News-Record* synopsized the reports by the Grunskys and Willis on the east side failure mode under the headline "Sixth Report on St. Francis Dam Offers New Theories."[188] In the same issue, the journal published an editorial that dismissively downplayed the Grunsky and Willis analyses as "speculation" and "ingenious theories" and criticized their reports for

> drawing far-reaching conclusions from the appearance of a wooden ladder caught in a crack in the remaining section of the dam...The report plays rather heavily on a single bit of circumstantial evidence—and circumstantial evidence, we know, must always be handled with great caution. But, taken in its entirety, the [Grunsky-Willis] report reinforces the findings of the earlier reports in saying that the foundation was thoroughly unsuitable and unsafe.[189]

The next month the journal published a brief rejoinder from the Grunskys, who complained that the *Engineering News-Record* editors had mischaracterized their analysis as somehow being in general accord with earlier west-side-first studies.[190] But after this minor dust-up, little discussion of the implications of the Grunskys' or Charles Lee's findings appeared in the engineering press.

In October 1929, the ASCE published a paper in its *Proceedings* entitled "Essential Facts Concerning the Failure of the St. Francis Dam: Report of the Committee of the Board of Direction." However, this was little more than an inconclusive review, presenting no analysis of its own and blandly declaring: "Some of the [investigating] reports express the belief that the break-up started at the west abutment, others at the east, and others reach no conclusion on this point."[191] The paper did not seek to evaluate the merits of the various investigations and, after members were given a chance to submit written comments, the ASCE editors opted not to publish the article in the far more prominent and widely available ASCE *Transactions*.[192]

The American civil engineering profession was greatly embarrassed that an engineer as well regarded as Mulholland had built such a deficient structure, and there seemed no reason to call any more attention to the disaster than absolutely necessary. Mulholland had publicly accepted blame and departed the professional arena. Never again would someone like him build a dam in California (and presumably in the nation as a whole) without outside review. So what was the harm in allowing the views of the Governor's Commission and others that ascribed the failure to the western abutment's dissolving conglomerate to predominate? What was wrong with declaring "defective foundations" the generic cause of the collapse and not parsing all the tiresome issues underlying the west side vs. east side failure debate? For both the engineering profession and parties interested in large-scale water development in the American West, there was little to be gained in revisiting the disaster. How else to say it? For many people, it was simply best to let sleeping dogs lie and not worry any more about what caused the collapse of the St. Francis Dam. That issue was yesterday's news. It was time to move on.

A Grim Reminder

Symbolically, one last task remained before the disaster could be relegated to the dustbin of history and, as far as possible, forgotten. In the immediate aftermath of the collapse, engineers seeking to defend the technology of concrete gravity dams had celebrated the surviving center section as evidence of stability and strength. This was proclaimed in the Governor's Commission *Report* and echoed by Governor Young in his telegram to Congressman Swing in late March. Through all the politicking that preceded passage of the Boulder Canyon Project Act, the notion of the center section as a symbol of strength could not be publicly questioned, at least not by city officials hoping to get a congressional green light for Boulder Dam. But once President Coolidge signed the Swing-Johnson Act into law on December 21, 1928, the St. Francis remnant took on new symbolic meaning.

As the anniversary of the disaster approached, plans to demolish the surviving center section were put forward, on the presumed grounds of public safety. Since the collapse, the site had become a tourist destination of sorts, a place where Angelenos could drive up in the morning,

FIGURE 6-18. An evocative snapshot of young visitors to the disaster site. In May 1928, a young man exploring the area with friends died when he fell from one of the concrete remnants. [DC Jackson]

clamber around, take some pictures, perhaps eat a picnic lunch, and then get back to the city by nightfall. On one such visit in late May 1928, eighteen-year-old Leroy Parker of Los Angeles fell to his death from a concrete block after a friend startled him with a thrown snake. He fell thirty feet to the canyon floor and later died from internal injuries at the San Fernando Hospital.[193] Through the remainder of the year no action was taken to eliminate the danger posed by the attractive nuisance of the concrete remains. But once the Boulder Canyon Project Act was enacted, official sentiment changed.

Of course, gravity dam proponents could have petitioned the city or the state to transform the dam site and surviving center section into a park, one dedicated to educating the public about the great, inherent strength of concrete gravity technology. Such a proposal would have been perfectly logical if the earlier pronouncements by the governor, the Governor's Commission, and gravity dam proponents like Walter Huber had been taken seriously. But in fact, the remains of the dam had come to symbolize something far different from structural strength. They stood as a relic of a disastrous city project, one that killed some 400 people and caused millions of dollars in damage. Why would anyone want to keep such a site intact, especially if some youngster might fall and die during an ill-advised visit? In an article describing the demolition work, the *Los Angeles Times* called the structural remains "a grim reminder of the disaster."[194] Why would anyone want to preserve that?

In April, cavernous shafts were excavated into the base of the center section. More than 300 holes were drilled into the concrete and packed

FIGURE 6-19. In the weeks and months after the disaster, the remnants of the St. Francis Dam became a tourist attraction for day-trippers from Los Angeles. [DC Jackson]

FIGURE 6-20. The surviving center section of the St. Francis Dam was heralded by many proponents of concrete gravity dams as a symbol of the technology's inherent strength. But interest in preserving remnants of the failed dam waned after the Boulder Canyon Project Act was signed into law by President Coolidge in December 1928. This view shows the center section being readied for demolition in spring 1929. [DC Jackson]

with five tons of dynamite. At 7:00 p.m. on May 11, 1929, "workmen and other spectators perched on the surrounding hills" to watch as "the electrically connected plunger exploded the dynamite...First the concrete wall bulged outward, then came a rumbling explosion followed by an ear-shattering roar as the concrete split into jagged blocks and crashed into the gigantic grave which had been dug at its base. And just as the night settled down the $15,000 job of demolition was done

FIGURE 6-21. Ruins of the St. Francis Dam in May 1929, leveled by five tons of dynamite. [*Engineering News-Record*, 1929]

and the last remaining portion of the ill-fated dam was gone."[195] Peace, or perhaps simply a pervasive desire to forget, had at last come to San Francisquito Canyon.[196]

CHAPTER SEVEN

Dam Building after the Flood

> The St. Francis Dam disaster in Southern California was a tragic event, which has created an hysteria on dam construction in California and the West. This catastrophe was of such an appalling extent that citizens have practically lost their heads on the subject of control of dam design and construction and have enlisted the aid of the legislature of California to pass a most drastic law covering the subject.
>
> M. M. O'Shaughnessy, ASCE Symposium on Public Supervision of Dams, 1933[1]

> In as much as dams of different types may be found feasible at certain sites...there is no good reason why the most expensive type, namely the gravity dam, should receive first and sometimes sole consideration.
>
> Fred Noetzli, ASCE Symposium on Public Supervision of Dams, 1933[2]

Within days after the St. Francis Dam collapsed, the press and public began clamoring for a new dam safety law, one that would eliminate the municipal exemption and presumably place all regulatory authority in one state office.[3] The final recommendation offered by the Governor's Commission gave voice to this demand, urging that all dams in California be "erected and maintained under the supervision and control of state authorities...with the police powers of the state...extended to cover all structures impounding any considerable quantities of water."[4]

Given that the St. Francis dam was built without state supervision, it was difficult for anyone to oppose, at least in public forums, such a seemingly common-sense proposal. Mulholland's failings appeared so egregious that, going forward, no right-minded person could support the municipal exemption to state regulation of dam construction. But behind the prospect of new legislation lay worries that, if carried too far, state regulation of dams could threaten economic growth: what if vital development of California's water resources were to be blocked by an unrealistic standard of safety? Bluntly stated, the safest dam is the dam that is never built because the cost of meeting conservative design standards proves prohibitively expensive. In the end, water projects—and the storage dams on which they so often depend—are always subject to budgetary constraints, no matter how much money may appear to be available to an irrigation district, a power company, a municipality, or a federal agency such as the Bureau of Reclamation.

Mindful that "the failure of the St. Francis Dam has greatly disturbed public confidence in the safety of all dams," State Engineer Hyatt also recognized the economic hardship that might result from overzealous regulation. Soon after the St. Francis collapse he warned that "for a time at least, proposals for the construction of new structures are going to face unmerited opposition no matter how carefully supervised by public authority. Even among competent engineers there will be a tendency toward undue conservatism."[5] In his view, the responsibility for safety vested in state authorities was politically perilous, not least because of possible financial impact: "We [in the California Department of Public Works] are thoroughly in sympathy with the feeling that public interest requires dams...to be made absolutely safe against failure and provided with adequate spillway capacity. At the same time, we feel that we must exercise great care to avoid insisting upon safeguards beyond the actual needs since many meritorious projects might be thereby rendered financially infeasible."[6]

As the issue of legislating dam safety played out through 1928, California's civil engineering community grew apprehensive about the possibility of unchecked governmental authority. Hoping to guide the development of new legislation, the four state chapters of the American Society of Civil Engineers (representing Sacramento, San Francisco, Los Angeles, and San Diego) formed a "Committee on Proposed Legislation on Design and Construction of Dams" in order to "assist in shaping and passing at the next session of the State Legislature suitable legislation governing the design, construction and operation of dams in the State of California."[7] In addition to advocating new legislation, the committee was charged with the more politically complicated task of protecting the interests of the professional engineering community in the face of authority wielded by state bureaucrats.

It was generally easy for the committee and ASCE members to support the idea of delegating the regulation of dams to a single state authority (that is, the state engineer and not the Railroad Commission). More problematic was whether new legislation should require that boards of engineers, presumably staffed by ASCE members, advise the state engineer on proposed dam projects.[8] During the committee's deliberations, Walter L. Huber (who later served as national president of the ASCE) forcefully advocated mandatory consulting boards, providing a justification that, while not specifically naming Mulholland, clearly referred to the St. Francis Dam disaster:

> I think that for any of the larger reservoirs the services of a Board of Consulting Engineers should be *mandatory* [emphasis in original]. Otherwise picture one of our powerful municipalities proposing a great reservoir to be built from plans by an engineer of that municipality who has much prestige but who alone considered the plans and the location. His reputation might far outweigh that of the State Engineer and might therefore make a critical review by the latter appear foolish. With political pressure reaching even to the Governor, would not a perfunctory review by the State Engineer, his early approval and a permit from the Department, be almost certain to follow without reference to consultants. Thus we would have failed to provide that protection to the public which we are seeking to accomplish.[9]

Samuel B. Morris, chief engineer of the Pasadena Water Department, shared Huber's contention that boards of consulting engineers should always review the judgment of the state engineer for major projects. He wrote to Hyatt and Leon Whitsell, chairman of the Railroad Commission, in September 1928: "Upon all major structures the State Engineer should employ a board of consulting engineers and geologists to review his analysis of dam design, foundation conditions, and any other details provided for in the [proposal]."[10]

In the end, the idea of de facto delegating a key police power of the state to boards of non-state employees proved unacceptable to the governor's office and the legislature. In early February 1929, Hyatt informed

Morris that "there have been several meetings of the various committees and parties interested in this subject and it appears that the differences will be satisfactorily adjusted, and a composite draft is in preparation at this time. The composite act [conjoining a Senate and an Assembly bill] eliminates federal dams, tanks, oil structures, [and] the mandatory [consulting engineer] board of review."[11] The final wording of the dam safety law, enacted in May 1929 and to take effect the next August, did not require use of consultants by the state engineer.[12] But lobbying by the ASCE committee and Morris did have an effect. As Hyatt proceeded to implement the new law, he sought input from consulting boards and welcomed their advice. Thus, the fears expressed by Huber—who served on at least one such board, and likely many more—were generally quelled, as long as Hyatt did not stray too far from the conservative orthodoxy that had rallied around massive gravity dam technology in the aftermath of St. Francis.[13]

An example of the way the new process worked occurred in the early 1930s, when the city of Los Angeles set out to build the Bouquet Canyon Dam above San Francisquito Power House No. 1. This dam was to replace the storage capacity lost at St. Francis and, of course, its construction required the approval of State Engineer Hyatt. But the earth embankment design was also reviewed and approved by a board of engineering consultants including J. B. Lippincott, L. C. Hill, and Charles T. Leeds, and geologists F. L. Ransome and Robert T. Hill, who advised Chief Engineer Harvey Van Norman. While Hyatt was not obligated to comply with the recommendations offered by the consulting board, he nonetheless accepted the city's plans and the project was approved. As state engineer, Hyatt held final say over the design, but he worked in collaboration with the city's engineering consultants in reaching a decision. The cost of the Bouquet Canyon Dam and related penstocks necessary to connect the reservoir to the aqueduct system totaled a reported $4 million—certainly more than the original St. Francis Dam (estimated at $1.25 million), but not more than the city considered affordable given its need for water storage. While the cost of dam-building increased in the post–St. Francis era, it did not rise so much as to seriously deter water development by large cities like Los Angeles or by federally financed agencies like the Bureau of Reclamation and the Army Corps of Engineers.[14]

California engineers generally accepted the necessity of the new dam safety law, but some continued to express concern about an all-powerful state bureaucracy dictating what could or could not be built and at what

FIGURE 7-1. Construction of the Bouquet Canyon Dam, 1933. This earth-fill embankment dam was built to replace the storage capacity lost when the St. Francis Dam collapsed. In accord with the state's 1929 dam safety law, it was built under the regulatory review of the state engineer. The design was also approved by a board of consulting engineers that included Mulholland's colleague J. B. Lippincott. [Los Angeles Department of Water and Power]

cost. At the extreme, M. M. O'Shaughnessy's comments to an ASCE–sponsored symposium on "Public Supervision of Dams" lambasted the new law as "most drastic" and declared that the St. Francis catastrophe had created "an hysteria" whereby citizens and legislators "have practically lost their heads on the subject of dam design and construction." O'Shaughnessy also raised the key question of who would exercise the

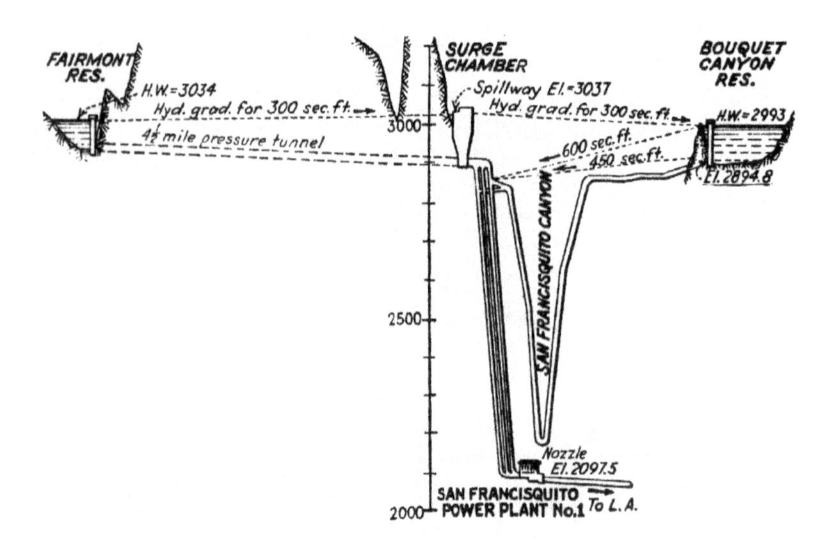

FIGURE 7-2. A schematic drawing illustrating how aqueduct water released from the Fairmount Reservoir in the Antelope Valley could either flow directly down to San Francisquito Power House No. 1 or pass over to the Bouquet Canyon reservoir via a steel siphon. The Bouquet Canyon reservoir site was far superior to St. Francis in terms of hydroelectric power generation, but the overall cost of the Bouquet Canyon system was much greater. In total, the Bouquet Canyon project cost over $4 million, while the St. Francis Dam was built for about $1.25 million. [*Engineering News-Record*, 1934]

state's regulatory authority: "The great problem is to find that 'some one' who has gathered enough wisdom from his experience and who has adequate force, technical knowledge, and authority to fill the job."[15] This important point likely struck a nerve in Sacramento, because Edward Hyatt was not a dam design engineer. He had risen through the state public works hierarchy based on his expertise in water rights adjudication and not in the design and construction of dams, highways, or other types of infrastructure.[16] As O'Shaughnessy no doubt knew, Hyatt possessed no direct experience as a dam builder and his success in the position of state engineer depended on his deftness as a bureaucrat.

A. H. Markwart, vice president in charge of engineering for the Pacific Gas and Electric Company, was more circumspect than O'Shaughnessy in his comments for the ASCE symposium, but he called attention to "the tendency...to require dams to be constructed stronger than actually necessary. Such excess strength can only be had from capital expenditures greater than have been required in the past."[17] Markwart did not mention the simplistic recommendation recently offered by Sibert's Colorado

River Board that would have increased the already ample dimensions of the Boulder Dam design by decreeing that the maximum compressive stresses be reduced from forty to thirty tons per square foot (see chapter 6). But it was this type of recommendation that would have spurred his concern about excessively conservative design standards. As it turned out, in the era of New Deal public works projects, fears that excessive costs might impede the financing of major dam projects proved overstated.[18] Nonetheless, Markwart accurately foresaw that capital costs for water development projects would escalate after St. Francis.

In perhaps the most revealing comment of the symposium, the European-trained engineer Fred Noetzli (who had immigrated to Southern California in the early 1920s) made a not-so-subtle objection to the growing dominance of massive-dam technology. Noetzli was a prominent advocate of thin-arch and multiple-arch dams who, in a 1924 ASCE *Transactions* article, had declared that "the gravity dam is an economic crime" because of its woefully inefficient use of concrete. In comments published as part of the dam supervision symposium, Noetzli warned that California's new law would encourage—perhaps even demand—construction of massive gravity dams at the expense of other potentially less expensive structures. Fearing that state authorities would automatically dismiss materially efficient and economically cost-effective alternatives, Noetzli protested that "in as much as dams of different types may be found feasible at certain sites...there is no good reason why the most expensive type, namely the gravity dam, should receive first and sometimes sole consideration."[19]

Amid all the inquiries and the bloviating that followed the St. Francis disaster, the issue of whether a multiple-arch buttress design might have been a better technology for the St. Francis site was never raised by any of the engineering investigations or by the press. Mulholland's choice of a curved concrete gravity design was always taken as a given, and after the failure it appeared—at least to Noetzli—that the hegemony of massive dams and the engineers who championed them would still prevail in the state's new regulatory regime. Notably, the St. Francis disaster did not prompt the ASCE to sponsor a session on a subject such as "What the St. Francis Disaster Can Tell Us about the Limitations of Gravity Dam Technology." Instead, the engineering society sponsored a symposium about how state regulation might be affected by the collapse of Mulholland's dam—far different from investigating and analyzing how the disaster might reveal deficiencies in the technology itself and prompt a search for alternatives.

Regulation and Innovation

Since 1929 California has cultivated a reputation for sustaining one of the most demanding dam safety bureaucracies in the world. Aside from the Baldwin Hills Reservoir collapse in 1963 (which killed five people and destroyed more than 250 homes), and the partial collapse of the Van Norman (Lower San Fernando) Dam during the Sylmar earthquake of 1971, the state has succeeded in preventing major dam disasters.[20] But, harkening back to fears of "undue conservatism," it may also be that efforts to ensure dam safety have suppressed innovation in designs and in construction techniques. In his article "The Evolution of the Arch Dam," Swiss engineer and dam historian Nicholas Schnitter noted that in California, "the number of [thin-]arch dams as well as their proportion in relation to other types decreased sharply in the 1930s."[21] Even more significant, the construction of multiple-arch dams built in the state dropped to virtually zero following passage of the new dam safety law. The reasons for this decline point to one of the most dramatic and lasting consequences of the St. Francis collapse for California dam builders.

As discussed in chapter 6, following passage of the 1917 law, State Engineer W. F. McClure was reluctant to approve any multiple-arch dams higher than 150 feet. The Littlerock Creek and Palmdale Irrigation Districts fought with McClure's office for almost four years before winning approval of what turned out to be a 170-foot-high dam. During this bureaucratic odyssey—which extended from 1918 through 1922—McClure sought advice from a variety of consultants, including Walter L. Huber of San Francisco, who largely dismissed the structural attributes of multiple-arch designs.[22] In 1928 Huber was allowed to inspect the St. Francis site soon after the failure and, even before release of the Governor's Commission *Report*, he described the surviving center section as a "great witness of the stability of a gravity section."[23] Huber had proven to be a proponent of gravity dam technology and, as apparent in recommendations made for the Littlerock design, he also had little interest in multiple-arch dams. So when in the spring of 1931 State Engineer Hyatt convened a special Multiple Arch Dam Advisory Committee to investigate the technology and appointed Huber to chair the group, it was not good news for advocates of large-scale buttress designs. In September 1932 Huber's committee presented its findings to the state engineer and, at best, damned multiple-arch designs with faint praise:

FIGURE 7-3. Murray Dam in San Diego County, a 117-foot-high multiple-arch dam, shortly after completion in 1917. The state's 1929 dam safety law effectively brought an end to multiple-arch dam construction in California. [DC Jackson]

> Although the multiple arch dam has its place...it should not be regarded as a cheap substitute for all other types. [The] committee was impressed that certain defects were common to a number of these structures [and] it is appropriate to call attention to these undesirable features...It is a naturally slender [type of] structure with integral members of relatively thin concrete... Some of them have been designed under competitive conditions resulting in structures successfully answering certain mathematical requirements, but hardly adequate from other points of view.[24]

Huber's committee offered a harsh assessment, especially for a technology that eliminated the threat of uplift—and uplift had comprised the essential cause of the St. Francis collapse. Imagine if post-disaster analysis of concrete gravity dams had been infused with the same dour skepticism that Huber brought to bear on multiple-arch dams, with

Mulholland's St. Francis Dam standing as a representative example of the technology. Viewed in this light, the importance placed by the Governor's Commission on the "defective foundations" of the St. Francis site, and not on gravity dam technology per se, becomes clearer.

The primary reason that gravity dam technology survived the St. Francis disaster with little direct criticism is that Mulholland's design was, in fact, egregiously *retardaire* in comparison with other major gravity designs of the 1920s. No great angst gripped the profession over the technology's possible weaknesses, and gravity dam design underwent no dramatic changes after the disaster. A simple illustration helps to make this point. The March 15, 1928, issue of *Engineering News-Record*—distributed the very week of the St. Francis disaster—carried a description of the East Bay Municipal Utility District's Lancha Plana Dam (soon to be renamed Pardee Dam). Construction of this 360-foot-high curved concrete gravity dam had started in late 1927. Designed by F. W. Hanna under the direction of Chief Engineer Arthur P. Davis, it was part of greater Oakland's water supply system. As drawings published with the article make clear, the design featured extensive foundation grouting, a drainage system running up both canyon walls, a drainage tunnel extending through the center of the dam, and a series of vertical contraction joints to control temperature cracks in the massive structure. All of these features were of course missing at St. Francis. But it was obviously not the St. Francis failure that prompted Davis and Hanna to include them in their Lancha Plana design.[25]

After St. Francis, gravity dam design did evolve, and concern about uplift continued to grow through midcentury.[26] And, while geologists were certainly consulted by the Bureau of Reclamation and by engineers such as John R. Freeman in the teens and early 1920s, such input became far more common in dam and water projects from the 1930s on. However, the profession never fundamentally questioned the suitability of gravity dam technology and, in fact, the technology that was transformed—or more accurately vitiated—by the St. Francis disaster was the multiple-arch dam. Gravity dam proponents first rallied around the "surviving center section" in order to protect the Boulder Dam project. After that success, the machinery of the state's new dam safety bureaucracy ensured that massive gravity dams would not be unduly criticized. And, as Noetzli feared, the same bureaucracy came to accept the advice of Huber's advisory committee and essentially brought an end to innovation in multiple-arch dam design. This was arguably the greatest technological legacy of the St. Francis disaster.[27]

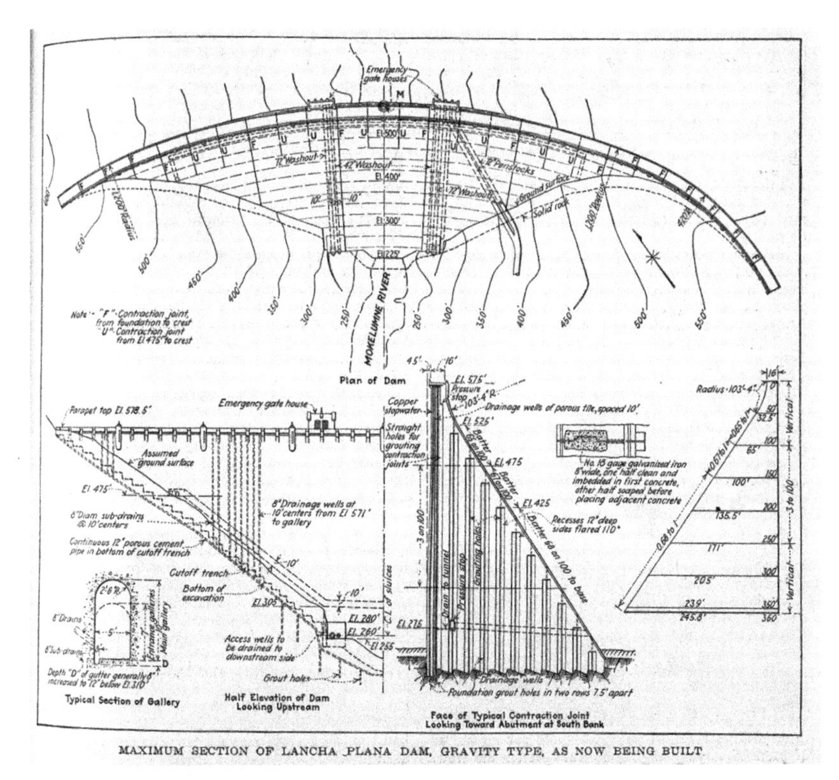

FIGURE 7-4. Drawings of the Lancha Plana Dam (soon to be renamed Pardee Dam), as published in *Engineering News-Record* on March 15, 1928, the same week as the St. Francis Dam disaster. Construction of this concrete curved gravity dam across the Mokelumne River east of Stockton had started the previous December, and its design bore scant resemblance to what Mulholland had used for St. Francis. Because Mulholland's design fell far short of what other dam engineers were already doing by the mid-1920s, the failure of St. Francis had only modest impact on concrete gravity dam design. [*Engineering News-Record*, 1928]

In a society that holds dear the benefits of competition and the expansion of free markets, state regulation is often portrayed as a deterrent to innovation and economic efficiency. Of course, the autonomy and latitude allowed to Mulholland by the municipal exemption in the 1917 dam safety law did not spur creative design at St. Francis. But neither should it be thought that freedom from government regulation invariably leads to inadequate, dangerous design—witness the gravity design that O'Shaughnessy built at Hetch Hetchy in the early 1920s. Dam politics (and some 400 lost lives) dictated that water storage technology in California would be much more closely regulated in the post–St. Francis era. In this context, some key yet seldom asked questions are prompted by the demise of multiple-arch dam technology. Huber's committee noted that multiple-arch dams were often "designed under competitive conditions resulting in structures successfully answering certain mathematical requirements, but hardly adequate from other points of view." So what standards beyond "mathematical requirements" should be employed in stringent government regulation, and how might such standards constrain technological innovation? Perhaps O'Shaughnessy was overly harsh in his view of California legislators who enacted the 1929 dam safety law, caving in to the "hysteria" of citizens who "practically lost their heads on the subject of dam design and construction."[28] But he saw clearly how the legacy of St. Francis might affect independent innovation in dam design. Many lessons can be learned from the St. Francis disaster, and some of the most important—specifically those relating to multiple-arch dams—are among the least obvious.

Resonating throughout the history of the St. Francis Dam is the simple truth that technology is not a force operating in isolation from the society that produces and sustains it. Technology and culture are inextricably intertwined, a connection we often wish to deny, or at least ignore. The St. Francis Dam was the product of human culture, one that allowed William Mulholland enormous freedom in selecting the dam's location, creating its design, and building it without outside review or interference. The structure of the St. Francis Dam took shape through the prism of a culturally and politically determined privilege. And through a related prism, the St. Francis Dam disaster became a part of history.

CHAPTER EIGHT

Postscript: William Mulholland

> He was 73 years of age on September 11 last and his resignation has been anticipated for several years...It is no secret that the collapse of the St. Francis Dam hastened Mr. Mulholland's resignation. The tragedy, which was keenly felt by the "Chief" as he is called at the water department, has aged him.
>
> *Los Angeles Times,* November 14, 1928[1]

> I suggest for the site of the memorial the location of an early home of Mr. Mulholland where he began his labors for the City Water Works near the corner of Los Feliz and Riverside Drive. A suggestion for the type of memorial–a water temple of suitable design through which large volumes of water from the Los Angeles River may flow and intermingle with water from the Owens River Aqueduct, and if feasible, Colorado River water.
>
> J. B. Lippincott, Mulholland Memorial proposal, July 30, 1937[2]

Given the pain, heartbreak, and death suffered by victims of the St. Francis disaster, William Mulholland was treated gently in the years following the dam's collapse.[3] His professional colleagues and California's political leaders did not publicly denounce him, once it appeared he had accepted responsibility for the disaster. Hailed as a "Big Man" in the engineering press, he was shielded from professional attacks that, for other engineers in similar circumstances, would have been considered fully justified.[4] Privately, however, at least a few of his peers criticized Mulholland's St. Francis design.

Shortly after the disaster the prominent engineer John R. Freeman advised J. B. Lippincott in a personal letter that "I have been careful...to say nothing [to newspaper reporters] regarding the Los Angeles dam which could come back to hurt Mulholland." But Freeman then candidly disparaged Mulholland's habit of not consulting independent experts: "he does not appreciate the benefit of calling in men from outside to get their better perspective and their independent point of view."[5] To another colleague, Freeman reinforced the point: "This [St. Francis Dam] site plainly required many precautions that were ignored, and while I have the highest personal regard for my good old friend William Mulholland, I can but feel that he trusted too much to his own individual knowledge, particularly for a man who had no scientific education."[6] Following a visit to the collapsed dam, Arthur P. Davis, another longtime associate of Mulholland and former director and chief engineer of the U.S. Reclamation Service, privately pointed out to Freeman the lack of a "deep cut-off trench," as well as the absence of "deep grouting" and "adequate drainage wells." Specifically referring to the Elephant Butte, Ashokan/Olive Bridge, and Arrowrock dams, Davis said, "Had provisions existed [at St. Francis], as established by recent practice [at] many other existing dams, the accident might have been avoided." But comments of this sort were expressed privately. The public did not hear such unvarnished criticism by Mulholland's peers.[7]

In the immediate aftermath of the disaster the Board of Water and Power Commissioners declined to relieve Mulholland of his duties, and he officially stayed on as chief engineer for several more months.[8] After the Coroner's Inquest absolved him of criminal negligence he kept a low profile. He avoided publicity, and nothing related to his ongoing work appeared in the press. In mid-May, his assistant chief engineer Harvey Van Norman, Bureau of Power and Light Chief Engineer Ezra Scattergood, City Councilman Peirson Hall, City Attorney Jess Stephens, and other city officials set off by automobile to inspect the proposed site for Boulder Dam. Mulholland was not included in the five-day caravan, which returned via the proposed route of the Colorado River Aqueduct.[9] Through the summer and into the fall he also remained personally distant from city-sponsored restoration work underway in the Santa Clara Valley. There is no evidence that Mulholland ever again journeyed to San Francisquito Canyon, apart from two visits to the site of the destroyed dam in the week after the flood; or that he ever traversed the disaster zone in the lower valley. Exactly what he did in his final months as chief engineer remains a matter of conjecture, if not mystery.

On November 14, 1928, the *Los Angeles Times* reported that he had "tendered his resignation to Board of Water and Power Commissioners"; as the *Times* further explained:

FIGURE 8-1. In the company of his chauffeur George Vejar, a weary Mulholland walked away from the St. Francis site in March 1928. The disaster weighed heavily on him—the *Los Angeles Herald* noted that he appeared to have aged some ten years in the traumatic days after the collapse. Mulholland lived on for another seven years, but after the flood he was no longer a dominant force in Los Angeles or in the world of hydraulic engineering. [Peirson Hall Papers, Huntington Library]

> He was 73 years of age on September 11 last, and his resignation has been anticipated for several years...It is no secret that the collapse of the St. Francis Dam hastened Mr. Mulholland's resignation. The tragedy, which was keenly felt by the "Chief" as he is called at the water department, has aged him.[10]

The exact timing of his departure was never precisely explained—his resignation officially became effective on December 1—but almost certainly it was spurred by the hopes of Angelenos that the Boulder Canyon Project Act would at last win Senate approval when Congress convened in early December (see chapter 6). Eight months on, the St. Francis disaster had faded from public view and lost its political potency. Nonetheless, Mulholland's departure and Van Norman's appointment as the new chief engineer were likely timed to assuage any lingering concern that Mulholland's continued leadership would discomfit advocates of Boulder Dam.[11]

In response to the announced retirement, R. F. del Valle, vice president (and former president) of the Board of Water and Power Commissioners, effusively proclaimed, "No engineering achievement in this country within the past half century has exceeded in difficulty or merit that of William Mulholland in keeping the water supply of this city with its rapid growth and development up to the requirements of its inhabitants." Maintaining a tone of goodwill, he assured Mulholland that he and his fellow commissioners were "fully appreciative of your sterling character and worth [and] we grant the request that you be relieved of the duties of your position only because you insist upon it." Del Valle did observe that "no ill fortune has been sufficient to shake the confidence of the people in this good and faithful servant." But other than this veiled reference, his testimonial offered scant reason for anyone to imagine what had occurred on Mulholland's watch only a few months before.[12] At the November 27 dinner marking Mulholland's retirement (held in the banquet hall of the City Club, like the dinner in honor of his twenty-fifth year of service to Los Angeles), none of the speakers mentioned the disaster, and even Ezra Scattergood chimed in with a gracious tribute that belied his fractious exchange with Mulholland over the siting of the St. Francis reservoir. If not a night of good cheer, it was certainly an occasion at which the spotlight fell exclusively on the glory and goodness of the long-serving Chief, a time for his accomplishments to be celebrated and his failings to pass unmentioned.[13]

A Man of History

After his resignation, the city retained Mulholland as a consultant at a salary of $500 a month, an assignment he held until his death. No specific duties were attached to this consultancy, and it served more as a pension or sinecure than as a way to keep him actively involved in departmental affairs or in the planning for the Colorado River Aqueduct.[14] His single known official chore came at the May 1930 trial of Ray Rising's civil suit against the city (for details, see chapter 5). Here, as the chief defense witness, he had the unenviable task of trying to raise doubt about the city's responsibility—and by extension his own—for the St. Francis disaster. Taking a very different tack from what might have been expected, given his inquest testimony and his prior willingness to accept blame, at the Rising trial he averred that an earth movement caused the collapse—in legal terms an "act of God"—and thus that the city should be absolved of liability for Rising's loss.[15] The jury was unswayed by the city's defense and Rising won a $30,000 judgment. Mulholland's reputation was hardly enhanced by his performance on the witness stand, but the trial did not attract much attention.[16]

Though he was no longer a paragon of the technological elite, the cachet of Mulholland's name lingered among at least part of the Los Angeles electorate. Three years removed from the St. Francis disaster and a year after his testimony at the Rising trial, he was deemed an appropriate booster for the bond election to finance construction of the Colorado River Aqueduct. His task: assume the role of a wise old sage familiar with the water problems of Southern California. At a July 1931 meeting of the Citizens' Colorado River Water Committee, he regaled the forty-seven attendees with some down-to-earth homilies: "I don't want to see the people ignore this need...For it is a need. You'll need it [water] when you get it; and if you don't get it—well you won't need it!"[17] The Metropolitan Water District's bond passed handily and in early 1932 he attended the inauguration of a district-sponsored groundwater replenishment program near San Bernardino.[18] His last known engagement as a public speaker was in May 1933 at the MWD ceremony marking the beginning of Colorado River Aqueduct construction at Cabazon. A crowd of 10,000 came to celebrate the start of tunnel blasting through the San Jacinto Mountains, where they heard MWD director John R. Richards praise Mulholland for "his work not only on the Owens Valley Aqueduct but in the preliminary work on this project [which] forms the basis upon which we have built." With the crowd giving him "the big hand of the day," Mulholland stepped to the microphone and spoke briefly: "Anything I might say would be pretty old stuff. I've tramped over these hills since '77, and I'm getting along. I'm glad to be of service to you and to this community now—and forever."[19]

In the aftermath of St. Francis, Mulholland was spared a professional purgatory where he would be denied ceremonial honors and personal respect. The September 1929 issue of *Southern California Business* overlooked his association with the St. Francis disaster, proclaiming him both a "Man of Broad Vision" and "Father of the Colorado Aqueduct project."[20] In 1931 he accepted an invitation from the Board of Water and Power Commissioners whereupon, on National Hospital Day, he was extolled "for producing a supply of water in the City of Los Angeles adequate for the uses of a population calculated on an unprecedentedly rapid basis of increase."[21] And in 1933 *Western Construction News* honored him "as a man of history and the maker of Los Angeles."[22]

But amid occasional accolades during his final years, Mulholland largely withdrew from public view. He fell into a melancholy that, despite time spent with his family, burdened him as he approached his seventy-ninth birthday.

The lingering sadness appears to have kindled a desire to reconnect with Fred Eaton, the visionary former mayor who introduced Mulholland to the possibilities offered by the Owens River, and who later became his great antagonist over rights to Long Valley. By the late 1920s Eaton's financial empire was a shambles, entangled in debt exacerbated by a run of bank failures in the Owens Valley. As his fortune spiraled downward, the city gained control of Long Valley and eventually was able to build Crowley Lake Dam—named in honor of a Catholic priest who long served the people of the eastern Sierra. The dam would be completed in 1941, concluding the Long Valley saga more than thirty years after Eaton and Mulholland first wrangled over control of the site. But long before this came to pass, on March 11, 1934, Fred Eaton died. Soon afterward Mulholland told a family member of a dream in which he and Eaton were walking together as young men, friends again. But the dream also brought a sense of impending death, where reconciliation could come only when both men were in the grave.[23]

Not long after Eaton's death Mulholland traveled with city officials to the site of the newly completed Bouquet Canyon Dam, a massive earth embankment built to replace the storage lost with the failure of St. Francis.[24] The *Los Angeles Times* article on the event did not mention any remarks made at the dedication, nor did it acknowledge that the dam replaced storage capacity lost when the St. Francis Dam collapsed. Some fifty city officials and guests (among them Los Angeles mayor Frank Shaw, Harvey Van Norman, Ezra Scattergood, MWD chairman William Whitsett, and Mulholland's old colleague J. B. Lippincott) made the trip to Bouquet Canyon on March 28, 1934, marking the last time Mulholland appeared at a public event. Perhaps it is fitting that the ceremony brought him back to the Los Angeles Aqueduct and to the upper reaches of the Santa Clara Valley.[25]

In October 1934 Mulholland fell and broke his arm. Two months later he suffered a stroke that brought him to the edge of death.[26] Paralysis gripped his body, and he lost the ability to speak or to eat solid food. On the morning of July 22, 1935, he died in his home on South St. Andrews Place, attended by nurses and surrounded by family. His granddaughter's account of his final months reveals a tragically incapacitated man who slowly wasted away: "Although the death certificate later stated cause of death as arteriosclerosis and apoplexy, his nurses agreed that he had simply starved to death."[27] Two days later his body lay in state in Los Angeles City Hall prior to a private burial at Forest Lawn Memorial Park. Longtime colleagues from the Bureau of Water Works and Supply—along with Ezra Scattergood—served as pallbearers. Fifty-eight years after arriving in Los Angeles as a humble laborer, he was laid to rest.[28]

Remembrances and tributes came quickly, with the city's flags lowered to half-staff and the Metropolitan Water District directing workers and contractors "that insofar as safety will permit all construction work [on the Colorado River Aqueduct] be halted for a period of one minute at 2:30 p.m. Thursday, July 25."[29] In the press, the collapse of the St. Francis Dam was rarely acknowledged and the main obituary in the *Los Angeles Times* did not mention the disaster; neither did a *Times* article with the subheading "Messages of Regret Pour in Voicing Unstinted Praise for Greatness of Mulholland" that summarized the eulogies extended by luminaries such as his former assistant, Van Norman; former Water and Power Board president R. F. Del Valle; W. A. Braunschweiger, president of the Los Angeles Chamber of Commerce; MWD chairman William Whitsett; and MWD chief engineer F. E. Weymouth.[30] Over the coming weeks and years, tributes to Mulholland almost always ignored the devastating flood that brought his career to a close.[31]

In 1938 a special issue of the nationally renowned *Engineering News-Record* was devoted to the Colorado River Aqueduct, with the editors heralding the "vision and foresight of the late William Mulholland," who more than a decade earlier "foresaw an ultimate need far exceeding any water supply that could be developed in the coastal basin." Praising his "indomitable courage," they further lauded his persistence in the "investigation and promotion of a plan for water importation that seemed almost impossible. Finally, as the results of his efforts, thirteen cities united to support the Colorado River aqueduct project [and] formed the Metropolitan Water District of Southern California." This passage fairly represents the way a multitude of engineers, politicians, and civic boosters in Southern California wanted to remember Mulholland and his legacy.[32]

Two years after his death, friends and colleagues initiated plans to erect a public memorial to him. It might seem that the Mulholland Dam, named in his honor in 1925 and prominently sited in the Hollywood Hills, would stand as a lasting monument to his service. After all, this was the structure in which he had expressed "a little pardonable pride," as his first concrete gravity dam.[33] But in the wake of the St. Francis

FIGURE 8-2. The Mulholland Dam above Hollywood featured a concrete curved gravity design similar to that used at St. Francis. In 1934, to address lingering discontent among the public, the city covered the downstream face of the structure with a blanket of earth fill to lessen its perceived height. The spillway crest had already been lowered thirty-one feet to reduce hydrostatic pressure acting against the dam. [*Engineering News-Record*, 1934]

disaster, the Mulholland Dam aroused concerns among downstream homeowners and businessmen, eventually addressed by reducing the allowable storage level of the dam's reservoir by thirty-one feet. While this preemptive action significantly decreased storage capacity (which the water department was not happy about), it also reduced the maximum hydrostatic pressure that could be exerted against the dam, thus enhancing its stability. This should have resolved all long-term safety issues, but scientific calculations affirming the structure's integrity were not enough to quiet public discontent. Mulholland Dam had been promoted as a symbol—a monument—attesting to both the city's civic stature and the chief engineer's technical prowess. After St. Francis the Mulholland Dam assumed a symbolic character very different from strength and durability.[34]

How to address this problem in public perception? The solution ultimately adopted called for blanketing the downstream face of the dam

FIGURE 8-3. After a few years, shrubbery and foliage covered the downstream face of the Mulholland Dam. These plantings helped sustain the illusion that the concrete dam was relatively small. [DC Jackson]

with a huge earth embankment—300,000 cubic yards of dirt that would forever shroud the concrete façade. In 1925 Mulholland had expressed disdain for the utilitarian character of such structures, likening them to "an old woman's apron—an object of utility but not of beauty." So it was no small step to take a once great symbolic structure and, through a literal cover-up, render the concrete facade largely invisible. The newly obscured Mulholland Dam could never serve as a monument celebrating The Chief.[35]

Over a period of several months in 1937, proposals for a memorial were discussed among a consortium of Mulholland's friends and associates. The group finally settled on a fountain to be located near Griffith Park and the Los Angeles River at the corner of Los Feliz Boulevard and Riverside Drive.[36] Here was a site that resonated with Mulholland's early years in the city, where he learned the practice of urban water supply and started his career as a hydraulic engineer. J. B. Lippincott conceived of the plan, telling his colleagues:

> I suggest for the site of the memorial the location of an early home of Mr. Mulholland where he began his labors for the City Water Works near the corner of Los Feliz and Riverside Drive. A suggestion for the type of memorial—a water temple of suitable design through which large volumes of water from the Los Angeles River may flow and intermingle with water from the Owens River Aqueduct, and if feasible, Colorado River water.[37]

Funds for the memorial were raised privately—no public monies were sought—and groups of youngsters were recruited to collect pennies, nickels, and dimes to support the cause. In October 1939 the Los Angeles school district sponsored a "Mulholland Memorial Week." In the words of the *Los Angeles Times*: "During the five-day period 300,000 students will study the life of the engineering genius who is hailed as the father of the municipal water system...Young Americans will find inspiration in the story of the youthful Irishman who came to these shores and found fame in the city of his adoption."[38] The fountain was completed at an estimated cost of $35,000 (students raised more than $1,000 of this sum).[39] Dedicated on August 1, 1940 (granddaughter Patricia Mulholland pushed the button to start the cascade), the memorial stood as a symbol of Mulholland's contribution to the burgeoning metropolis of twentieth-century Los Angeles. A small plaque adjacent to the fountain paid tribute to this legacy:

> A penniless Irish immigrant boy, who rose by the force of his industry, intelligence, integrity and intrepidity, to be a sturdy American citizen, an engineering genius, a whole-hearted humanitarian, the father of his city's water system and the builder of the Los Angeles Aqueduct: This memorial is gratefully dedicated by those who are the recipients of his unselfish bounty and the beneficiaries of his prophetic vision.[40]

Residents of the Santa Clara Valley or Owens Valley may have held a different view of the "unselfish bounty" brought to the city, but this was not a memorial designed to reflect their experiences with The Chief.

FIGURE 8-4. The Mulholland Memorial Fountain at Riverside Drive and Los Feliz Boulevard, August 1940. [Los Angeles Department of Water and Power]

FIGURE 8-5. A large boulder was brought from the Owens Valley to adorn the Mulholland Memorial Fountain. [Los Angeles Department of Water and Power]

Legacy and Image

The Mulholland Memorial was dedicated at a time when America hovered on the brink of war. Peace still reigned in the United States, but France had fallen to the Nazis and the German Blitz darkened Britain's skies. Only after the Japanese attack on Pearl Harbor in December 1941 did neutralist sentiments evaporate and America embark on a military mission to save the world from fascism. The socioeconomic impact of World War II on Los Angeles was tremendous, with Southern California serving as both a vital entryway to the Pacific Theater and a center of airplane production. Once the war ended in 1945, regional growth continued unabated through the remainder of the century.

Mulholland's postwar legacy as a hydraulic engineer found expression in both the operation of the Metropolitan Water District of Southern California and in the growth of Los Angeles' city-owned water supply system. But direct memories of his personality and professional demeanor diminished as the years passed. Public knowledge of the St. Francis disaster also ebbed, at least until Santa Paula native son Charles Outland published *Man-Made Disaster: The Story of St. Francis Dam* in 1963. Outland's book did not attract a mass readership, but it sold well enough to justify a second edition in 1977, and it brought the disaster to the attention of anyone with a serious interest in the history of Los Angeles.

The Owens Valley controversy retained a hold on the public imagination and Carey McWilliams' *Southern California Country: An Island on the Land* (1947) and Remi Nadeau's *The Water Seekers* (1950) helped ensure that Mulholland's work in building the Los Angeles Aqueduct would remain a staple tale in Southern California history.[41] Treatments that were more academic in character were published in the 1980s—both William Kahrl's *Water and Power* and Abraham Hoffman's *Vision or Villainy* offer scholarly analysis of the city's reach into the Owens Valley—but Mulholland's imprint on popular culture was most forcefully made by the 1974 film *Chinatown*.

A film noir classic, *Chinatown* does not claim to precisely depict early twentieth-century Los Angeles water politics. But through its dramatization of water department skullduggery, hapless farmers dispossessed of rural orchards, and a newly proposed dam and aqueduct conceived to foster urban growth, the film vividly brought to life the imperious character of Los Angeles as a civic enterprise. Most notably, the figure of Noah Cross personified the domineering deportment of a visionary who, enthralled with "the future," sought to bring ever more water to the city.[42] The film's particulars may be historical distortions,

FIGURE 8-6. Downtown Los Angeles in the late 1930s, not long after Mulholland's death. Perhaps more than any other person, Mulholland transformed the city from a nineteenth-century regional backwater into a major American city. [DC Jackson]

but John Huston's portrayal of Cross captured a sense of the determination, if not arrogance, that Mulholland brought to his labors. It is this cinematic rendering of Mulholland's personality that will live on in the public mind. And residents of the Santa Clara Valley in the 1920s certainly experienced this determination and arrogance during the construction of the St. Francis Dam—a time when Los Angeles sought to draw over 5,000 acre-feet of water annually out of the Santa Clara River watershed, and Mulholland created a 12-billion-gallon reservoir without outside review or counsel.

Mulholland is regarded in twenty-first-century Los Angeles as the man who created the hydraulic infrastructure for a world-class city. Once he is extolled for accepting responsibility for the St. Francis disaster, however, interest in how his design failed to meet gravity dam standards practiced by his professional contemporaries generally recedes. The authors of *Heavy Ground* have taken no joy in presenting a history that foregrounds the magnitude of Mulholland's deficiencies. But the disaster is as much a part of his life story as the building of the Los Angeles Aqueduct or the planning of the Colorado River Aqueduct. Not to confront his responsibility for the St. Francis tragedy is to deny the full trajectory of his career as an engineer. It may not be a happy story but, to borrow Mulholland's most famous invitation: There it is. Take it.[43]

APPENDIX A

How Fast Could the Reservoir Have Been Lowered?

In the late morning of March 12, 1928, William Mulholland and his assistant, Harvey Van Norman, visited the St. Francis Dam to investigate leakage that had been reported to them by dam keeper Tony Harnischfeger. As recounted in chapter 3, Mulholland determined that the leaks were not as serious as Harnischfeger had feared. After their inspection, he and Van Norman returned to Los Angeles in the early afternoon without taking any action to lower the reservoir and thus reduce the water pressure acting on the dam and its foundations. Some twelve hours later the dam collapsed, releasing over 12 billion gallons of water into the Santa Clara Valley.

During the spring of 1927, the St. Francis reservoir was filled within ten feet of the spillway crest for almost two months (April 10–June 5), and during this time it remained at three feet below the spillway for more than two weeks (May 10–May 27). Throughout this period the dam safely withstood the hydrostatic pressure exerted by the reservoir. It was not until March 5, 1928 that, for the first time, the reservoir level came within three inches of the spillway crest at elevation 1,835 feet. It stayed at this level for a week—and then the dam failed.[1] So it is reasonable to wonder: if the reservoir had been lowered three feet (back to a level that the dam had safely withstood for more than two weeks during the spring of 1927), could this action have perhaps averted the collapse? As a corollary: how long would it have taken to lower the reservoir three feet if Mulholland had taken decisive action at noon on March 12?

As a witness at the Los Angeles County Coroner's Inquest following the collapse, Mulholland downplayed the possibility that anything he might have done that morning could have had an appreciable effect in lowering the reservoir. He declared that water could have been released at a rate of only 1,000 cubic feet per second (cfs) and, as a result, that the reservoir could have been lowered about "a foot and a half a day—not that much—about a foot a day." And he repeated this allegation, emphasizing to the Coroner's Jury, "there is no means of lowering it more than a foot a day."[2] Van Norman agreed, testifying, "the facilities for releasing [water] were so limited, [the reservoir level] couldn't be reduced but a very little in that period."[3] Neither during the inquest nor at any later time did anyone question Mulholland's claim that it would have taken a full day to lower the reservoir one foot—or by extension, three days to lower it three feet. But is this contention reasonable?

Taking Mulholland's assertion that the maximum outflow through the outlet pipes was 1,000 cfs, and then modifying this rate by assuming that in March 1928 the upper San Francisquito Creek was delivering a flow of 50 cfs into the reservoir (thus reducing the de facto outflow rate to 950 cfs), the following analysis provides an answer that stands in stark contrast to what Mulholland told the Coroner's Jury:

An acre-foot is equivalent to 43,560 cubic feet. At a discharge rate of 950 cfs it would have taken 45.852 seconds to drain one acre-foot from the reservoir (950 cfs × 45.852 seconds = 43,560 cf). This is equal to about 1.31 acre-feet per minute or about 78.6 acre-feet per hour. Assuming that at elevation 1,835 feet the St. Francis reservoir had a surface area of 615 acres, a one-foot drop in elevation would have required a discharge of 615 acre-feet.[4] At a 950 cfs release rate it would have taken a little more than seven and a half hours to reduce the depth of the reservoir by one foot (615 acre-feet ÷ 78.6 acre-feet per hour = 7.82 hours). As the reservoir dropped, the rate of outflow would have decreased slightly over time, but this decrease would have been relatively negligible during the first few feet of drop. Thus it is reasonable to believe that the reservoir could have been lowered three feet in less than a day (7.82 hours × 3 = 23.46 hours). Put mildly, Mulholland's testimony at the Coroner's Inquest was far off the mark.

A further question should be asked: was Mulholland's supposition of a maximum outflow of 1,000 cfs a reasonable estimate? To check this figure it is necessary to calculate the discharge of the five outlet pipes under the pressure of a full reservoir. The outlet pipes were all 30 inches (2.5 feet) in diameter.[5] Use of Torricelli's Theorem to calculate discharge through a pipe was standard practice in the early twentieth century.[6] It involves calculating an idealized maximum flow ($A \times V$ or AV, where A equals the pipe's cross-sectional area and V equals velocity) and then multiplying this by a "coefficient of discharge" that reflects how much the maximum theoretical flow (AV) would need to be reduced because of friction losses and other inefficiencies created by the shape of the orifice. The basic flow formula derived from Torricelli's Theorem is:

$A \times V$ = Theoretical maximum flow
$V = \sqrt{2hg}$, where
g = gravitational constant = ~32.2 feet per second per second
h = the head (or depth of water) above the level of the outlet pipe
A = area of pipe cross section ($A = \pi R^2$)
For a pipe 2.5 feet in diameter, $A = \pi\ (1.25\ \text{feet})^2 = 4.90$ sqft

The actual estimated flow Q is calculated by multiplying $A \times V$ by the coefficient of discharge:

$Q = A \times V \times .60$ (using a standard coefficient of discharge = .60)

Assuming the five outlet pipes extending through the St. Francis Dam emptied at depths h = 36 feet, 72 feet, 108 feet, 144 feet, and 177 feet below the elevation of the full reservoir, the outflow from the five pipes under pressure from a full reservoir can be readily calculated (with minor rounding off).[7] For h = 36 feet the calculation for outflow, Q, is as follows:

$A = 4.90$ sqft
$V = \sqrt{2 \times 36 \times 32.2} = \sqrt{2318.4}$ ft per second = 48.15 ft per second
$A \times V = 4.90 \times 48.15 = 235.93$ cfs
$Q = (235.93 \times .60) = 141.5$ cfs

For the other pipes extending through the St. Francis Dam, a similar calculation can be made, providing the following results:

h = 36 feet, Q = 141.5 cfs
h = 72 feet, Q = 200.2 cfs
h = 108 feet, Q = 245.2 cfs
h = 144 feet, Q = 283.1 cfs
h = 177 feet, Q = 313.9 cfs

Adding up and rounding off, the total outflow from all five pipes = 1,184 cfs

From this calculation, it appears that Mulholland significantly underestimated the rate of outflow possible if all five outlet pipes had been opened to full capacity (1,000 cfs compared to 1,184 cfs, a difference approaching 20 percent).

Subtracting 50 cfs to account for inflow from upper San Francisquito Creek, the adjusted discharge rate with all outlet pipes open comes to 1,134 cfs. Thus it would have been possible to lower the reservoir at a rate of one acre-foot every 38.4 seconds, or 1.56 acre-feet per minute—equivalent to a one-foot drop in approximately 394 minutes, or about six and a half hours. This indicates that in thirteen hours the reservoir could have been lowered two feet, and in twenty-four hours close to four feet. Consequently, it appears that, in about two and a half days, the reservoir could have been lowered to a level that the dam had safely withstood for almost two months during the spring of 1927.[8]

Was Mulholland's testimony at the Coroner's Inquest a purposeful misstatement, one intentionally designed to obfuscate public understanding of how quickly the reservoir could have been lowered? Or was it an honest mistake, one derived from a flawed understanding of how quickly the reservoir could have been drained if all the outlet pipes had been opened to full capacity? Parsing his motives is difficult at a remove of over eighty years. Nonetheless it appears clear that Mulholland wished to preemptively rebut any contention that, had he acted aggressively at midday on March 12 to start lowering the reservoir, the disaster could have been averted.

Based on the above analysis, it is reasonable to question any arguments postulating that—had Mulholland opened all the outlet gates to full capacity and sent a tremendous, sustained surge into the Santa Clara Valley—the reservoir could not have been lowered quickly enough to have a meaningful impact on the dam's stability. Of course, lowering the reservoir some three feet to preclude imminent disaster would have had long-term significance only if the reservoir had never again been allowed to rise to the level of the spillway crest for any appreciable period of time. And that would have required enormous discipline on the part of Mulholland, in forgoing the prospect of storing an additional 1,800 acre-feet of water in the upper three feet (between elevations 1,832 and 1,835) of the St. Francis reservoir.

APPENDIX B

Was Failure Inevitable?

Were conditions at the St. Francis site such that, no matter what Mulholland might have done, any major structure for water storage built at that location in San Francisquito Canyon was inevitably destined to collapse? This issue was raised in the Governor's Commission *Report* on the disaster, which asserted:

> With such a formation [that is, the red conglomerate forming the west abutment], the ultimate failure of this dam was inevitable, unless water could have been kept from reaching the foundation. Inspection galleries, pressure grouting, drainage wells and deep cut-off walls are commonly used to prevent or remove percolation, but it is improbable that any or all of these devices would have been adequately effective, though they would have ameliorated the conditions and postponed the final failure.[1]

Offering only a sketchy description of how flow emanating from the west side might have triggered a massive slide of the east side mica schist formation, the commission blandly acknowledged that failure might have been delayed if "water could have been kept from reaching the foundation [under the dam]." And in the quoted passage, they point out a variety of technological features that Mulholland might have employed to block subsurface percolation. Nonetheless, they held fast to the view that disintegration of the west side conglomerate—and the excessive leakage that would have been precipitated by such disintegration—comprised the essential and unavoidable cause of the disaster.

If disintegration of the west side conglomerate had truly undermined the structure and caused the collapse, it would perhaps be possible to accept the fatalistic conclusion of the Governor's Commission. But the west side did not collapse first. Failure was initiated on the east side of the canyon, with uplift pressure acting on the dam through the fractured mica schist and a "high velocity orifice outflow" driven by water from the reservoir pressing on and through the lower depths of the east side canyon wall (see chapters 3 and 6).[2] The Governor's Commission was so fixated on the west-side-first failure hypothesis—based upon a presumed excessive leakage caused by sustained disintegration of the red conglomerate—that they dismissed any possibility that disaster could have been averted (not simply postponed) by efforts to ameliorate the effect of uplift.

A significant problem for arguments espousing "inevitable collapse" is the fact that the dam was in operation and successfully impounding water for almost two years prior to failure. Storage behind the dam began in March 1926 and, after the summer of that year, the reservoir level was never allowed to drop more than 75 feet below the spillway crest (at elevation 1,835 feet above sea level).[3] In the spring of 1927, the reservoir was filled to above 1,825 feet, where it stayed for almost two months (April 10–June 5). Most significantly, during this time it stayed at a height of 1,832 feet (just three feet below the spillway crest) for over two weeks (May 10–May 27). Throughout the remainder of 1927 the reservoir never dropped lower than 1,813 feet. Refilling began at the end of the year and about January 25, 1928, the reservoir reached 1,825 feet. It continued to rise through February and on March 5 came within three inches of the spillway crest (1,835 feet). It stayed at that level until the dam's collapse a few minutes before midnight on March 12.

Consider if the reservoir level had never been allowed to exceed an elevation of 1,825 feet above sea level—ten feet below the spillway crest at 1,835 feet. Could failure have been avoided if the maximum storage level had been reduced ten feet? Recall that water had been stored at a level of 1,832 feet for over two weeks in May 1927 and no collapse ensued. In addition, the dam had successfully operated for almost two months with the reservoir at 1,825 feet or higher. So what would have happened if storage had never been allowed to exceed the 1,825 foot level?

Positing 1,825 feet as a storage limit is not an arbitrary suggestion. The original design of the St. Francis Dam had featured a spillway crest of 1,825 feet, but in the midst of construction Mulholland had raised the dam ten feet. This change (discussed in chapters 2 and 3) placed the spillway crest at an elevation of 1,835 feet. And because of this alteration, substantially greater pressure was exerted against the dam and foundations under a full reservoir. Failure came because of water percolating into and through the east abutment; this created uplift forces of sufficient magnitude to dislodge the dam from its foundation and set loose a massive landslide of fractured schist. The extent of uplift forces is directly proportional to the water pressure exerted by a reservoir. A lower reservoir level equates to lower water pressure, and this necessarily diminishes the destabilizing effect of uplift. This is not a radical insight—as a safety precaution engineers commonly reduce the level of spillway crests or limit the height of water that can be stored in a reservoir as a way to enhance structural stability.

In addition, what if the St. Francis foundation had been pressure grouted and drainage wells and deep cutoff walls (or trenches) had been built up both canyon walls? And what if the dimensions of the concrete profile adopted by Mulholland and his staff had been more amply proportioned, creating a thicker and more massive, heavier structure? And what if the integrity of the east side foundation had not been compromised by the excavation of a tunnel some thirty to forty feet long and large enough to allow passage of a man pushing a wheelbarrow?[4] This tunnel, no matter how diligently (or not) it had been filled in with concrete, disturbed the "heavy ground" of the east side schist foundation at the location where failure of the dam was initiated. What if no tunnel of this sort had ever been excavated and Mulholland had implemented a full range of technologies to counteract the effect of uplift? Wouldn't all of the above measures to counter the effect of uplift—if implemented at St. Francis—have significantly enhanced the stability of Mulholland's concrete gravity design?

Taking an objective view, it is fair to conclude that a decision to limit storage to an elevation of 1,825—combined with an amply dimensioned gravity profile and enhanced efforts to vitiate the effect of uplift along the full length of the foundation and up each canyon wall—could have fostered construction of a stable structure at the St. Francis site. We cannot wind the clock back to determine with absolute certainty whether a redesigned St. Francis Dam would have safely stood for decades or more. But it appears unreasonable to assert that any dam built at the site was doomed to collapse.[5] Clearly, Mulholland's "standard of care" fell egregiously short of what other engineers followed in the design and construction of gravity dam projects of the 1920s, and even a few years earlier (see chapter 2, especially the description of San Francisco's Hetch Hetchy/O'Shaughnessy Dam). It was not preordained that Mulholland would design or build the St. Francis Dam the way he did. He could have done it differently.[6]

If the failure was not inevitable, then why did it happen? How could Mulholland have acted with such impunity and arrogance in adopting a design that, as his dam-designing contemporaries well understood, paid only minimal heed to the danger posed by uplift? And how could he have raised the dam's height ten feet in the midst of construction and blithely presumed that stability would be assured? He certainly had no desire to endanger the lives of those living downstream from the St. Francis Dam. He was no monster. But he also could not imagine that whatever dam he chose to build would not function the way he wanted. Hubris may seem too easy an explanation for Mulholland's actions in forgoing outside review or counsel when creating a 38,000 acre-feet storage reservoir high in the Santa Clara River watershed. But how else to account for it? Mulholland famously took "blame" for the disaster, but the muddled narrative presented at the Coroner's Inquest depicting the Bureau of Water Works and Supply's development of the St. Francis design (recounted in chapter 2) speaks to his relatively nominal involvement in the actual planning of the project and its details. Vigilantism in Owens Valley and grand plans for a new and wondrous Colorado River Aqueduct dominated his attention in 1924–25. He simply assumed that whatever was built at St. Francis would work, because it was *his* dam, no matter what the final design might be. A harsh assessment? No doubt. But therein lies the root cause of the disaster.

Acknowledgments

Many people and organizations helped the authors in researching *Heavy Ground.* Their assistance is gratefully acknowledged: Paul Soifer, consulting historian for the Los Angeles Department of Water and Power; Fred Barker of the Los Angeles Department of Water and Power; Charles Johnson of the Museum of Ventura County; Matt Roth and Morgan Yates of the Automobile Club of Southern California; Murray McEachron, Senior Hydrologist for the United Water Conservation District in Santa Paula; Alan Pollack of the Santa Clarita Valley Historical Society, Newhall; John Nichols of the John Nichols Gallery, Santa Paula; the staff of the state Division of Safety of Dams in Sacramento; Linda Vida of the Water Resources Center Archives, University of California, Riverside; Randall Brandt of the Bancroft Library, University of California, Berkeley; the Huntington Library, San Marino, including Peter Blodgett, Bill Frank, Jennifer Goldman, Daniel Lewis, Jennifer Watts, and the Reader Services Department. Special thanks to Bill Deverell, director of the Huntington-USC Institute on California and the West, and to Ann Stansell. Thanks also to Janet Fireman, Jenn Rossmann, Art Kney, Richard Wiltshire, Henry Petroski, and David P. Billington.

Research for *Heavy Ground* was generously supported by a Trent R. Dames Fellowship in Civil Engineering History that allowed Donald C. Jackson to be in residence at the Huntington Library during the summer of 2012. The Academic Research Committee of Lafayette College and the endowment supporting the Cornelia F. Hugel Professor of History also provided valuable travel and logistical support over the course of the research and writing and thanks to Tammy Yeakel, History Department secretary.

Last but not least, thanks to Richard K. Anderson and to Michael V. Carlisle of InkWell Management.

◆ ◆ ◆

Many thanks to the University of Nevada Press team that helped bring the paperback edition of *Heavy Ground* to publication, including Joanne Banducci, Sara Vélez Mallea, Sara Hendrickson, Jinni Fontana, Iris Saltus, and Clark Whitehorn. Their hard work and dedication is much appreciated.

Thanks also to Jean Patterson; Bill Deverell; Michael Green at UNLV; Alan Pollock, President of the Santa Clarita Valley Historical Society and forceful advocate of the St. Francis Dam Disaster Memorial; and Fred Barker, longtime engineer with the Los Angeles Department of Water and Power, for identifying John R. Haynes and Reginald Del Valle (both members of the Board of Water and Power Commissioners) as companions of William Mulholland in the 1927 photo appearing on page 103.

—DC Jackson, May 2020

About the Authors

Norris Hundley jr. (1935–2013)
Norris Hundley received a BA from Mt. San Antonio College and a Ph.D. from UCLA. In 1964 he joined the UCLA History Department, becoming a leader in the history of the American West and in the nascent field of water history. He also served for twenty-nine years as editor of the *Pacific Historical Review*. His first book, *Dividing the Waters: A Century of Controversy between the United States and Mexico* (1966), focused on the dispute between Mexico and the United States over rights to the Rio Grande and the Colorado River. Then came *Water and the West: The Colorado River Compact and the Politics of Water in the American West* (1975), his landmark investigation of the most important watercourse in the arid Southwest. The trilogy culminated with *The Great Thirst: Californians and Water, 1770s–1990s* (1992), a volume that gave stature to the study of water as a driving force in the political economy of California. After retiring from UCLA in 1994, Hundley served as president of both the Western History Association and the Pacific Coast Branch of the American Historical Association. In 2001 he began collaborating on the research underlying *Heavy Ground.*

Donald C. Jackson (1953–)
Donald C. Jackson received a BS in Engineering from Swarthmore College and a Ph.D. from the University of Pennsylvania. In 1989 he joined the History Department at Lafayette College in Easton, Pennsylvania, where he is presently the Cornelia F. Hugel Professor of History. A "dam historian" of long standing, his books include *Building the Ultimate Dam: John S. Eastwood and the Control of Water in the West* (1995); *Big Dams of the New Deal Era: A Confluence of Engineering and Politics* (2006, co-authored with David P. Billington); and *Pastoral and Monumental: Dams, Postcards, and the American Landscape* (2013). In 2010, his article "Structural Art: John S. Eastwood and the Multiple Arch Dam" was awarded the Overseas Prize by the London-based Institution of Civil Engineers; that same year he gave the keynote address at the Hoover Dam 75th Anniversary History Symposium, sponsored by the American Society of Civil Engineers. He first visited the site of the St. Francis Dam in 1979, recognizing early in his career its importance in the hydraulic history of the American West.

Notes

PREFACE

1 There are two Santa Clara Valleys in California, one north of Los Angeles and the other south of San Francisco Bay; all references to the Santa Clara Valley in this book are to the former. Santa Clarita is a municipal entity that was created in 1987 through the merging of the communities of Saugus, Newhall, Valencia, and Canyon Country. The names Santa Clarita and Santa Clarita Valley (as opposed to Santa Clara Valley) are artifacts of the late twentieth century and are never used in historical sources relating to the St. Francis Dam disaster. See Alan Pollack, E. J. Stephens, and Kim Stephens, *Then and Now: Santa Clarita Valley* (Charleston, S.C.: Arcadia Publishing, 2014). For more on the number of flood victims, see chapter 4.

2 David McCullough, *The Johnstown Flood* (New York: Simon & Schuster, 1968), Outside of the United States, other major dam failures that preceded St. Francis include the Puentes Dam in Spain that killed 600 people in 1802 and the Dale Dyke Dam in Britain that killed almost 250 people in 1864; see Norman Smith, *A History of Dams* (Secaucus, N.J.: Citadel Press, 1972), 122–25, 213–16. Almost 140 people died in the 1874 Mill River Dam disaster in western Massachusetts; see Elizabeth Sharpe, *In the Shadow of the Dam: The Aftermath of the Mill River Flood* (New York: Free Press, 2004).

3 Donald C. Jackson and Norris Hundley jr., "Privilege and Responsibility: William Mulholland and the St. Francis Dam Disaster," *California History* 82 (2004): 8–47.

4 Charles Outland, *Man-Made Disaster: The Story of St. Francis Dam*, rev. ed. (Glendale, Calif.: Arthur H. Clark Co., 1977), 237. Unless otherwise noted, all subsequent references to *Man-Made Disaster* are to the 1977 edition.

5 J. David Rogers, "A Man, a Dam, and a Disaster: Mulholland and the St. Francis Dam," *Southern California Quarterly* 77 (Spring/Summer 1995): 82. Rogers claims that "many engineers were just beginning to appreciate the destabilizing effects of uplift pressures in the late 1920s" (30), a statement that was contested by the authors in "Privilege and Responsibility." "A Man, a Dam" was reprinted in *The St. Francis Dam Disaster Revisited* (pp. 1–109), published in

1995 by the Historical Society of Southern California and the Ventura County Museum of History and Art.

6 J. David Rogers, "Reassessment of the St. Francis Dam Failure," in Bernard W. Pipkin and Richard J. Proctor, eds., *Engineering Geology Practice in Southern California*, Special Publication no. 4 (Belmont, Calif.: Association of Engineering Geologists, Southern California Section, 1992), 639–66.

7 For Mulholland's supposed exoneration or apparent vindication, see "The Night the Dam Broke: Geological Look 64 Years Later Clears Mulholland and His Engineering Marvel in Tragedy That Killed 450," *Los Angeles Times*, October 25, 1992; Margaret Leslie Davis, *Rivers in the Desert: William Mulholland and the Inventing of Los Angeles* (New York: Harper Collins, 1993), 263–64, which credits Rogers for Mulholland's possible "exoneration" and for the belief that the chief engineer was "innocent of professional negligence"; Ruth Pittman, *Roadside History of California* (Missoula, Mont.: Mountain West Publishing Co., 1995), 233, which states that "more than 50 years after his death, Mulholland was exonerated"; Kim Weir, *Southern California Handbook* (Chico, Calif.: Moon Publications, 1998), 230, claims that Rogers "largely exonerated him"; and Catherine Mulholland, *William Mulholland and the Rise of Los Angeles* (Berkeley: University of California Press, 2000), 382 n. 4, heralds Rogers as "masterfully" analyzing the disaster and acknowledges "his apparent vindication" of her grandfather.

8 After "Privilege and Responsibility" was published in late 2004, the authors began research for the manuscript that has become *Heavy Ground*. Much archival work was carried out in 2005–6 and the drafting of various chapters began soon thereafter. Professional demands and health issues slowed work on the project, but in the summer of 2012 author Jackson was awarded a Trent R. Dames Fellowship in Civil Engineering History at the Huntington Library and progress accelerated. Unfortunately, Hundley died in April 2013 after a long period of declining health, but much work on *Heavy Ground* had already been done. Although Jackson was responsible for completing the book and preparing it for publication, *Heavy Ground* stands as a true collaboration.

9 For example, the Joseph B. Lippincott Papers held by the Water Resources Center Archives at the University of California, Riverside, contain a multitude of files related to Lippincott's work on western water projects. Lippincott served under Mulholland as assistant chief engineer during construction of the Los Angeles Aqueduct and remained a friend and colleague until Mulholland's death in 1935. Lippincott subsequently published a widely read tribute to his former boss ("William Mulholland: Engineer, Pioneer, Raconteur," *Civil Engineering* 2 [February–March 1941]: 105–6), but there is nothing surviving in his papers at the WRCA relating to the St. Francis Dam.

10 Transcript of "Testimony and Verdict of the Coroner's Jury in the Inquest over Victims of St. Francis Dam Disaster" (hereafter Coroner's Inquest), book 26902, box 13, folder 2, Richard Courtney Collection, Huntington Library. This voluminous transcript documents more than three weeks of inquest testimony given in late March and early April 1928. As a municipal employee responsible for maintaining city records, Richard Courtney facilitated the transfer of the

transcript to the Huntington Library; the Coroner's Inquest transcript was accessioned as part of the Richard Courtney Collection.

PROLOGUE

1 Lillian Curtis recalled the events of that night nearly fifty years later, in 1978 (by then she had remarried as Lillian Eilers), in a newspaper interview; see Marianne Tyler, "The Day the Dam Broke," *Los Angeles Herald-Examiner,* February 9, 1978. Pre-flood photographs of the Curtis family are posted on the Santa Clarita Valley Historical Society website; see http://www.scvhistory.com/scvhistory/id2101.htm and http://www.scvhistory.com/scvhistory/id2801.htm.

2 Tyler, "The Day the Dam Broke"; H. B. Gardett and A. R. Arledge, "Reconstruction of San Francisquito No. 2 Power Plant" (November 14, 1928), 3–6, file WP23–8:17, Historical Records Program, Los Angeles Department of Water and Power Archives (hereafter LADWP Archives), Los Angeles, Calif.

3 Lillian Curtis' memory seems to have been faulty when she contended, "We'd lived there [the employee housing compound] two years...and watched the dam being built." Her claims against the city of Los Angeles for the deaths of her husband and children and for property damage state that her husband's length of service had been fourteen months, therefore beginning about January 1927; the dam was completed on May 4, 1926. Perhaps Lyman Curtis and his family had been nearby before his employment with the Bureau of Power and Light began in January 1927. See Claims of Lillian Curtis, and Claims of Ray E. Rising, file WP23–8:15, LADWP Archives; Tyler, "The Day the Dam Broke"; Los Angeles Board of Water and Power Commissioners, "Twenty-Fifth Annual Report for the Fiscal Year Ending June 30, 1926," 32; Outland, *Man-Made Disaster,* 44, 46, 51. Ray Rising's memories of the night the dam collapsed were recounted to a reporter thirty-five years later; see Ted Thackrey Jr., "LA's Worst Disaster 35 Years Ago: 450 Killed," *Los Angeles Herald-Examiner,* March 10, 1963. Portions of Rising's testimony at a civil trial brought against the city of Los Angeles in 1930, and recorded by the *Los Angeles Times,* also describe what occurred when the flood hit his frame bungalow; see "Broken Dam's Roar Echoes: St. Francis Dam Disaster Survivor Describes Flood at Trial of $175,000 Suit Against City," *Los Angeles Times,* May 26, 1930.

4 Tyler, "The Day the Dam Broke"; Thackrey Jr., "LA's Worst Disaster"; Gardett and Arledge, "Reconstruction of San Francisquito No. 2 Power Plant," 4; Outland, *Man-Made Disaster,* 82, 98, 222.

5 Exactly when Curtis and Rising encountered each other after the flood hit is unclear: Curtis implies that it happened soon after she ascended the knoll; Rising later reported that they "met in a fire break near the flooded area at dawn." Curtis further describes the night of the flood: "When we realized the water was receding a little and we were safe, we decided to pull some brush and make a little nest for ourselves because it was very cold. About that time our big white spotted dog came along and I pulled him over Danny and me to help us keep warm. After that we just sobbed and cried and thought and kept calling for help." It appears that the "we" in this quotation does not include Rising.

See Tyler, "The Day the Dam Broke," on Curtis' experiences, and Thackrey Jr., "LA's Worst Disaster," for Rising's recollection.

6 Rising's description of the flood, including quotations, is drawn from "Broken Dam's Roar Echoes."

7 An early listing of those who died in the Power House No. 2 compound appears in "Roll of the Identified and Missing," *The Intake* 5 (April 1928): 3; this list mistakenly includes D. C. Mathews, who was in Newhall the night of the flood and survived, and does not include either the sister-in-law or the infant nephew of Power House No. 2 employee William Weinland, who were visiting in San Francisquito Canyon on the night of March 12. *The Intake*, published by the Employees Association of the Department of Water and Power, usually provided upbeat news of employee activities.

CHAPTER ONE

1 "Tribute to Master Mind of Aqueduct," *Los Angeles Times*, November 6, 1913.

2 W. W. Hurlbut, "The Man and the Engineer," *Western Construction News* 1 (April 25, 1926): 44. Hurlbut's name is sometimes incorrectly spelled "Hurlbert" (or Hurlburt, as in the Los Angeles County Coroner's Inquest transcript).

3 Numerous sources treat Mulholland's early years; see Elisabeth Mathieu Spriggs, "The History of the Domestic Water Supply of Los Angeles" (M.A. thesis, University of Southern California, 1931), 63–68, 65; Catherine Mulholland, *William Mulholland*, 6–13; Richard Prosser, "William Mulholland: The Maker of Los Angeles," *Western Construction News* 1 (April 25, 1926): 43; J. B. Lippincott, "William Mulholland: Engineer, Pioneer, Raconteur," *Civil Engineering* 2 (February–March 1941): 105–6; Harvey A. Van Norman, "William Mulholland," *Transactions of the American Society of Civil Engineers* 101 (1936): 1604–5; Robert W. Matson, *William Mulholland: A Forgotten Forefather* (Stockton, Calif.: Pacific Center for Western Studies, University of the Pacific, 1976), 5–6.

4 Prosser, "William Mulholland," 43; Catherine Mulholland, *William Mulholland*, 17; William L. Kahrl, *Water and Power: The Conflict over Los Angeles' Water Supply in the Owens Valley* (Berkeley and Los Angeles: University of California Press, 1982), 19.

5 Catherine Mulholland, *William Mulholland*, 17; Lippincott, "William Mulholland," 106; Van Norman, "William Mulholland," 1605.

6 Spriggs, "History of the Domestic Water Supply of Los Angeles," 29–45; Kahrl, *Water and Power*, 11–12; Catherine Mulholland, *William Mulholland*, 20–22. In some parts of the American West, forty miner's inches equals a flow of one cfs, but in Southern California it is fifty.

7 Vincent Ostrom, *Water and Politics: A Study of Water Policies and Administration in the Development of Los Angeles* (Los Angeles: Haynes Foundation, 1953), 45.

8 Matson, *William Mulholland*, 7; Lippincott, "William Mulholland," 106; Catherine Mulholland, *William Mulholland*, 28.

9 Catherine Mulholland, *William Mulholland*, 30–32; Prosser, "William Mulholland," 43; "William Mulholland, Los Angeles Water Supply Engineer, Dies," *Engineering News-Record* 115 (July 25, 1935): 136.

10 Catherine Mulholland, *William Mulholland*, 32, 35–37.

11 Ibid.

12 Glenn S. Dumke, *The Boom of the Eighties in Southern California* (San Marino, Calif.: Huntington Library, 1944); Catherine Mulholland, *William Mulholland*, 38–40, 42, 45.

13 Spriggs, "History of the Domestic Water Supply of Los Angeles," 46–59; Kahrl, *Water and Power*, 9–18; Catherine Mulholland, *William Mulholland*, 75–80. Blake Gumprecht, *The Los Angeles River: Its Life, Death, and Possible Rebirth* (Baltimore: Johns Hopkins University Press, 1999), 92–95.

14 J. B. Lippincott, "Mulholland's Memory," *Civil Engineering* 9 (March 1939): 199; Lippincott, "William Mulholland," 161; *Constructive Californians: Men of Outstanding Ability Who Have Added Greatly to the Golden State's Prestige* (Los Angeles: Saturday Night Publishing Co., 1926), 132; Ostrom, *Water and Politics*, 1–25.

15 Ostrom, *Water and Politics*, 90–91. Because of special financial circumstances consequent on the construction of the Los Angeles Aqueduct, beginning in 1906 Mulholland also reported to a special advisory committee within the Board of Public Works on his role as chief engineer of the aqueduct.

16 Kahrl, *Water and Power*, 23–25; Norris Hundley, jr., *The Great Thirst: Californians and Water: A History*, rev. ed. (Berkeley: University of California Press, 2001), 143–44.

17 *Vernon Irrigation District v. Los Angeles*, 106 Cal. 237, 250, 251 (1895).

18 Los Angeles Department of Public Works, Bureau of the Los Angeles Aqueduct, "Third Annual Report" (1908), 23; Ostrom, *Water and Politics*, 9, 144–46.

19 "Titanic Project to Give City a River," *Los Angeles Times*, July 29, 1905. Like many others in city government, Mulholland eventually became convinced that the city would outrun its current supply, but he did not come to this position until much later than did Fred Eaton; see Catherine Mulholland, *William Mulholland*, 120–21. Eaton had regarded the Owens River as a potential source of water supply for Southern California by the early 1890s and explored a variety of schemes for implementing it. On the evolution of Eaton's plans to tap the Owens River, see Kahrl, *Water and Power*: Eaton "had been promoting the [aqueduct] idea with Mulholland and anyone else who would listen since 1892, but no one had paid much attention" (47–50).

20 Los Angeles Department of Public Works, "First Annual Report of the Chief Engineer of the Los Angeles Aqueduct to the Board of Public Works" (March 15, 1907), 18; Los Angeles Aqueduct Investigation Board, "Report of Aqueduct Investigation Board of the City of Los Angeles" (August 31, 1912), 10, 35–37, 42, 53; *Los Angeles Examiner*, August 25, 1905; Abraham Hoffman, *Vision or Villainy: Origins of the Owens Valley–Los Angeles Water Controversy* (College Station, Tex.: Texas A&M Press, 1981), 64–66; Kahrl, *Water and Power*, 48–49, 54–55, 248; Hundley, *Great Thirst*, 146–47.

21 William Kahrl describes Eaton's maneuvering over Long Valley and options to the Rickey Ranch in *Water and Power*, 64–69.

22 Los Angeles Aqueduct Investigation Board, "Report," 10, 36–37, 41, 42; Los Angeles Department of Public Works, "First Annual Report of the Chief Engineer," 23, 46, 84; *Los Angeles Times*, August 5, 1905; *Los Angeles Examiner*, August 5, 1905; Los Angeles Department of Public Service, "Complete Report on Construction of the Los Angeles Aqueduct" (1916), 13; Hundley, *Great Thirst*, 148–50.

23 Los Angeles Aqueduct Investigation Board, "Report," 35.

24 Kahrl, *Water and Power*, 63–77, 108–26; Hoffman, *Vision or Villainy*, 67–86, 79; Los Angeles Aqueduct Investigation Board, "Report," 53–54.

25 Hoffman, *Vision or Villainy*, 80–90.

26 "Titanic Project to Give City a River," *Los Angeles Times*, July 29, 1905; also see *Inyo Register*, August 3, 1905.

27 "Los Angeles Plots Destruction," *Inyo Register*, August 3, 1905.

28 Hundley, *Great Thirst*, 154–55.

29 Ibid., 155.

30 Kahrl, *Water and Power*, 126; Hundley, *Great Thirst*, 163.

31 William Kahrl notes Mulholland's relative lack of experience with big construction projects prior to the aqueduct in *Water and Power*, 148–49.

32 Ibid., 149. Each member was approved by the City Council, the Board of Water Commissioners, the Board of Public Works, the Chamber of Commerce, the Municipal League, and the Merchants and Manufacturers Association—a who's who of the city's public and business leadership. As Kahrl observes, this was a clear sign that "the aqueduct was regarded as a joint venture of Los Angeles's public officials and its business leaders" and that "the prosperity of the city was at stake."

33 "First Annual Report of the Chief Engineer of the Los Angeles Aqueduct to the Board of Public Works," 59–60.

34 John R. Freeman, Frederic P. Stearns, and James D. Schuyler, "Report of the Board of Consulting Engineers on the Project of the Los Angeles Aqueduct from Owens River to San Fernando Valley" (December 22, 1906), Appendix E, in Los Angeles Department of Public Works, "Annual Report of the Chief Engineer," 117–32, 118, 119, 123.

35 Los Angeles Board of Public Service Commissioners, "Complete Report on the Los Angeles Aqueduct" (1916), 9–29, 75–81, 98–110, 149–56, 199–209, 250–59, 271; Los Angeles Board of Public Works, "Los Angeles Aqueduct: Report of Progress for the Fiscal Year 1912–1913," 5; Water Service (Aqueduct) Business Unit Historical Records, Los Angeles Aqueduct Construction Records, file WP19–1:1A, Historical Records Program, Los Angeles Department of Water and Power; Burt A. Heinly, "The Longest Aqueduct in the World," *Outlook* 93 (September 25, 1909): 215–20; J. B. Lippincott, "Tufa Cement, as Manufactured and Used on the Los Angeles Aqueduct," *Transactions of the American*

Society of Civil Engineers 76 (December 1913): 520–81; J. Gregg Layne, *Water and Power for a Great City* (Los Angeles: Los Angeles Department of Water and Power, 1957), 122, 125–48; Hoffman, *Vision or Villainy*, 145–54; Kahrl, *Water and Power*, 159.

36 Los Angeles Board of Public Service Commissioners, "Complete Report on the Aqueduct," 31; Los Angeles Board of Public Service Commissioners, "Twenty-First Annual Report for the Fiscal Year Ending June 30, 1922," 8; "Silver Torrent Crowns the City's Mighty Achievement," *Los Angeles Times*, November 6, 1913. For more on the opening ceremony and the chief engineer's complete remarks, see Catherine Mulholland, *William Mulholland*, 244–46.

37 William Mulholland to Arthur Powell Davis, February 5, 1912, file WMOF-3-13, LADWP Archives.

38 Matson, *William Mulholland*, 2; W. W. Hurlbut, "The Man and the Engineer," *Western Construction News* 1 (April 25, 1926): 44.

39 Ostrom, *Water and Politics*, 98–99.

40 R. F. del Valle testimony, Coroner's Inquest, 615.

41 Los Angeles Aqueduct Investigation Board, "Report," 3, 4, 48; Hoffman, *Vision or Villainy*, 126–28, 144–45, 274–75; Kahrl, *Water and Power*, 96–99, 133–34, 193–94, passim; Hundley, *Great Thirst*, 160–62.

42 Nordskog wrote a scathing letter to the California Legislature on the activities of Lippincott, Eaton, and Mulholland in the Owens Valley, "Letter of Transmittal to the Owens Valley Investigating Committee of the State Senate," *Journal of the California Assembly* (April 27, 1931): 2713–26. See also Hoffman, *Vision or Villainy*, 142–45, 208–43; Kahrl, *Water and Power*, 155–56, 320–21, 328–29; Catherine Mulholland, *William Mulholland*, 149.

43 Gervaise Purcell, W. H. Sanders, F. C. Finkle, and Chester B. Loomis, *Report on Municipally Manufactured Cements Used on Los Angeles Aqueduct from Owens River to the San Fernando Valley, California* (Philadelphia: Association of American Portland Cement Manufacturers, 1912); F. C. Finkle, "Los Angeles Aqueduct Mistakes," *Journal of Electricity, Power and Gas* 34 (January 9, 1915): 25–28; F. C. Finkle, "Los Angeles' $40,000,000 White Elephant," *Irrigation Age* 30 (May 1915), 200–202, 216; Kahrl, *Water and Power*, 160, 164, 191–92. Edward Johnson and Edward S. Cobb reinforced Finkle's criticism of the cement and the purchase of the tractors in "Report on the Los Angeles Aqueduct: After an Investigation Authorized by the City Council of Los Angeles" (1912), 8–13.

44 "Tribute To Master Mind of Aqueduct," *Los Angeles Times*, November 6, 1913.

45 Kahrl, *Water and Power*, 245–47; Freeman, Stearns, and Schuyler, "Report of the Board of Consulting Engineers," 125.

46 Los Angeles Aqueduct Investigation Board, "Report," 4; Kahrl, *Water and Power*, 193–94.

47 Eventually the city would build a large dam and reservoir in Long Valley, but only after both Mulholland and Eaton had died. Los Angeles acquired the

Long Valley site after Eaton's holdings went into foreclosure after financial reversals in 1932. The city bought the land for $650,000 and subsequently purchased downstream water rights as well, opening the way to the construction of Long Valley Dam. Completed in 1941, it forms the reservoir known as Crowley Lake; Kahrl, *Water and Power,* 246–50, 334, 340–41; Hoffman, *Vision or Villainy,* 178, 261.

48 Kahrl, *Water and Power*, 252–54, 259–62.

49 Los Angeles Board of Public Service Commissioners, "Twenty-First Annual Report," 9.

50 Los Angeles Board of Public Service Commissioners, "Twenty-Second Annual Report for the Fiscal Year Ending June 30, 1923" (1923), 7; Hundley, *Great Thirst*, 164–65.

51 The final dynamite attack in this period occurred in November 1931, but the violence had a lasting place in the valley's collective memory. As late as September 15, 1976, an attack was made against the aqueduct near Lone Pine. See Hoffman, *Vision or Villainy*, 181, 191–92, 196, 259–60, 268.

CHAPTER TWO

1 Letter from Edward Godfrey on gravity dam design, *Engineering News* 70 (August 21, 1913): 371.

2 Bailey Willis, "Report on the Geology of St. Francis Damsite, Los Angeles County, California," *Western Construction News* 3 (June 25, 1928): 411.

3 Los Angeles Department of Public Service, "Complete Report on the Aqueduct." In this widely distributed report of 1916, no future dams or reservoirs are mentioned other than Long Valley (273–74).

4 Irrigation for the San Fernando Valley and the needs of the city's hydropower plants during winter are given as reasons to develop sizable storage capacity at the lower end of the aqueduct; see William Mulholland to Board of Public Service Commissioners, July 1, 1922, in Los Angeles Board of Public Service Commissioners, "Twenty-First Annual Report," 9.

5 For a list of various dams (with reservoir capacities) built by the Bureau of Water Works and Supply in the 1920s, see Rogers, "A Man, a Dam," 20.

6 Ezra F. Scattergood, "Respecting St. Francis Reservoir" (August 31, 1933), 1, in John R. Haynes to A. J. Mullen, February 21, 1934, William Mulholland Collection, Historical Records Program, LADWP Archives; Lippincott, "William Mulholland: Engineer, Pioneer, Raconteur," 163–64.

7 Mulholland to Board of Public Service Commissioners, July 1, 1922, in Los Angeles Board of Public Service Commissioners, "Twenty-First Annual Report," 9.

8 In September 1922, the Board of Water Commissioners authorized preliminary surveys of the site for the dam and reservoir; James P. Brome (secretary of the Board of Water and Power Commissioners); Coroner's Inquest testimony, 592.

9 The name of St. Francis Dam is derived from San Francisquito Creek, and the creek's name was derived from the Mexican-era Rancho San Francisco that encompassed much of the land in the upper Santa Clara Valley. There is no religious significance attached to the name.

10 Use of U.S. Forest Service lands for dams, reservoirs, canals, power plants, transmission lines, etc., was authorized in the federal 1901 "Right-of-Way Act" (February 15, 1901, chap. 372, 31, stat. 790). In contrast to national parks under the protection of the U.S. Department of the Interior, land under the control of the Forest Service was to be developed to serve a broad public interest. A municipally owned dam and reservoir located within an otherwise unremarkable canyon was readily accommodated by Forest Service officials in the early twentieth century. See James P. Brome testimony, Coroner's Inquest, 590; and George E. Cecil, Angeles Forest supervisor to William Mulholland, August 3, 1925, in "San Francisquito Reservoir, March 1925–August 1926," file WP04–6:18, LADWP Archives. The city-owned property at the site is described as "the St. Francis Ranch and the La Brea Ranch, totaling approximately 480 acres, all other land in the site belongs to the United States Government, the city having filed on same up to the 1825 contour." See Los Angeles Board of Public Service Commissioners, "Twenty-Second Annual Report," 23.

11 See H. W. Dennis (construction engineer, Southern California Edison Company) to William Mulholland, May 14, 1925; William Mulholland (by Assistant Engineer J. E. Phillips) to H. W. Dennis, August 11, 1925; and H. W. Dennis to H. A. Van Norman, September 28, 1925; all in "San Francisquito Reservoir, March 1925–August 1926," file WP04–6:18, LADWP Archives. The cost of the line relocation was $10,350, paid by the city to Southern California Edison.

12 Los Angeles Board of Public Service Commissioners, "Twenty-First Annual Report," 6, 8–9; "Twenty-Second Annual Report," 5–9, 23, 24, 25; Scattergood, "Respecting St. Francis Reservoir," 2.

13 Scattergood's memorandum, "Respecting St. Francis Reservoir," cited in full in n. 6, above, indicates that a decade before he had "strenuously opposed the construction of the St. Francis Reservoir" because of its impact on electric power generation.

14 Scattergood, "Respecting St. Francis Reservoir" (August 31, 1933), 1. At the Coroner's Inquest, Mulholland acknowledged that the St. Francis Dam/reservoir "was rather a disadvantage to [the] power [bureau] than otherwise"; Mulholland testimony, 26. Ostrom documents the financial importance of the city's San Francisquito power plants in *Water and Politics*: "The [Southern California] Edison Company was obligated to sell the city whatever quantity of power was needed to meet the demands of the city's distribution system that were not met by its own hydroelectric generators. Rapidly rising power demands soon absorbed the full capacity of the hydroelectric plants along the Los Angeles Aqueduct [i.e., the San Francisquito power plants], and Edison Company was providing the municipal power system with 55 percent of its electrical energy at rates substantially higher than the cost for the hydroelectric energy generated by the municipal plants along the aqueduct" (64). It

is often assumed, incorrectly, that the St. Francis Dam was built as part of the city's hydroelectric power system. For example, Sarah Elkind, in *How Local Politics Shape Federal Policy: Business, Power & the Environment in Twentieth-Century Los Angeles* (Chapel Hill: University of North Carolina Press, 2011), refers to the city's "purchase of the municipal power grid [from private power companies] in 1919 and the construction of the ill-fated St. Francis Dam and power plant" (146), claiming that "power bonds paid for construction of the Saint Francis Dam" (220). Elkind cites Jackson and Hundley, "Privilege and Responsibility," as the apparent source for the latter statement; in fact, the authors explain that the dam was detrimental to the city's generating system.

15 A reservoir at Big Tujunga would have been downstream from both San Francisquito Power House No. 1 and Power House No. 2, and presumably would have had less effect on the city's power generation system than deployment of the St. Francis reservoir.

16 For the Hollywood and, later, the St. Francis design, Rogers speculates that "concrete was likely chosen over earth because of the relative paucity of the clayey material within the abutment materials, basic requisites for any earth fill embankment" ("A Man, a Dam," 24).

17 Allen Hazen and Leonard Metcalf, "Middle Section of Upstream Side of Calaveras Dam Slips into Reservoir," *Engineering News-Record* (April 4, 1918): 679–81; "Failure of Part of Calaveras Dam," *Western Engineering* 9 (May 1918): 173–74.

18 "City Honors Mulholland—Dedicates New Dam in Name of Engineer," *Los Angeles Times,* March 18, 1925.

19 "Building for Future in Huge Dam Project: Weid Canyon Reservoir to Impound Billions of Gallons of Water for Hollywood's Use," *Los Angeles Times,* October 21, 1922. See Los Angeles Board of Public Service Commissioners, Minutes of Board Meeting, August 11, 1922; G. Gordon Whitehall to Department of Public Service, October 31, 1924; memo to Mulholland, December 29, 1924, all from file WP01–98:1A, LADWP Archives; and "Twenty-First Annual Report," 28.

20 "Condemns Projected Reservoir: Hollywood Chamber Says Weid Canyon Dam Would be Menace to City," *Los Angeles Times,* August 5, 1922; "Building for Future in Huge Dam Project: Weid Canyon Reservoir to Impound Billions of Gallons of Water for Hollywood's Use," October 21, 1922; also see "Weid Canyon Dam Opposed by Hollywood," *Los Angeles Times,* December 24, 1922. For more on how fears about the Mulholland Dam were spurred by the St. Francis collapse, see "Board Intends to Prove Mulholland Dam Safe: Hollywood Wants Fears Quieted," *Los Angeles Times,* March 28, 1928; "Dam Appraisal to Allay Fears: Weid Canyon to Undergo Rigid Test," *Los Angeles Times,* March 29, 1928; "Abandonment of Dam Urged," *Los Angeles Times,* February 28, 1935; and Crete Cage, "Hollywood Reservoir Seen Like 'Sword of Damocles,'" *Los Angeles Times,* March 26, 1936.

21 For a recent geological discussion of the Pelona schist present in the east abutment formation, see Gordon B. Haxel et al., "The Oracopia Schist in Southwest

Arizona: Early Tertiary Oceanic Rocks Trapped or Transported Far Inland," in *Contributions to Crustal Evolution of the Southwestern United States*, ed. Andrew Barth, Special Paper no. 365 (Geological Society of America, 2002).

22 Willis, "Report on the Geology of St. Francis Damsite," 409–12, 410.

23 Ibid., 409, 410.

24 Ibid., 411.

25 Rogers' comments are taken from an interview published in Ann Lucius, "Cause of Dam Failure Discovered...Almost 70 Years Later," *Water Operation and Maintenance Bulletin* (June 1997): 33–36. Publication of this journal was sponsored by the Bureau of Reclamation.

26 "Sixth Annual Report of the Bureau of the Los Angeles Aqueduct to the Board of Public Works" (1911), 42.

27 Joseph B. Lippincott to John R. Freeman, March 26, 1928, box 54, John R. Freeman Papers (MC51), Institute Archives and Special Collections, Massachusetts Institute of Technology, Cambridge, Mass.

28 During the Coroner's Inquest, Mulholland was repeatedly questioned about how carefully he had studied the geologic conditions at the dam site. Subsurface drilling or coring into the foundation was mentioned, but no logs were kept as a record, and the cores were stored below the dam and washed away during the flood; see R. R. Proctor, Coroner's Inquest testimony, 360; and Edgar Bayley testimony, 335. Joseph B. Lippincott notes his lack of involvement in the design of the dam in a letter to John R. Freeman, March 26, 1928, box 54, Freeman Papers, MIT.

29 The authors have not found any evidence that either an embankment design (utilizing either earth fill or rock fill) or a concrete buttress design (featuring either a flat-slab or multiple-arch upstream face) was considered for the Weid Canyon or the St. Francis dam sites. Neither of Mulholland's two concrete gravity dams received much notice in the engineering and technical press. Brief descriptions of Weid Canyon/Hollywood/Mulholland appear in "Air and Water Jet Cleans Concrete Surface for Next Pour," *Engineering News-Record* 93 (August 28, 1924): 351; and "Truck Hauls Plums Directly on to Dam," *Engineering News-Record* 93 (October 23, 1924): 680. Notices of the construction, only a paragraph long, appeared in "Los Angeles Building New Dam in San Francisquito Canyon," *Engineering News-Record* 94 (March 19, 1925): 497; and "St. Francis Dam Added to Los Angeles Water System," *Engineering News-Record* 96 (May 6, 1926): 743.

30 As engineers would put it, the force created by adding the horizontal water pressure to the vertical weight of masonry must intersect the "middle third" of the base in order to ensure safety in a gravity dam. If the weight of the dam is reduced (or the height of the dam is increased without widening the base), the resultant force will necessarily fall closer to the downstream face. The closer it comes to the downstream face, the greater the dam's instability. For discussion of nineteenth-century gravity dam design, see N. A. F. (Norman) Smith, *A History of Dams* (Secaucus, N.J.: Citadel Press, 1972), 195–206; and Smith, "The Failure of the Bouzey Dam in 1895," *Construction History* 10 (1994): 47–65.

31 Describing Liverpool's 136-foot-high Vyrnwy Dam, completed in 1892, Smith states that "concern about the effects of water percolating through a dam [explains] why it was fitted with a network of drainage tunnels, apparently the first dam to feature such equipment"; *History of Dams*, 206.

32 Early sources for discussion of the ways to mitigate uplift include C. L. Harrison, "Provision for Uplift and Icethrust in Masonry Dams," *Transactions of the American Society of Civil Engineers* 65 (1912): 142–25; and Charles E. Morrison and Orrin L. Brodie, *Masonry Dam Design Including High Masonry Dams* (New York: John Wiley, 1916), 1–15.

33 For an insightful critique of gravity dam designs, see George Holmes Moore, "Neglected First Principles of Masonry Dam Design," *Engineering News* 70 (September 4, 1913): 442–45.

34 Frederick H. Newell, first director of the U.S. Reclamation Service, expressed such a perspective: "We [the Reclamation Service] have been inclined to adhere to the older, more conservative type of solid [gravity] dam, largely perhaps because of the desire to have them appear so and recognized by the public as in accord with established practice"; Newell to A. H. Dimock, April 16, 1912, entry 3, "Discussion Related to Dams" file, Denver Federal Records Center, Record Group 115 (U.S. Reclamation Service).

35 "City Honors Mulholland: Dedicates New Dam in Name of Engineer," *Los Angeles Times*, March 18, 1925.

36 Mulholland testimony, Coroner's Inquest, 30.

37 See Los Angeles Department of Public Service, "Complete Report on the Construction of the Los Angeles Aqueduct," for a description of the aqueduct's organizational administration (250).

38 Edgar Bayley testimony, Coroner's Inquest, 219–21. Responding to the question, "Can you produce Mr. Wilkinson here [at the inquest]?" Edgar Bayley stated: "Yes sir, I think Mr. Wilkinson is here" (221). There is no record that Wilkinson testified. Because he functioned as the draftsman under Bayley, who inked the drawings, it was quickly deemed inappropriate to refer to Wilkinson as the designer, and Bayley subsequently offered testimony as to what occurred in the drafting room. It is puzzling nonetheless that Wilkinson did not testify.

39 Bayley testimony, Coroner's Inquest, 220. 247. (In the inquest transcript, Bayley's name is sometimes incorrectly spelled "Bailey"). Bayley likely consulted Morrison and Brodie, *Masonry Dam Design Including High Masonry Dams* (see n. 32, above); and William Creager, *Masonry Dams* (New York: John Wiley, 1917). For evidence that Morrison and Brodie's book was, at least nominally, used to develop a profile for the St. Francis Dam, see Governor's Commission *Report*, 31.

40 Bayley testimony, Coroner's Inquest, 225. Bayley's subsequent testimony (239) emphasized that his (and presumably Wilkinson's) role was subordinate to Mulholland's authority: "Q. Then the examination into the [shrinkage] stress, which would take part in that kind of analysis, wouldn't be the function of you or Mr. Wilkinson? A. No, our function was the cross sectional profile. Taking

care of the stress was entirely taken care of by the Chief Engineer." In addition, Bayley directly credited Mulholland with the "stepped" sloping of the downstream face (345).

41 Hurlbut testimony, Coroner's Inquest, 319–20.

42 Ibid., 321–22.

43 Ibid., 332–33.

44 Ibid., 334.

45 In the aftermath of the disaster, the Association of General Contractors criticized the city's decision not to use an independent contractor for St. Francis. At the time, the president of the association was W. A. Bechtel, whose firm would in a few years join the Six Companies, Inc., consortium responsible for building Hoover/Boulder Dam. In a telegram to California governor C. C. Young, Bechtel commented: "This association has repeatedly recorded its judgment that the use of contractors on such work is necessary in the public interest." The self-interest is apparent, but it is undoubtedly also true that our understanding of the design of St. Francis would be far more complete if Mulholland had engaged an outside contractor. See telegram from W. A. Bechtel to Governor C. C. Young, March 22, 1928, St. Francis Dam file, Division of Safety of Dams, Sacramento.

46 "St. Francis Dam Added to Los Angeles Water System," *Engineering News-Record* 96 (May 6, 1926): 743.

47 Governor's Commission *Report*, 29.

48 Outland, *Man-Made Disaster*, 29–30.

49 Los Angeles Board of Public Service Commissioners, "Twenty-Second Annual Report," 5, 24.

50 "Twenty-Third Annual Report of the Board of Public Service Commissioners of the City of Los Angeles for the Fiscal Year ending June 30, 1924," July 1, 1924, 23.

51 "Twenty-Fourth Annual Report of the Board of Public Service Commissioners of the City of Los Angeles for the Fiscal Year ending June 30, 1925," August 1, 1925. A height of 205 feet with a storage capacity of 38,000 acre-feet is mentioned in "Los Angeles Building New Dam in San Francisquito Canyon," *Engineering News-Record* 94 (March 19, 1925): 497. This brief notice represents the earliest published reference to a capacity of 38,000 acre-feet or a height exceeding 200 feet.

52 Outland, *Man-Made Disaster* (1977), 230.

53 Mulholland testimony, Coroner's Inquest, 28–29.

54 Governor's Commission *Report*, 9–10. A small concrete cutoff wall was built across the streambed at the dam's upstream face to facilitate dewatering and clearing of the foundation. However, this wall was not a trench that extended into the schist foundation; it also did not extend up either canyon wall. A dike or "wing wall section" was built atop the ridge forming the west abutment to reach the dam's final height. This approximately ten-foot-high dike required the excavation of a shallow trench, but the dike—which played no role in causing the dam to collapse—was separate from the main curved gravity structure.

55 Mid-nineteenth-century gravity dam design is described by Smith in "Failure of the Bouzey Dam," 47–65.

56 I. Davidson, in "George Deacon (1843–1909) and the Vrynwy Works," *Transactions of the Newcomen Society* 59 (1987–88): 81–95, discusses how uplift figured into the design of Vrynwy Dam. Early theorizing on uplift by German professors Lieckfeldt, Keil, and Link is noted in Serge Leliavsky, *Uplift in Gravity Dams* (London: Constable & Co., 1958), 2–7. In the United States, a prominent article published in 1895 cautioned engineers that upward pressure is not "an imaginary or remote danger." See John D. Van Buren, "Notes on High Masonry Dams," *Transactions of the American Society of Civil Engineers* 24 (1895): 493–520.

57 Edward Godfrey, "The Failure of the Reservoir Wall at Winston," *Engineering News* 50 (December 3, 1904): 672.

58 Edward Godfrey, "Failure of the Hot Spring Dam," *Engineering News* 58 (July 11, 1908): 55; Morrison and Brodie, in *High Masonry Dam Design*, state that "no account is made [by Wegmann] of the condition of uplift due to water penetrating the mass of masonry, nor of the ice thrust acting horizontally" (1–3).

59 "The Failure of a Concrete Dam at Austin, Pa. on Sept. 30, 1911," *Engineering News* 66 (October 5, 1911): 419–22; Marie Kathern Nuschke, in *The Dam That Could Not Break* (Coudersport, Pa.: The Potter Enterprise, 1960), lists seventy-eight victims by name (40).

60 For biographical information, see "John Ripley Freeman," *Transactions of the American Society of Civil Engineers* 98 (1933): 1471–76.

61 John R. Freeman, "Some Thoughts Suggested by the Recent Austin Dam Failure Regarding Text Books on Hydraulic Engineering and Dam Design in General," *Engineering News* 66 (October 19, 1911): 462–63, 462. Mulholland was aware of the Austin Dam failure and referred to it during the planning for the Lower San Fernando Dam in 1912, but did not specifically discuss uplift as a cause of failure; see Mulholland to Board of Public Service Commissioners, January 12, 1912, file WMOF-3-13, LADWP Archives.

62 Alfred D. Flinn, "The New Kensico Dam," *Engineering News* 67 (April 25, 1912): 772–79, 773; Wegmann, *Design and Construction of Dams*, 431–33. Freeman served as consulting engineer for the Catskill Aqueduct system from 1906 until his death in 1932.

63 Arthur P. Davis to L. C. Hill, October 11, 1912, entry 3, box 793, Rio Grande Project, General Administration and Projects 1902–1919, Record Group 115, Federal Records Center, Denver.

64 In 1917, in *Irrigation Works Constructed by the United States Government*, Davis described the "variety of precautions...adopted to prevent percolation under the [Elephant Butte] dam, and to relieve any upward pressure that might develop there." See E. H. Baldwin, "Excavation for Foundation of Elephant Butte Dam," *Engineering News* 73 (January 14, 1915): 49–55; Baldwin, "Grouting the Foundation of the Elephant Butte Dam," *Engineering News-Record* 78 (June 28, 1917): 625–28; Arthur. P. Davis, *Irrigation Works Constructed by the United States Government* (New York: John Wiley, 1917), 243–45.

65 Davis, *Irrigation Works,* 117. After the service was renamed the Bureau of Reclamation in 1923, concern about uplift continued. For example, Black Canyon Dam in southern Idaho, a 184-foot-high concrete gravity structure completed by the bureau in 1924, featured two rows of grout holes "drilled into the bedrock along the upstream edge of the dam along its entire length. [A] row of drainage holes was drilled 8 feet downstream from the second row of grout holes." In case anyone missed the point, *Engineering News-Record* declared: "The purpose of this drainage system is to collect and lead off any water that might accumulate and to prevent an upward pressure under the dam"; see Walter Ward, "Building Black Canyon Irrigation Dam in Western Idaho," *Engineering News-Record* 93 (November 20, 1924): 818–23, 818. After leaving the Reclamation Service in 1923, Davis became chief engineer of Oakland's East Bay Municipal Utility District, where his concern about uplift became manifest in the Llana Plancha (later renamed Pardee) Dam. This curved concrete gravity structure featured foundation grouting and an extensive drainage system running up both canyon walls. Construction started in 1927 and the design was illustrated in *Engineering News-Record* the same week that the St. Francis Dam collapsed. See F. W. Hanna, "Designing a High Storage Dam for the Mokelumne Project," *Engineering News-Record* 100 (March 15, 1928): 444–46.

66 Harrison, "Provision for Uplift and Icethrust in Masonry Dams," 142–225, 221.

67 C. R. Weidner, "Experiments on Uplift Pressure in Masonry Dams," *Engineering News* 70 (July 31, 1913): 202–5; Letter from Edward Godfrey, *Engineering News* 70 (August 21, 1913): 371.

68 Chester W. Smith, *The Construction of Masonry Dams* (New York: McGraw-Hill, 1915), 100–109.

69 Morrison and Brodie, *Masonry Dam Design,* 8–9.

70 References to uplift appear throughout the book; see Creager, *Masonry Dams,* 12, 13, 16–18, 25–33, 46, 48–59, 61, 63, 68, 69, 79–98, 98–104, 117–119, 120–30, 135, 172. In the preface, Creager advised readers: "The methods of design described and the assumptions recommended [in *Masonry Dams*] represent conservative practice and correspond to a proper degree of safety for the average enterprise...where considerable damage to property and loss of human life would result if failure occurred" (vii).

71 Savage's designs for Lower Otay Dam (completed in 1917) and Barrett Dam (1922) are described in "Arched Gravity Dams to Be Built at Lower Otay and Barrett Dam Sites," *Engineering News* 74 (August 12, 1916): 195–96.

72 "Concrete Dam on Eel River Built on Shale Foundation," *Engineering News-Record* 86 (May 5, 1921): 750–54, 750.

73 Planning and construction of the Don Pedro Dam is described in chap. 12 (114–26) of Dwight H. Barnes, *The Greening of Paradise Valley: The First Hundred Years of the Modesto Irrigation District* (Modesto, Calif: Modesto Irrigation District, 1987). See "Exchequer Dam Construction Plant Built in Narrow Canyon," *Engineering News-Record* 94 (May 28, 1925): 880–84. A photograph of Exchequer Dam showing drainage pipes extending up the canyon

walls is available in folder 79, Charles Derleth Papers, Water Resources Center Archives (hereafter WRCA), UC Riverside.

74 "Plant and Program on the Hetch Hetchy Dam," *Engineering News-Record* 89 (September 21, 1922): 464–68, 467. The extensive drainage system devised for O'Shaughnessy Dam is documented in "Drainage System Showing Galleries, Wells, and Contraction Joints-Sheet No. 3," signed by Michael M. O'Shaughnessy in May 1919. A copy of this drawing is available in folder 88, Derleth Papers, WRCA, UC Riverside

75 See M. H. Slocum testimony, Coroner's Inquest, 487–88.

76 Fred A. Noetzli, "Types of Storage Dams and Their Adaptation to Western Conditions," *Modern Irrigation: The Magazine of Applied Hydraulics* 3 (June 1927): 21. Other references to the ways engineers countered uplift prior to the disaster include Ezra B. Whitman, "The New Loch Raven Dam at Baltimore, Md.," *Engineering News* 72 (August 13, 1914): 331–37; "Upward Water Pressure Test Pipes Constructed in Concrete Dam," *Engineering News-Record* 82 (May 15, 1919): 954; "New Dam Will Double Water Supply of Portland, Ore.," *Engineering News-Record* 98 (May, 26, 1927): 842–46; and B. E. Torpen, "The Bull Run Storage Dam for Portland, Ore.," *Engineering News-Record* 103 (August 8, 1929): 204–8; construction of the 200-foot-high, curved gravity Bull Run Dam started in May 1927, along with grouting and a large drainage system, "a cut-off trench...excavated 6 feet below the finished foundation extend[ed] the entire length of the dam" (204).

77 See Kahrl, *Water and Power*, 21.

78 *California Statutes*, chap. 337, sec. 2 (1917): 517–18. The text of the 1917 law is available in *California Irrigation District Laws as Amended 1921* (Sacramento: California State Printing Office, 1922): 89–90. Specifically, the state engineer was given no authority over dams built by any "municipal corporation maintaining a department of engineering."

79 The phrase "municipal exemption" has been devised by the authors of *Heavy Ground* to refer to the provision in California's 1917 dam safety law that precluded the state engineer from supervising dam building undertaken by large cities.

80 Michael M. O'Shaughnessy to Edward Hyatt, October 3, 1928, Supervision of Dams file, 1928, Public Utility Commission Records, California State Archives, Sacramento.

81 For example, in the period 1907–16, New York City relied on a panel of engineering consultants to help design the Catskill water supply system (including the Ashokan/Olive Bridge Dam; see Wegmann, *Design and Construction of Dams*, 433). In 1913 the Miami Conservancy District (a model for the Tennessee Valley Authority) engaged a group of consulting engineers to review designs for flood-control dams in central Ohio; see Arthur E. Morgan, *The Miami Conservancy District* (New York: McGraw-Hill, 1951), 152–53, 161–74. On the federal level, the Reclamation Service (later Bureau) initiated a policy in 1903 requiring dam designs and other projects to be reviewed by "engineering boards." See U.S. Reclamation Service, "Third Annual Report of the Reclamation Service, 1903–4" (Washington, D.C.: Government Printing

Office, 1905), 41. The bureau announced, in April 1928 in *Engineering News-Record:* "The recent unfortunate failure of the St. Francis Dam in California… justifies special mention of the extensive geological and engineering investigations that preceded the approval of the site and designs for the Owyhee Dam," which included input from three geologists and three engineers not on the bureau staff. See John L. Savage, "Design of the Owyhee Irrigation Dam," *Engineering News-Record* 100 (April 26, 1928): 663–67.

82 Donald C. Jackson, *Building the Ultimate Dam: John S. Eastwood and the Control of Water in the West* (Lawrence: University Press of Kansas, 1995), 112, 114–15. In 1911, the Great Western Power Company also engaged the highly respected engineers James D. Schuyler and Alfred Noble to review John Eastwood's design for the Big Meadows Dam.

83 Mulholland to A. P. Davis, February 8, 1912. The nine-page "Report of A. P. Davis on Plans for San Fernando Dam" was incorporated into a report Davis made to Board of Public Service Commissioners, April 5, 1912, where he acknowledges that "in accordance with your request I have examined the location, materials of construction, and plans for the high dam near San Fernando to form a reservoir for the storage of water as it comes from the power plants of the Los Angeles Aqueduct." Davis concluded by advising the Board that "all the local conditions…are favorable at the proposed San Fernando dam site, and a dam constructed upon the lines planned will be as safe and secure against possible accident as any works constructed by human agency"; both reports are in file WMOF-3-13, LADWP Archives.

84 Mulholland testimony, Coroner's Inquest, 621.

85 Jackson, in *Building the Ultimate Dam*, provides a detailed history of early twentieth-century buttress dams, with a focus on multiple-arch dams in California.

86 John R. Freeman to A. P. Davis, September 26, 1912, box 63, Freeman Papers, MIT. Also see Jackson, *Building the Ultimate Dam*, 121–26.

87 Sylvester Q. Cannon, "The Mountain Dell Dam," *Journal of Utah Society of Engineers* 3 (September 1917): 223–31.

88 For biographical information, including his work as a leader in the Church of Jesus Christ of Latter-day Saints, see "Sylvester Quayle Cannon," *Transactions of the American Society of Civil Engineers* 109 (1944): 1472–74.

89 For more on the history of the Mountain Dell Dam, see Jackson, *Building the Ultimate Dam*, 146–54.

CHAPTER THREE

1 W. W. Hurlbut to William Mulholland, July 1, 1925; Los Angeles Board of Public Service Commissioners, "Twenty-Fourth Annual Report," 24.

2 Stanley Dunham testimony, March 21, 1928; Coroner's Inquest, 44.

3 Hurlbut to Mulholland, July 15, 1923; Board of Public Service Commissioners, "Twenty-Second Annual Report," 23.

4 "General Historical Data from January 1, 1921 to February 1, 1926," in "Historical Data: General & St. Francis Dam, 1921–1926," file WP04–3, LADWP Archives. This typescript provides short entries for activities involving the BWWS; the first mention of St. Francis Dam is "August, 1923 St. Francis Reservoir[:] Lindsey started work on road around the east side of the St. Francis Reservoir to take the place of the present county road." The report contains no entries for the period October 1923 to March 1924, and despite the date range in its title, the final entry is dated October 8, 1925.

5 April 1924 entry, "General Historical Data from January 1, 1921 to February 1, 1926."

6 An estimated 72 percent of Weid Canyon/Hollywood Dam was complete on July 18, 1924; entry dated July 18, 1924, "General Historical Data from January 1, 1921 to February 1, 1926." For a brief description of how concrete forms were used during construction of the Weid Canyon/Hollywood Dam, see "Convenient Scheme for Forms on Face of Concrete Dam," *Engineering News-Record* 93 (August 14, 1924): 271.

7 Short notices regarding the St. Francis Dam appeared in "Los Angeles Building New Dam in San Francisquito Canyon," *Engineering News-Record* 94 (March 19, 1925): 497; and "St. Francis Dam Added to Los Angeles Water System," *Engineering News-Record* 96 (May 6, 1926): 743. The latter included a photograph showing the dam nearing completion.

8 "Historical Data on Construction of the St. Francis Dam from May 1, 1924 to date of completion," in "Historical Data: General & St. Francis Dam, 1921–1926," file WP04–3, LADWP Archives. This three-page typescript, which lists events during construction of St. Francis Dam, ends abruptly in May 1925.

9 The clearing of foundations using pressurized water, frequently termed "hydraulicking," was a common feature of early twentieth-century dam construction. In his inquest testimony, Harold Hemborg described how water was taken from the aqueduct "about three hundred feet" above the dam site and then directed under pressure against the canyon walls. There are no known photographs documenting this process; Coroner's Inquest, 137.

10 Other entries from late January 1925 through April 1925 include:

> January 30. Old tower hole filled with concrete.
>
> March 3. Upstream forms for elevation 1660 were set.
>
> March 9. First concrete poured from new [steel] tower and mixing plant (7 trucks hauling sand—6 cement).
>
> March 18. Upstream forms [set] for elevation 1670. Cement conveyor belt installed.
>
> April 20. Set forms on upstream side for elevation 1684.60 Down stream points [set] for elevation 1685.00.

11 Governor's Commission *Report*, 9.

12 The "Historical Data on Construction of the St. Francis Dam" memo does not explain how concrete was placed in the foundation for six weeks (October 1–November. 12) prior to completion of the wooden tower.

13 William Lindsey testimony, Coroner's Inquest, 362.

14 Mulholland testimony, Coroner's Inquest, 373–74.

15 Dunham testimony, Coroner's Inquest, 44–45.

16 Bennett testimony, Coroner's Inquest, 162–65.

17 In later testimony at the inquest concerning a visit to the dam site by State Engineer Wilbur F. McClure, Mulholland refers to the "tunnel that Dunham had driven near the east end of this dam." When Coroner Frank Nance asks, "Was that on the east side?' Mulholland responds, "Yes sir, into the rock"; Coroner's Inquest, 360. Harold Hemborg also refers to the east side tunnel in his testimony: "Q. Did you witness any preliminary borings or tunneling into the site of the dam to determine the nature of the rock before the dam was placed there? A. [Hemborg]: I witnessed the tunneling into the east side"; Coroner's Inquest, 136.

18 Hurlbut to Mulholland, July 1, 1923; Los Angeles Board of Public Service Commissioners, "Twenty-Second Annual Report," 23–25.

19 Hurlbut to Mulholland, July 1, 1924; "Twenty-Third Annual Report," 22–24.

20 For the first public notice of a planned reservoir capacity of 38,000 acre-feet, see "Los Angeles Building New Dam in San Francisquito Canyon," *Engineering News-Record* 94 (March 19, 1925): 497

21 Hurlbut to Mulholland, July 1, 1925; Los Angeles Board of Public Service Commissioners, "Twenty-Fourth Annual Report," 24.

22 To the east, the topography of the St. Francis dam site is defined by a steep canyon wall of gray mica schist. To the west, and generally perpendicular to the east abutment, lies a ridge (at times referred to as a spur) of red sandstone conglomerate. The gorge between the west side ridge and the east abutment forms the main dam site. Because the top of the west ridge lies at an elevation of about 1,825 feet, it presented a natural limit for the dam height; any dam with a spillway level significantly higher would require special measures to ensure that the reservoir would not overtop the western ridge. See Hurlbut to Mulholland, July 1, 1926; Los Angeles Board of Water and Power Commissioners, "Twenty-Fifth Annual Report," 32.

23 Kahrl, *Water and Power,* 287–88; Catherine Mulholland, *William Mulholland,* 287.

24 Kahrl, *Water and Power,* 292–93; Catherine Mulholland, *William Mulholland,* 294–96.

25 Kahrl, *Water and Power,* 292–93.

26 Although surveying was briefly discussed at the inquest, no one questioned Mulholland about the decision to raise the height of the dam to 1,838 feet, and exactly when the decision was made remains uncertain. BWWS surveyor Hemborg testified that he arrived at the dam site on July 2, 1924, and soon "set the first stakes of the dam." In response to the question, "What were the dimensions of the dam as you laid it out there?" he replied, "We brought the dam up to an elevation of 1838 [feet] from the bedrock" (Hemborg testimony, Coroner's Inquest, 127–28). If he was directed to set the top elevation at 1,838 feet at the time

he started staking the dam in early July 1924, then why did the Board of Public Service Commissioners annual report, publicly distributed at the same time, not provide the same information?

27 "Auxiliary Dam Nearing Completion," *Los Angeles Times,* August 30, 1925.

28 The mixing and placement of concrete in the dam is described in the testimony of Edward Hendrik (Coroner's Inquest, 144–48) and David S. Menzies (Coroner's Inquest, 148–54).

29 Governor's Commission *Report,* "Report of Compression Tests of Concrete," 21. Three samples tested by the Raymond G. Osborne Laboratories exhibited compressive strength of 2,007 to 2,717 pounds per square inch. These samples had densities of 137.4, 142.8, and 141.7 pounds per cubic foot. The commission considered the "quality of concrete" to be "satisfactory" (*Report,* 10).

30 There are three typescript reports documenting concrete operations at St. Francis Dam in "Historical Data: General & St. Francis Dam, 1921–1926," file WP04-3, LADWP Archives: "Daily Concrete Placement St. Francis Dam from November 13, 1924 to conclusion on May 12th, 1926"; "Report on Monthly Concrete Placement at St. Francis Dam and Monthly Labor Costs from March 1, 1925 to the conclusion on May 12, 1926"; and "Report of Weekly Placement...at St. Francis Dam from March 30, 1925 to conclusion May 12, 1926." When operation of the steel hoisting tower began in March 1925, approximately 30,000 cubic yards of concrete had been cast in the prior seven months. Concrete placement averaged about 15,000 cubic yards per month from April through August 1925. It then gradually tapered off through the next spring. The "Report of Weekly Placement" indicates that concrete placed in the dam came to a total of 177,540 cubic yards (without "factors") and 173,783 (with "factors"); what these "factors" were is unknown. Published reports that 175,000 cubic yards of concrete were used in building the dam are compatible with these data.

31 A cost of $1.25 million for construction is given in "New Reservoir Put in Use," *Los Angeles Times,* March 13, 1926; "Water Pouring into Reservoir," *Los Angeles Times,* March 20, 1926; and "Huge Dam is Completed," *Los Angeles Times,* July 25, 1926. "Report on Monthly Concrete Placement" (WP04-3) indicates that about $900,000 was spent on the dam from April 1925 through May 1926. That would mean that about $350,000 was expended from May 1924 through March 1925, a reasonable figure given that during this time extensive excavation occurred, two hoisting towers were erected (one wood, one steel), and about 30,000 cubic yards of concrete were cast.

32 "St. Francis Dam Given Inspection," *Los Angeles Times,* June 6, 1926.

33 "Huge Dam is Completed," *Los Angeles Times,* July 25, 1926.

34 Robert H. Wright testimony, Coroner's Inquest, 546–47. Identified as "Chief Criminal Deputy, Los Angeles," Wright, called to testify regarding "rumors" that a dynamite attack had caused the dam's collapse, stated: "About a year ago, an officer received a telephone message from some unknown party stating that there were men on the way from Inyo County [i.e., the Owens Valley] for the purpose of dynamiting St. Francis Dam, told us to be sure to get some of-

ficers on the way as quick as possible, which we did, kept them there for about ten days. During that time nothing transpired in any way, shape or form that would cause us to believe this was true." Dynamite attacks on the aqueduct from May through July 1927 are described by Catherine Mulholland in *William Mulholland,* 312–14.

35 William T. Hoke Jr. testimony, Coroner's Inquest, 246. About a trip to the dam and reservoir on March 10, 1928, Hoke testified: "I went there for a little social fishing. It was a common custom for employees around the power house to go fishing in the lake, which, of course, was against the existing ordinance. We cheated a little bit, but it was a common practice among we boys there."

36 Hurlbut to Mulholland, July 1, 1923, Los Angeles Board of Public Service Commissioners, "Twenty-Second Annual Report," 24.

37 Louis C. Hill, J. B. Lippincott, and A. L. Sonderegger, "Summary of Report on the Water Supply for the City of Los Angeles and the Metropolitan Area," August 14, 1924, 7; "Water Supply—Special Reports, May 1924–October 1927, file WP04–7:31, LADWP Archives (also discussed in Kahrl, *Water and Power,* 289–90). After the disaster, the city estimated that the 38,000 acre-feet of water lost in the flood amounted to a loss of $1.25 million. With this figure as a gauge, the 5,240 acre-feet of San Francisquito Creek flood flow claimed by Los Angeles was worth about $170,000 per year to the city.

38 J. W. Stewart (attorney for the Newhall Land and Farming Company) to the City of Los Angeles...and to its Board of Public Service Commissioners, September 4, 1924; Aqueduct Division Reference Library Collection, "Dams and Reservoirs: St. Francis Dam, 1921–1926," file WP19–16:4, LADWP Archives.

39 "Supervisors Oppose Dam Working Harm to Santa Clara," *Ventura Post,* January 17, 1925; "Damming of Headwaters of Santa Clara River Menace to Water Users of Valley," *Ventura Free Press,* January 26, 1928.

40 The origins of the Santa Clara River Protective Association are recounted in Vernon M. Freeman, *People—Land—Water: Santa Clara Valley and Oxnard Plain, Ventura County, California* (Los Angeles: Lorrin L. Morrison, 1968), 83–87. The archives of the United Water Conservation District in Santa Paula contain the early records of the Protective Association.

41 By the summer of 1925, Los Angeles had filed a claim for 1,000 cubic feet per second of surplus (or flood) water within San Francisquito Creek; see Freeman, *People—Land—Water,* 88.

42 For example, see William Mulholland to Mayo Newhall, March 5, 1925; J. E. Phillips (BWWS assistant engineer) to Harmon S. Bonte (attorney for Newhall Ranch), June 26, 1925; W. B. Mathews to William Mulholland, telegram, April 9, 1926; W. Mayo Newhall to William Mulholland, June 9, 1926; all in "San Francisquito Reservoir, March 1925–August 1926," file WP04–6:18, LADWP Archives.

43 Mulholland made this prediction at an August 17, 1926, hearing in Los Angeles convened by Edward Hyatt, chief of the state Division of Water Rights. Attorneys representing the Newhall Ranch and the Santa Clara River Protective Association were in attendance, as was consulting engineer C. E. Grunsky

on behalf of the Protective Association; see Freeman, *People—Land—Water,* 90–92; and Outland, *Man-Made Disaster,* 38–42.

44 Results of the percolation test are described in the California Department of Public Works, Division of Water Rights report, "San Francisquito Experiment, Harold Conkling Engineer-in-Charge, September 1926." There is a copy in the WRCA, UC Riverside (WRCA G46029 E6).

45 Mulholland testimony, Coroner's Inquest, 13.

46 The *Los Angeles Times* photograph shows him visiting the dam when the reservoir was about thirty-five feet below the spillway crest, dating it to about February 1927. The photograph was apparently never published, but survives in the *Los Angeles Times* Photographic Archive, Charles E. Young Research Library, Department of Special Collections, UCLA; photo ID# uclamss_1429_1919.

47 "Program of Silver Anniversary Banquet tendered to William Mulholland in Commemoration of the Twenty-Fifth Anniversary of the Los Angeles Water Works and Supply," in box 2, R. F. del Valle Papers, HM 43836, Huntington Library. Toasts were also offered by Ezra Scattergood and Harvey Van Norman, and William Hurlbut presented Mulholland with a "silver set" in honor of his service. Although Harry Chandler attended and offered a toast, no notice of the banquet appeared in the *Los Angeles Times.*

48 All data on the reservoir level is drawn from Governor's Commission *Report,* 7, and "St. Francis Reservoir: Daily Record of High Water Elevations," plate 6, p. 33.

49 Mulholland's inquest testimony reflected a flawed understanding of the reservoir's operational history, as he claimed that "it was full or nearly full then within a foot and a half of last year—not all last year—but six or seven months." The reservoir never came within a foot and a half of the spillway crest until late February 1928; Coroner's Inquest, 13.

50 In the era of wooden ships, builders commonly used tarry rope called oakum to fill spaces between wooden planks and make hulls watertight.

51 Robert Atmore testimony, Coroner's Inquest, 165–72.

52 Henry Ruiz testimony, Coroner's Inquest, 289; the court reporter erroneously spelled his name Reiz in the transcript. Ruiz spent the night of March 12–13 at Power House No. 1 and so survived the flood, but eight members of his family perished.

53 Chester Smith testimony, Coroner's Inquest, 177–83.

54 A photograph of Tony Harnischfeger, his former wife, Gladys, their son Coder (who died in the flood), and their daughter Gladys Antoinette (who was with her mother and not in San Francisquito Canyon the night of the flood), is posted on the Santa Clarita Valley Historical Society website: scvhistory.com/scvhistory/lw2403.htm.

55 At the inquest, Mulholland's chauffeur, George Vejar, described the visit, reporting where he drove and parked, and where Mulholland and Van Norman walked. Vejar makes no mention of any photographer or of any photographs

taken; Coroner's Inquest, 722–26. In recent years, Mulholland and Van Norman have sometimes been identified (erroneously) in a photograph of three men in which the reservoir appears very full (confirming that it was taken in early March 1928); see Outland, *Man-Made Disaster,* 65, and Paul Rippens, *The Saint Francis Flood* (Alhambra: The Copy-Rite Press, 2003), 38. At the Coroner's Inquest, Paul Caughren and Ben Allen of Glendale testified that they visited St. Francis Dam on Sunday, March 11, and mentioned that Allen took photographs, likely including the one in question; Paul Caughren testimony, Ben Allen testimony, Coroner's Inquest, 256–58.

56 Mulholland testimony, Coroner's Inquest, 13–14. Also see Outland, *Man-Made Disaster,* 94–68, for a description of the March 12 trip.

57 Harvey Van Norman testimony, Coroner's Inquest, 91.

58 Ibid., 100.

59 Ibid., 92.

60 Mulholland testimony, Coroner's Inquest, 33.

61 Ibid.,34.

62 Ibid., 35–36.

63 Van Norman testimony, Coroner's Inquest, 93.

64 In response to a juror's question, "After you found this new leak [that Harnischfeger drew his attention to], what happened, what did you do?" Mulholland responded, "I had no plan in mind yet." The juror followed up: "I presume you planned to lower the flow carefully?" and Mulholland said, "Yes, draw the water out slowly. There is no means of lowering it more than a foot a day." He seemed to want to have it both ways: he wanted to claim that any purposeful drawdown of the reservoir would have been too slow to avert the disaster; but he also wanted to make it sound as though he intended to "draw the water out slowly." But given that it would be a slow process, why didn't he simply start lowering the reservoir as soon as possible? Perhaps he was concerned about the leaks but wanted to avoid taking any action that might draw public attention; Coroner's Inquest, 35. See also Appendix A, "How Fast Could the Reservoir Have Been Lowered?"

65 There would have been a sizable cost to the city if large quantities of water had been allowed to flow unimpeded down the length of San Francisquito Creek. Water rates for domestic consumers in Los Angeles were 13 cents per 100 cubic feet, or about $56 per acre-foot. If the reservoir level had been drawn down three feet this would have meant a loss of about 1,800 acre-feet, equal to a loss to the city of about $101,000. The city's domestic water rates as of March 1928 are given in "Los Angeles Water Rate Increase," *Wall Street Journal,* April 19, 1928.

66 For Archibald's service as Los Angeles Fire Chief and later career and interests, see lafire.com/fire_chiefs/FireChief-index.htm.

67 "Report of Death Claim Leona Johnson #2538," St. Francis Dam Claims Records, LADWP Archives.

68 Work was under way to build a new road to the top of the west side abutment, and during the afternoon of March 12 Harnischfeger talked by phone with

BWWS construction superintendent J. H. Bouey. Bouey later recalled that Harnischfeger had mentioned "Mr. Van Norman and Mr. Mulholland being up there, and things were the same as they were the last report, had nothing new to report to me"; J. H. Bouey testimony, Coroner's Inquest, 441.

69 See Outland, *Man-Made Disaster*, 68–70, on the reasons the adit to the aqueduct was closed (or blocked) during the afternoon of March 13.

70 The release of thirty cubic feet per second of aqueduct water down Drinkwater Canyon to San Francisquito Creek is described in Outland, *Man-Made Disaster*, 62–63. This release added to the flow in the lower San Francisquito Creek, as noted by Harvey Van Norman, Coroner's Inquest, 125–26.

71 Geologist Bailey Willis described the basic character of this phenomenon in *Western Construction News* shortly after the collapse: "The east abutment is located on a landslide. It was an old landslide which had become inactive... When it had become soaked by water standing in the reservoir against its lower portion, it became active and moved"; "Report on the Geology of St. Francis Damsite," 409–12.

72 Ibid., 411.

73 Governor's Commission *Report*, "St. Francis Reservoir: Daily Record of High Water Elevations," plate 6, p. 33. Had the reservoir level never been allowed to surpass 1,825 feet (the spillway height stipulated by the BWWS in 1923–24), it is likely that the structure could have maintained stability, as it had during the spring of 1927 when the reservoir stood above 1,825 feet for seven weeks (and above 1,830 feet for over two weeks). See Appendix B for further discussion.

74 Charles Lee described the challenges of after-the-fact investigations: "The failure of the St. Francis Dam occurred without human witnesses and the only available information regarding it lies in the mute evidence of broken concrete blocks, eroded foundations, and the portions of the dam which still remain in position"; "Theories of the Cause and Sequence of Failure of the St. Francis Dam," *Western Construction News* (June 25, 1928): 405.

75 Outland, *Man-Made Disaster*, 74, 79–80, 242.

76 Dean Keagy testimony, Coroner's Inquest, 425. Also see Outland, *Man-Made Disaster*, 74.

77 Exactly where Leona Johnson's body was discovered is not indicated in any records produced in the immediate aftermath of the disaster. However, Outland reported in *Man-Made Disaster* that he had been informed by Otto Steen (a Bureau of Power and Light employee involved in recovery efforts) that her clothed body had been found close to the dam site, indicating that she had been near the dam at the time of failure. Had she been in the Harnischfeger cabin when the flood hit, it would be impossible to explain how her body could have been carried upstream a quarter mile to the dam site (*Man-Made Disaster*, 74). Don Ray also mentions the discovery of Johnson's body near the dam in "1928 St. Francis Dam Disaster: Reunion Honors Victims and Survivors," *Los Angeles City Historical Society Newsletter* 10 (March 1988): 1–3.

78 Katherine Spann testimony, Coroner's Inquest, 434, and Elmere E. Steen testimony, 846–49; his first name, Helmer (Outland, *Man-Made Disaster,* 79–80), was presumably misspelled by the court reporter. Steen worked for the Bureau of Power and Light at Power House No. 2.

79 Ace Hopewell testimony, Coroner's Inquest, 425–31. Also see Outland, *Man-Made Disaster,* 80.

80 Analysis of the dam's collapse is drawn from key reports and studies documenting the "east side first" failure mechanism. These include C. E. Grunsky and E. L. Grunsky, "The St. Francis Dam Failure," *Western Construction News* (May 25, 1928): 314–20; Willis, "Report on the Geology of St. Francis Damsite"; Charles H. Lee, "Theories of the Cause and Sequence of Failure of the St. Francis Dam," *Western Construction News* (June 25, 1928): 405–8; and Rogers, "A Man, a Dam," 1–110. In chapter 6, below, the "east side first" hypothesis is compared to the once commonly held but erroneous "west side first" hypothesis.

81 Rogers, "A Man, a Dam," 42–43, 58.

82 Dunham testimony, Coroner's Inquest, 44–45; Lee, in "Theories of the Cause and Sequence," makes the point that concrete remnants from the lower east side abutment (specifically blocks 35 and 32) were among those deposited farthest downstream from the dam (408).

83 Rogers, "A Man, a Dam," 59.

84 Ibid., 68. In discerning how the massive east side landslide temporarily impeded the flood (for "30 to 90 seconds"), Rogers expands upon the insights of the Grunskys, Bailey Willis, and Charles Lee, and makes an important contribution to our understanding of the collapse and the initial stages of the flood.

85 Lillian Curtis is quoted by Marianne Tyler in "The Day the Dam Broke," *Los Angeles Herald-Examiner,* February 9, 1978.

86 Lee notes the direction of the bent pipe, and its significance for the location of initial failure, in "Theories of the Cause and Sequence," 408.

87 Lee describes the end of the failure sequence (ibid.): "The final step was the undermining of the slab represented by blocks No. 2, 3, and 4, and its toppling over [into the east side gap]. The edges of these pieces are fresh and sharp, showing no signs of erosion, so that the reservoir must have been nearly empty at this time."

88 For discussion of the center section's displacement and its relation to the failure of the dam's west side, see Rogers, "A Man, a Dam," 69–70.

89 Pulverized schist suspended in the floodwater enhanced its capacity to transport large, heavy remnants of the dam. The density of the liquid muck was increased by this sediment, which provided greater buoyancy. Had the flood been made up of silt-free water, the blocks would not have been transported as far.

90 For more on the rate of outflow, see Rogers, "A Man, a Dam," 75. Rogers believes that 1 million cubic feet per second is a conservative estimate and that, for at least a short time, maximum flow approached 1.7 million cfs.

CHAPTER FOUR

1 Chester Smith testimony, Coroner's Inquest, 180.

2 "Fillmore Funeral Chapel Stacked with Bodies of Victims of the Flood," *Santa Paula Chronicle*, "Extra," March 13, 1928.

3 H. L. Tate testimony, Coroner's Inquest, 459, 667–70. In *Made-Made Disaster*, Outland sometimes gives the impression that the flood hit places relatively close to the dam later than those farther downstream (87–94).

4 Governor's Commission *Report*, 8.

5 Harnischfeger apparently remained in the canyon even though he feared for his safety; at least, that was a story expressed soon after the collapse. A few days before the disaster he reportedly told D. C. Mathews that the dam "looks bad" because of leaks on the west side. "If she [the dam] is here Wednesday, why I will come down and we [he and his family] will go out." Why Harnischfeger planned to wait until Wednesday (March 14) before acting on his professed fears is unknown; see D. C. Mathews testimony, Coroner's Inquest, 198, 247, 389. In "Roll of the Identified and the Missing," *The Intake* 5 (April 1928), Leona is described as "missing" (3). But Outland indicates that her body was found close to the dam and well upstream from the cabin she shared with Harnischfeger and his son (*Man-Made Disaster*, 74).

6 The deaths of Aaron J. Ely, his wife, Margaret, and their sons, Jack and Roy, are noted in "Roll of the Identified and the Missing," *The Intake* 5 (April 1928), along with Earl Pike (and his wife, Selda, and son, Richard). Pike's work as a dam attendant is noted in "First Claim in Flood Filed," *Los Angeles Times*, March 21, 1928. For detailed information on flood victims, see the spreadsheet attached to Ann Stansell, "Memorialization and Memory of Southern California's St. Francis Dam Disaster of 1928" (M.A. thesis, California State University Northridge, 2014; hereafter cited as Stansell, "Memorialization and Memory," victim spreadsheet).

7 The speed of the flood wave is documented in the Governor's Commission *Report*, 8.

8 H. L. Tate testimony, Coroner's Inquest, 459.

9 T. A. Pantes to E. F. Scattergood, "Memorandum Concerning Destruction at San Francisquito Power Plant No. 2 and St. Francis Dam—March 13, 1928" (March 27, 1928), file WP28–28:11A, Historical Records Program, LADWP Archives; "Memorandum to E. F. Scattergood from Electrical Engineer in Charge of Operation, April 24, 1928," file WP23–8:15, LADWP Archives; "Failures of Power Supply from Southern California Edison, Jan. 1, 1928–March 31, 1928," file WP23-8:15, LADWP Archives; Coroner's Inquest, 459–69. The March 27 "Scattergood Memorandum" is presented in considerable detail in Coroner's Inquest testimony by Earl Martindale, assistant operating engineer with the city's Bureau of Power and Light; see Martindale testimony, 468–74. When the flood hit Power House No. 2, only one of the two generators (No. 2) was in operation. Because generator No. 1 was off line, it suffered significantly less damage to its electrical components and proved much easier to repair. The Scatter-

good memorandum reported that hours after the flood had passed, "Generator No. 2 [was] still rolling and burning" (472).

10 E. H. Thomas Testimony, Coroner's Inquest, 446, 447; "Corpses Flung in Muddy Chaos by Tide of Doom," *Los Angeles Times,* March 14, 1928. Cited henceforward as "Tide of Doom."

11 D. C. Mathews testimony, Coroner's Inquest, 197–99.

12 "Bodies Found of Mrs. Imus and Her Baby," *Banning Herald,* March 19, 1928: "The body of Olive Kabar Imus was found quite a distance below the St. Francis Dam site. Mrs. Imus had been visiting her sister Mrs. Weinland at the Weinland home near Power House No. 2 when it was destroyed by the flood. A short time previous to the finding of Mrs. Imus' remains, the body of her eight-month-old baby was recovered." (The *Banning Herald* article is available in California Scrapbook No. 8, Huntington Library.) Cecelia M. Small, teacher at the schoolhouse near Power House No. 2, also died in the flood, but she lived farther down the canyon. She was not an employee of the Bureau of Power and Light or the BWWS and was not listed as a victim in the April 1928 issue of *The Intake*. The Coroner's Inquest lists Cecelia Small as a victim (5).

13 Frank R. Webb testimony, Coroner's Inquest, 6. Webb also stated that Julia Rising's autopsy was carried out in the W. G. Noble mortuary in San Fernando on March 15, 1928.

14 D. C. Mathews testimony, Coroner's Inquest, 199. For a fuller description of Rising's and Curtis' miraculous escape, see the Prologue, above.

15 Chester Smith testimony, Coroner's Inquest, 177–87, 178, 180.

16 Ibid., 180, 181. Stansell, "Memorialization and Memory," victim spreadsheet, identifies Alva Kennedy, his wife, Reba, their two-year-old son, Charles, and their five-year-old daughter, Evelyn, as having lived at the Chester Smith Ranch. The entire Kennedy family was buried in Santa Ana, California. Guy Lundy, an electrician who worked for Famous Players Studio (Paramount) in Hollywood, maintained a cabin near the Smith ranch that was also wiped out in the flood. Lundy was not staying at his rural retreat on March 12, but came up from his house in Los Angeles the morning after the flood to aid in rescue work. See Guy D. Lundy testimony, Coroner's Inquest, 386–87.

17 James J. Erratchuo testimony, Coroner's Inquest, 298–99. Outland also describes Erratchuo's struggle to survive in *Man-Made Disaster* (89), but cites no source; he apparently interviewed Erratchuo (263).

18 "Statement of B. W. Hunick, taken at Sheriff's Office, Newhall, California" (March 17, 1928), 6–7, Charles F. Outland Collection, Museum of Ventura County, Ventura, California.

19 Los Angeles Citizens' Restoration Committee, "Report on Identified Dead: Victims of St. Francis Dam Disaster" (May 4, 1928). See the listing under "Hunick" in the section "Survivors and Members of Family Lost," St. Francis Dam Claims Records, LADWP Archives. The recovery of Jeff Hunick's body and its removal to the Moorpark morgue is also noted in "Inquest This Morning of Flood Victims Here," *Moorpark Enterprise,* March 15, 1928. His sister appears

in claims records as Ellen (Hunick) Crosno; see Stansell, "Memorialization and Memory," victim spreadsheet.

20 Several members of the Halen family resided at the Price ranch. Edward Price, his sister, Carrie Halen, her sons Leon John Halen and Kenneth Halen, and her five-year-old granddaughter, Jane Halen, all died in the flood; see Stansell, "Memorialization and Memory," victim spreadsheet.

21 "Tide of Doom"; "200 Dead, 300 Missing, $7,000,000 Loss in St. Francis Dam Disaster," *Los Angeles Times*, March 14, 1928. Cited henceforward as "200 Dead, 300 Missing." Myra Nye reported Carey's absence from the valley because of the premiere of *The Trail of '98* in "Society of Cinemaland," *Los Angeles Times*, March 14, 1928. Also see "Harry Carey Trading Post," brochure, n.d., History Section, Los Angeles Public Library; Michele E. Buttelman, "St. Francis Dam Disaster of March 12, 1928, Remembered," *The* [*Newhall*] *Signal*, March 12, 2000. See also scvhistory.com/scvhistory/carey-bunse-5.htm.

22 "Tide of Doom."

23 "274 Perished, 700 Missing, in Torrent Loosed by California Dam; Flood Engulfs Victims as They Sleep," *New York Times*, March 14, 1928.

24 Governor's Commission *Report*, 8.

25 The original topographic alignment of San Francisquito Creek as it passed to the north of the Round Mountain knoll is illustrated in "Whole Story of St. Francis Dam Disaster Pictorially Told by 'Times' Staff Artist," *Los Angeles Times*, March 25, 1928.

26 A. M. Newhall and George A. Newhall Jr., "Report on St. Francis Dam Flood for the Newhall Land and Farming Company" (March 24, 1928), 1–9, file 627.82, Museum of Ventura County; available online through the Santa Clarita Valley Historical Society at scvhistory.com/scvhistory/nlf-stfrancis.htm#1.

27 Ibid.

28 Modern-day Route 126, which extends through the Santa Clara Valley, was historically known as Telegraph Road, a designation it retains in Ventura County. In Los Angeles County the road is named Henry Mayo Drive in honor of Henry Mayo Newhall, the patriarch of the Newhall Land and Farming Company.

29 "Deaths Reach Total of 243 as Final Hunt for Missing Continues," *Los Angeles Times*, March 17, 1928.

30 "Wild Race with Flood Described," *Los Angeles Times*, March 17, 1928.

31 "Southern Pacific Among Sufferers from St. Francis Dam Disaster," *Railway Engineering & Maintenance* 24 (May 1928): 200–201. (Stansell gives the names of the Southern Pacific victims at Castaic Junction in "Memorialization and Memory," victim spreadsheet.)

32 Ibid.

33 Newhall and Newhall, "Report on St. Francis Dam Flood," photograph no. 12 and caption.

34 "Tragic Survivors of Catastrophe: Younger Rivera Trio See Rest of Family Swept to Death," *Los Angeles Times*, March 17, 1928. Also see "New Bonds Link Flood

Survivors: Elder Sister Appointed Guardian of Family in Place of Lost Parents," *Los Angeles Times*, June 6, 1928.

35 Newhall and Newhall, "Report on St. Francis Dam Flood," 9; Outland, *Man-Made Disaster*, 95. "St. Francis Dam Claim and Demand for Damages of the Newhall Land and Farming Company to City of Los Angeles and Board of Water and Power Commissioners of the City of Los Angeles" (March 30, 1928), Newhall Land and Farming Company Collection, Huntington Library; Ruth Waldo Newhall, *The Story of the Newhall Land and Farming Company* (San Marino, Calif.: Huntington Library, 1958), 83–84.

36 Newhall and Newhall, "Report on St. Francis Dam Flood," 7.

37 Thanks to Ann Stansell, a list of the thirty-two automobiles and one motorcycle recovered at Kemp (giving owners and license plate numbers) is available at scvhistory.com/scvhistory/as2804.htm.

38 Governor's Commission *Report*, 8–9; Newhall and Newhall, "Report on St. Francis Dam Flood," 6–7; H. C. Stinchfield to R. H. Ballard, March 14, 1928, 2; "Correspondence with Southern California Edison Co.," folder 17, Outland Collection, Museum of Ventura County. The disaster at Kemp was noted blandly in the Southern California Edison in-house magazine, *Busy Button Bulletin* (April 1928). Little was said in the *Bulletin* about the circumstances that led to the death of eighty-four employees (or why employees at the Saugus substation did not get word of the flood to people downstream in Ventura County); but the magazine proudly noted that "Edison Lines Carry Entire City Load With Only 2 Minutes Darkness"; see box 406, Southern California Edison Collection, Huntington Library. The authors of *Heavy Ground* are not aware of any first-person accounts of the Kemp tragedy.

39 "Death and Destruction Carried by Great Flood Wave When Big Dam Breaks," *Ventura County News*, March 16, 1928.

40 Frank Chandler to Mrs. W. B. Chandler, March 16, 1928, in the *Maryville Enterprise*, March 26, 1928, John Nichols Papers, Museum of Ventura County.

41 Sometime around 12:30 a.m. the Edison dispatcher in Los Angeles had been warned by company employees in Saugus about "lots of water." That information apparently led the Edison dispatcher and the Los Angeles Bureau of Power and Light dispatcher to suspect problems at either St. Francis Dam or the Los Angeles Aqueduct. But if that was the case, why did the Edison dispatcher fail to alert—or at least attempt to alert—the men at Kemp and law enforcement officials in the lower Santa Clara Valley? A key memo prepared by T. A. Pantes, a Bureau of Power and Light electrical engineer for Ezra Scattergood (the "Scattergood Memorandum"), drew from the logbook of the bureau's dispatcher to document actions taken by bureau and Edison employees in the early hours of March 13. In his research for *Man-Made Disaster*, Charles Outland came to question the accuracy of parts of this memo, believing that it had "serious flaws." Outland also noted that after the Borel-Lancaster transmission line went dead shortly before midnight, no effort was made by personnel in Saugus to journey up San Francisquito Canyon to locate the site of the line break, which would have been the first step in restoring service (for an electric power company, transmission line failures are to be rectified as soon as

possible because they threaten the lifeblood of the enterprise: money). Did Edison employees in Saugus suspect that the line break had resulted from the dam collapse and thus did not want to embark on a suicide mission up the canyon? Outland hoped that these questions might be answered if he could gain access to Edison Company records, but there is no evidence that he did so; he died in 1988. In 2012 many records related to the history of the Southern California Edison Company were made available through the auspices of the Huntington Library. These records include several photographs taken of the flood zone in March 1928 but they do not include any material related to actions taken, or not taken, by Saugus-based personnel on March 13. For an extended rumination on the difficulty of reconciling various accounts of the way flood warnings were disseminated, see Outland, *Man-Made Disaster,* 95–120; he quotes extensively from a letter he received in 1976 from Raymond Starbard, an Edison employee stationed in Saugus in March 1928. Also see T. A. Pantes to E. F. Scattergood, "Memorandum Concerning Destruction at San Francisquito Power Plant No. 2 and St. Francis Dam—March 13, 1928," March 27, 1928, LADWP Archives.

42 Despite his skepticism about official reports, Outland was willing to attribute to general confusion the delay in getting a warning to Kemp and the lower valley: "The contradictions in the evidence here are of interest only to the extent of illustrating the confusion that was generated by events following the failure of the dam." Moreover, he noted, "given the same knowledge and circumstances, it is very doubtful if the writer [Outland] or the reader [of his book] would have done better"; Outland, *Man-Made Disaster,* 102, 120.

43 Pantes to Scattergood, "Memorandum Concerning Destruction," 1–5; *Santa Paula Chronicle,* March 24, 1928. Outland cites this article in *Man-Made Disaster,* 101.

44 Pantes to Scattergood, "Memorandum Concerning Destruction," 3; "Ventura County Sheriff's Office Log of Incoming Phone Calls on Night of March 12–13, 1928," Outland Collection, Museum of Ventura County.

45 Telephone operators and police officers in the lower reaches of the Santa Clara Valley had just begun to issue alarms when the flood wave hit Kemp. There is no evidence that these telephone operators had the means to contact the Edison work camp.

46 "In the Matter of the Inquisition upon the Bodies of Those Who Lost Their Lives by Breaking of the St. Francis Dam, Moorpark, Calif., March 15, 1928" (hereafter Ventura County Coroner's Inquest, Moorpark), 2; St. Francis Dam Claims Records, LADWP Archives; "In the Matter of the Inquisition upon the Bodies of Those Who Lost Their Lives by Breaking of the St. Francis Dam, Oxnard, Calif., March 15, 1928" (hereafter Ventura County Coroner's Inquest, Oxnard), 4–7; Pantes to Scattergood, "Memorandum Concerning Destruction," 3; see "Ventura County Sheriff's Office Log of Incoming Phone Calls on Night of March 12–13, 1928," Outland Collection, Museum of Ventura County, for evidence of calls from Los Angeles County Sheriff's Office to Ventura County Sheriff's Office beginning at 1:20 a.m. on March 13. Hearne rose to the occasion in getting the word out, but later he could not recall clearly what had

transpired on the night of the flood. Two days later, when questioned by the Ventura County deputy coroner as to when he received the call from Los Angeles, Hearne answered "about midnight" or "about eleven fifty" on Tuesday, March 12. "March 12th? March 12th, is that right? Instead of being the 13th?" asked the coroner, who knew that calls like the one made to Hearne were not placed until after 1:15 a.m. on March 13. "Yes sir," came the reply. See Ventura County Coroner's Inquest, Moorpark, 2. Others also had difficulty recalling the timing or character of the disaster. Raymond Ransdell, another Ventura County deputy sheriff, testified that the call to him came at "about ten minutes of one"—also a faulty recollection, since the order to make warning calls to Ventura was not issued for another twenty-five minutes. See "In the Matter of the Inquisition upon the Bodies of Those Who Lost Their Lives by Breaking of the St. Francis Dam, Santa Paula, Calif., March 15, 1928" (hereafter Ventura County Coroner's Inquest, Santa Paula), 23.

47 Ventura County Coroner's Inquest, Moorpark, 2–4; Governor's Commission *Report*, 8.

48 Ventura County Coroner's Inquest, Santa Paula, 23–24; "In the Matter of the Inquisition upon the Bodies of Those Who Lost Their Lives by Breaking of the St. Francis Dam, Fillmore, Calif., March 15, 1928" (hereafter Ventura County Coroner's Inquest, Fillmore), 23–24; "In the Matter of the Inquisition upon the Bodies of Those Who Lost Their Lives by Breaking of the St. Francis Dam, Ventura, Calif., March 16, 1928" (hereafter Ventura County Coroner's Inquest, Ventura), 3–5.

49 "Recollections of Thornton Edwards Re St. Francis Dam," folder 16, Outland Collection, Museum of Ventura County; Outland, *Man-Made Disaster,* 128–31.

50 "Flood Ride Hero Cited for Daring," *Los Angeles Times,* March 25, 1928. Also see "Recollections of Thornton Edwards Re St. Francis Dam."

51 "Flood Revere Gets Citation," *Los Angeles Times,* March 24, 1928; "Flood Ride Hero Cited for Daring," *Los Angeles Times,* March 25, 1928; "Modern Paul Revere Honored: Medal Given a Man Who Warned of Dam," *Los Angeles Times,* May 2, 1928. Edwards later became Santa Paula's chief of police, a position he held until his retirement in the 1960s. There is a photograph of Edwards in his chief of police uniform in Outland, *Man-Made Disaster,* 129.

52 The vital contribution made by telephone operators was recognized immediately by the local press. See "Hello Girls Stick to Job to Warn: Many Saved by Heroism of Telephone Girls in S.P. Telephone Office and at Fillmore," *Santa Paula Chronicle,* "Extra," March 13, 1928. Also see "Honors Sought for Heroines," *Los Angeles Times,* March 15, 1928; and "Crews Working at Night to Speed Up Rebuilding," *Los Angeles Times,* March 15, 1928.

53 "Crews Working By Night to Speed Up Rebuilding," *Los Angeles Times,* March 15, 1928, reported that "two telephone operators, Mrs. Carrie Johnson at Fillmore, and Louise Gipe at Santa Paula, remained at their posts and transmitted warnings to the operators in smaller communities although they expected the crest of the disaster to reach their offices at any minute." Basolo's service is noted in "Roll of Flood Heroes Called," *Los Angeles Times,* March 19, 1928; also see Outland, *Man-Made Disaster,* 128.

54 Governor's Commission *Report*, 8.

55 Outland, *Man-Made Disaster*, 149.

56 On the night of March 12–13 Outland was bunked in the Santa Paula High School, hoping to apprehend a nocturnal miscreant who had been stealing from the school. The school lay in the flood plain of the surge but suffered no serious damage. Outland experienced the flood close up, but for him it did not constitute a life-threatening event; see Outland, *Man-Made Disaster*, 139.

57 Zelma Wilson Oral History, November 9, 1988, tape 2, side 1; UCLA Center for Oral History Research, transcript available at oralhistory.library.ucla.edu/Browse.do?descCvPk=27453.

58 Ibid.

59 Outland, *Man-Made Disaster*, 139–42.

60 "Santa Paula Hardest Hit Town in Valley," *Fillmore Herald*, March 23, 1928.

61 "Flood Dead in Santa Paula Now Sixteen," *Santa Paula Chronicle*, April 6, 1928. The roll of the identified dead reveals that half had Hispanic surnames (three Torres, three Perez, and two Samaniego) and four more had the likely Hispanic surname Luna. The part of town closest to the river had a greater concentration of Hispanic residents, and this is reflected in the casualty list.

62 Governor's Commission *Report*, 8.

63 Outland, *Man-Made Disaster*, 149.

64 Memorandum from E. T. Scott, Asst. Dist. Maintenance Engr. to [S. V.] Cortelyou, District Engineer, Los Angeles, March 15, 1928, St. Francis Dam File, Division of Safety of Dams, Sacramento.

65 Outland, *Man-Made Disaster*, 146–49; John Nichols, *St. Francis Dam Disaster* (Chicago, 2003), 6; "Search Sea for Bodies Believed Carried There," *Fillmore Herald*, March 23, 1928.

66 As Outland put it, once the flood passed into the Pacific, "the curtain had come down on the main act" (*Man-Made Disaster*, 150).

67 By Wednesday, March 14, "the emergency hospital [in Newhall] which was opened by Mrs. Louise Burnell, director of nursing activities of the Los Angeles chapter of the Red Cross, and Mrs. Edna Karcofe, the first nurses dispatched to the scene of the disaster, [was] closed. After the treatment of about twenty-five refugees for minor cuts and bruises there was no necessity for its maintenance." See "Organized Relief Brings New Hope for Survivors," *Los Angeles Times*, March 15, 1928.

68 "Old Mattress Saves Lives of Four in Family," *Los Angeles Times*, March 14, 1928.

69 "Thousands Rush to Aid in Work of Rescue and Relief," *Los Angeles Times*, March 14, 1928.

70 "200 Dead, 300 Missing."

71 Earl Martindale testimony, Coroner's Inquest, 472.

72 "Tide of Doom"; "Rebuilders Follow Rescuers in St. Francis Dam Tragedy," *Los Angeles Times*, March 15, 1928; cited henceforward as "Rebuilders Follow Rescuers." By the end of the first week "more than 300 bodies of horses and

cattle found in the bed of the river" were destroyed. See "Call Forty in Inquest," *Los Angeles Times,* March 20, 1928.

73 "Dead Listed at 296 in California Flood," *New York Times,* March 15, 1928.

74 "St. Francis Dam Inquest Set Here on Wednesday: Over 1500 Paid Workers Now Aid Rebuilding of Flood Zone; Victims Placed at 441," *Los Angeles Times,* March 18, 1928.

75 "200 Dead, 300 Missing."

76 "Thousands Rush to Aid in Work of Rescue and Relief," *Los Angeles Times,* March 14, 1928. This article reports that "two hours after Mary Grogan, field worker for the Catholic Welfare Bureau of the Los Angeles and San Diego diocese, sent the call to bureau headquarters here for clothing for destitute flood suffers, a truck loaded with supplies, was speeding northward yesterday afternoon. Rev. Thomas O'Dwyer, director of the bureau, conferred with the Queen's Daughters, women's volunteer relief organization, and the St. Vincent de Paul Society, men's organization. The salvage department of the latter was cleaned out of all men's and women's overcoats and additional coats were purchased. The Queen's Daughter supplie[d] warm children's garments, while underclothing for men, women, and children was purchased and added to the allotment. The supplies were turned over to the Red Cross at Newhall."

77 "Inquest on Dam Set Tomorrow," *Los Angeles Times,* March 20, 1928. Because the charter of the Red Cross prohibited placing children who had lost their parents in disasters in orphanages, a search later ensued for adoptive families. The Catholic Church sought out Catholic families and eighty responded to the initial call. The American Legion announced its readiness "to care for any ill or destitute children" of any former servicemen or—women. See "Deaths Reach Total of 243 as Final Hunt for Missing Continues," *Los Angeles Times,* March 17, 1928; and *Los Angeles Times,* March 29, 1928.

78 Memo, undated, in box 1, St. Francis Dam Disaster Papers, Red Cross and American Legion, folder HM 70397, Huntington Library. Much of the newsreel "Destruction of a Dam," produced within a few weeks of the disaster, consists of a simulated flood and generic river flood scenes. It also includes some images of food being served at a Red Cross relief center. But toward the end there are authentic scenes of rescue work in the Santa Clara Valley and destruction in the Bardsdale and Santa Paula districts. This public domain film is available through the Prelinger Archives in San Francisco; see archive.org/details/destruction_of_a_dam.

79 "200 Dead, 300 Missing"; "Rebuilders Follow Rescuers."

80 Governor Young's visit to Santa Paula is described in "200 Dead, 300 Missing." "Nation Offers its Sympathy," *Los Angeles Times,* March 15, 1928. This article featured the subheading "Governor Promises Early Road Repair."

81 "Roads and Highways Present Huge Problem," *Los Angeles Times,* March 16, 1928.

82 "Flood Quiz Under Way," *Los Angeles Times,* March 19, 1928.

83 "Local People Return from Flood Scenes," *Banning Herald,* March 19, 1928, in California Scrapbook No. 8, Huntington Library.

84 "200 Dead, 300 Missing."

85 "Tide of Doom."

86 Letter from Walter Stephens to the *Marysville Enterprise* and friends in Blount County, dated March 28, 1928, printed in the *Marysville Enterprise,* April 18, 1928. The authors are grateful to John Nichols of Santa Paula for making this source available.

87 "Persons Handling Bodies Should Now Wear Rubber Gloves," *Fillmore Herald,* March 23, 1928.

88 Quotation from untitled report on the impact of the St. Francis Dam disaster on Ventura County, p. 8, attached to a letter from George B. Travis to C. C. Teague, dated August 22, 1929, Charles C. Teague Papers, Bancroft Library, UC Berkeley.

89 Receipt from Lester B. Tozier dated March 18, 1928, with typed certification by Sheriff Robert Clark, ca. September 1, 1928, in St. Francis Dam Disaster Papers, box 2, folder HM 70405, Huntington Library. Efforts to induce the American Red Cross to reimburse Clark for the whiskey purchase ultimately came to naught, because no prescription issued by a medical doctor authorized the sale by druggist Tozier. The Red Cross was concerned that it might be criticized for expending funds on a transaction that, in the Prohibition Era, was illegal. In the end, Clark was required to reimburse Tozier $10.50 from his own funds. See C. C. Teague to J. W. Richardson, September 21, 1928, in HM 70405. The work at Bardsdale appears to have been particularly onerous and, three days after the flood, a call from the local American Legion commander went out for "men in old clothes willing to give their labor in the salvage work and the search for bodies still believed to be somewhere in the huge mats of debris." This effort was energized by earlier success, "where bodies of four living persons were uncovered in the debris." See "Workers Needed in Bardsdale District," *Ventura County News*, March 16, 1928.

90 Quotation from an untitled report on the impact of the St. Francis Dam disaster on Ventura County, p. 8, attached to a letter from Travis to Teague, Teague Papers, Bancroft Library, UC Berkeley.

91 C. C. Teague to George Eastman, May 31, 1928, St. Francis Dam Disaster Papers, box 2, Death Claims Committee, folder HM 70406, Huntington Library.

92 A week after the flood, it was reported that "six crews of twenty-five men each, working under the Red Cross, are now combing the river bottom east of the Saticoy bridge. Huge tractors tackle piles of debris, pull the rubbish and heavy trees apart, and then move on while the crews search every bit of rubbish. After a thorough search, trucks haul away the rubbish to a different spot, where a burning crew destroys it. In this manner, it is pointed out by Red Cross officials, there is no chance for any bodies to be missed." See "Call Forty in Inquest," *Los Angeles Times,* March 20, 1928. There was at least one instance, however, in which the charred bodies of victims were discovered later. In researching Ventura County coroner's records, Patricia Allen came across evidence that in August 1928 a ranch crew "found the charred bones of three victims, a man, woman, and child in the ashes of a fire they had built to burn

the trash deposited by the flood water"; "The Death Toll," *Ventura County Historical Society Quarterly* 29 (Winter 1984): 17–23.

93 "Steam Shovels Grope About," *Washington Post,* March 15, 1928.

94 "Scores More Thought Buried in Debris of Wild Water," *Los Angeles Times,* March 14, 1928.

95 "Flood Deaths Reach 400: City is Blamed," *The World,* March 15, 1928.

96 "Fillmore Funeral Chapel Stacked with Bodies of Victims of the Flood," *Santa Paula Chronicle,* "Extra," March 13, 1928.

97 Telegram from Dr. Sarah L. Murray to Elizabeth Wineland [*sic*], March 13, 1928, California Scrapbook No. 8, Huntington Library. The telegram, which misspelled the Weinland family name, was sent from Newhall at 2:15 p.m. and received in Banning at 2:50.

98 Cynthia Grey, "Wife of 'Tony' Seeks Body of Little Boy," *Los Angeles Record,* March 26, 1928.

99 "Search Sea for Bodies Believed Carried There," *Fillmore Herald,* March 23, 1928.

100 "Body of Mrs. William Weinland Identified," *Banning Record,* March 22, 1928, in California Scrapbook No. 8, Huntington Library. This article reports on the search for the body of Lloyd Weinland in the Santa Clara Valley morgues.

101 At the start of the Los Angeles County Coroner's Inquest, Nance entered into the record the names of sixty-nine victims whose bodies had been recovered in Los Angeles County by the time the inquest commenced on March 21. He also selected one of them, Julia Rising, to stand for all of the victims. "This testimony," the coroner explained, "is to establish the record of the death of Julia Rising. The information adduced at the hearing will cover all those persons whose deaths occurred as the result of the breaking of the St. Francis Dam" (see chapter 6, below); Frank R. Webb testimony, Coroner's Inquest, 1–5, 3, 6.

102 "Inquest This Morning of Flood Victims Here," *Moorpark Enterprise,* March 16, 1928.

103 Ventura County Coroner's Inquest, Fillmore (March 15, 1928), 1–2, 3. Photographs of the deceased, identified and unidentified, were included with the transcript of each inquest.

104 Ventura County Coroner's Inquest, Fillmore (March 15, 1928), 2–32; Ventura County Coroner's Inquest, Santa Paula (March 15, 1928), 4, 7. The cover page of the Fillmore Inquest apparently errs in stating that there were thirty-six dead (twenty-three identified and thirteen unidentified) in Fillmore. According to the official tally as recorded in the inquest, the dead numbered forty-six (thirty-three identified and thirteen unidentified), 4–35. However, even these numbers may be suspect. When Ventura County District Attorney Hollingsworth began to list the unidentified dead, he included a "white" female; he was challenged by S. A. Wagner, an assistant who participated in attempts to identify the dead as they were brought into the morgue: "That body was identified as Mary Ruiz of San Francisquito Canyon," declared Wagner. "Of San Francisquito Canyon, Los Angeles County, California?" asked Hollingsworth. "Yes sir,"

replied Wagner. "White race?" queried Hollingsworth. "No," replied Wagner, "she was Mexican." The supposition when this memorandum was made was that she was white, but she was identified as Mexican. Questions of race and ethnicity aside, Hollingsworth, by mistake or for reasons he never disclosed, continued to list this woman among the Santa Paula Inquest's "Unknown Persons, Deceased"; see 18, 32.

105 For example, six of the victims found in the Santa Paula area were from Los Angeles County: one described as having lived "near" St. Francis Dam, two from Saugus, and three from Castaic Junction. Those from Ventura County included ten workmen at the Edison camp at Kemp, three from Fillmore, and nine whose residences were not disclosed, with only one victim from Santa Paula. Twenty of the victims could not be identified, with eight described as "Mexican." See Ventura County Coroner's Inquest, Santa Paula (March 15, 1928), 2–33.

106 "Find Body of Mrs. Cowden at Bardsdale," *Santa Paula Chronicle,* April 11, 1928.

107 "Weinland Family Stricken by Flood," *Banning Express,* March 14, 1928.

108 "Body of Mrs. William Weinland Identified," *Banning Record,* March 22, 1928, in California Scrapbook No. 8, Huntington Library.

109 "New Body Found in Flooded Area," *Ventura Free Press,* March 23, 1928.

110 "Three More Bodies Are Recovered," *Santa Paula Chronicle,* April 6, 1928.

111 "New Bodies Found in Devastated Area," *Ventura Free Press,* April 7, 1928. The article also reported that "descriptions of these bodies will be broadcast over KFI in an attempt to aid identification." There is no evidence that the announcement proved effective in either case.

112 "Water Body Asks Loan," *Los Angeles Times,* March 30, 1928.

113 Margaret Gilbert (executive secretary, Los Angeles Chapter, American Red Cross), to J. W. Richardson, July 13, 1928, St. Francis Dam Disaster Papers, box 1, Red Cross and American Legion, folder HM 70397, Huntington Library.

114 Richard Bard, Chairman Sub-Committee on Death and Injury, to Charles A. Storke, Santa Barbara, July 10, 1928, St. Francis Dam Disaster Papers, box 2, Death Claims Committee, folder HM 70406, Huntington Library.

115 C. A. Storke to Richard Bard, July 11, 1928, St. Francis Dam Disaster Papers, box 2, Death Claims Committee, folder HM 70406, Huntington Library.

116 C. C. Teague to J. W. Richardson, July 21, 1928, St. Francis Dam Disaster Papers, Red Cross and American Legion Folder, HM 70397, Huntington Library. Relations between Mexicans and Americans (as the two groups were commonly identified at the time) in the Santa Clara Valley were comparable to interactions between the two ethnic groups throughout Southern California. What would be deemed unacceptably racist in twenty-first-century America was the norm in the 1920s. Nonetheless, there were aspects of Santa Clara Valley life that exhibited less contentious relations between the two groups. For example, "Mexican Refugees Grateful: Tender Banquet to People of Santa Paula Who Aided in Flood Relief," *Los Angeles Times,* April 24, 1928, describes a "Spanish dinner" in late April at which "leading citizens of [Santa Paula], the Mayor, president of the Chamber of Commerce, bankers, merchants and Red

Cross executives sat down at the banquet board at the [Red Cross] refugee camp. Later, in addresses given in the evening, they expressed their appreciation and goodwill. Mexican spokesmen reciprocated by also extending the goodwill and thanks of their people...at the center entrance three flags were flying—the Mexican, Red Cross, and Old Glory."

117 "200 Dead, 300 Missing"; "Flood Deaths Near 1,000: Wall of Water Sweeps Many California Towns," *Chicago Tribune*, March 14, 1928. The body of the article was slightly more circumspect than the headline, reporting, "The list of known dead as the result of the St. Francis dam disaster stood at 231, with the number of missing estimated at from 300 to 600 early today. All the missing were presumed to have perished." The Associated Press story was published across the nation; for example, see "Giant California Dam Bursts: 76-Foot Water Wall Devastates Canyon," *Omaha Bee-News*, March 14, 1928, which reported, "A rapidly mounting death toll...Tuesday night showed 275 had lost their lives, while upwards of 700 persons were reported missing."

118 "Rebuilders Follow Rescuers"; "Flood Debris Gives Up Bodies of 305 Victims," *Chicago Daily Tribune*, March 15, 1928.

119 Ventura County Coroner's Inquest, Fillmore (March 15, 1928); Ventura County Coroner's Inquest, Moorpark (March 15, 1928); Ventura County Coroner's Inquest, Santa Paula (March 15, 1928); Ventura County Coroner's Inquest, Oxnard (March 15, 1928); Ventura County Coroner's Inquest, Ventura (March 16, 1928).

120 Los Angeles County Coroner's Inquest, 1.

121 "Flood Quiz Under Way," *Los Angeles Times*, March 19, 1928; a subheadline announced "St. Francis Dam Dead and Missing Total 450; 46 Unidentified."

122 Governor's Commission *Report*, 9.

123 "Rehabilitation Forces Move In," *Los Angeles Times*, March 26, 1928.

124 "Dam-Flood Death List Set at 378," *Los Angeles Times*, May 12, 1928.

125 "Report on Identified Dead," May 4, 1928, Los Angeles Citizens Committee, St. Francis Dam Claims Records, LADWP Archives, which reports: "Total—231—identified dead," but in fact only 223 victims are listed. The total for the "Missing" also seems inaccurate; it is given as 172, but the list enumerates only 161.

126 "List of Missing and Dead St. Francis Dam Disaster," attached to letter from A. J. Ford to City Attorney Jess Stephens, May 28, 1928, in St. Francis Dam Claims Records, LADWP Archives. The totals of identified and unidentified victims were calculated by Hundley and Jackson from the report's raw data.

127 Citizens' Restoration Committee, Subcommittee on Death and Disability Claims, July 15, 1929, file 979.49, Museum of Ventura County.

128 Ann Stansell, "Memorialization and Memory." These three groups (identified bodies, unidentified bodies, and individuals reported missing) total 403—the maximum number of fatalities. But this would mean that the total of 195 still missing did not include any of the 68 unidentified victims—not impossible, but highly unlikely. The number of actual victims thus lies somewhere between 335 and 403. If half of the unidentified bodies belong among the missing,

the total would be 369; if only 20 were among the missing, the fatalities would amount to 383. At a remove of more than eighty years, calculating the extent of overlap between the two categories is fraught with uncertainty, but the authors believe that Stansell's careful research shows that any figure either notably lower or higher than 400 is untenable.

129 "Report of A. J. Ford After Inspection of Flood Area, March 14, 15, 16, 17, 1928," 4; St. Francis Dam Claims Records, Reports of Flood Damage Areas, A. J. Ford, March 14-September 14, 1928, LADWP Archives.

130 "Service of Ruiz Family Read by Catholic Pastor," *Los Angeles Times,* March 20, 1928.

131 "Eight of One Family Laid into Grave," *Los Angeles Times,* March 19, 1928. Later in the spring, movie star William S. Hart paid for a "memorial monument dedicated to those who lost their lives in the St. Francis Dam disaster," which was placed in the Ruiz cemetery; see "Shaft Raised to Memory of Flood Victims," *Los Angeles Times,* May 11, 1928. Hart and a group known as the Newhall Cowboys were also involved in efforts to provide a burial for a young boy whose body remained unidentified at the Newhall morgue for more than a week after the flood. Shortly before the boy was to be buried at the Ruiz cemetery, he was identified as the son of a woman (referred to in sources as either Mrs. Trexler or Mrs. Prixler) who apparently had been staying at the McIntyre Cabins at Castaic Junction; he was subsequently buried next to the woman identified as his mother in the Oakwood Cemetery in the San Fernando Valley. This story was recounted by Leon Worden in "Requiem to a Little Soldier," *The [Newhall] Signal,* May 17, 2003.

132 "Eight of One Family Laid into Grave."

133 "City Will Vote One Million to Aid," *Los Angeles Times,* March 19, 1928.

134 "Flood Quiz Under Way"; "Homage Paid to Dead in Valley," *Los Angeles Times,* March 20, 1928.

135 "Federal Board Will Fix Blame in Dam Disaster," *Los Angeles Times,* March 16, 1928, reported that "H. C. White, representing the water and power bureau, yesterday told authorities he is empowered to advance $150 for the burial of each unidentified body." In "Organized Relief Brings New Hope for Survivors," *Los Angeles Times,* March 15, 1928, it was reported that the Board of Water and Power Commissioners "agreed to finance and make possible family burials for all victims of the disaster that desire this service." During the week after the flood, A. J. Ford of the Department of Water and Power made extensive visits to morgues in Ventura County and established a "uniform scale" for burial services that the city would pay to local undertakers: "Local burial, including plot $150 for adults, $135 for children of ten and under; Body prepared for shipment by rail $175 for adults, $160 for children of ten and under; Body ready for delivery without casket $50 for adults, $25 for children of ten and under; Body for delivery with casket $90 for adults, $75 for children of ten and under; Hearse delivery, single body, to Los Angeles $35." See "Report of A. J. Ford After Inspection of Flood Area," entry under "Santa Paula, 8:00 A.M., March 17, 1928," LADWP Archives.

136 "Homage Paid to Dead in Valley." The victim spreadsheet included in Stansell, "Memorialization and Memory," notes the burial of Leona Johnson and the Lyman family at Forest Lawn.

137 "Homage Paid to Dead in Valley."

138 "Funeral Services Held for the Victims of the St. Francis Dam Disaster," *Banning Record*, March 22, 1928, in California Scrapbook No. 8, Huntington Library.

139 The May 28, 1928, list of the identified dead compiled by the Department of Water and Power gives the burial sites for many of the victims handled by morgues in Ventura County, but does not include that information for bodies taken to the Newhall morgue in Los Angeles County. Most flood victims were buried in Southern California, but more than a few were interred elsewhere, for example: Lodi, California; Walnut Grove, Missouri; Warren, Ohio; Knoxville, Tennessee; Galesville, Wisconsin; and Kalispell, Montana. See "List of Missing and Dead St. Francis Dam Disaster," attached to letter from A. J. Ford to City Attorney Jess Stephens, May 28, 1928, in LADWP Archives. Stansell, in "Memorialization and Memory," provides an extensive list of burial sites.

140 "Eight of One Family Laid into Grave."

141 Restoration of aqueduct service is noted in "Grand Jury Dam Inquiry to Depend on Coroner," *Los Angeles Times*, March 25, 1928; generating unit B at Power House No. 2 was back online by mid-June; see "Small Army of Men Work Night and Day to Rebuild Power House No. 2," *Los Angeles Times*, May 14, 1928, and "St. Francis Power Unit Turned On," *Los Angeles Times*, June 15, 1928. Reconstruction of unit A (which was in operation at the time the flood hit) was completed by early November, bringing the plant to full generating capacity of 44,000 kW; see "Power Plant Now Restored," *Los Angeles Times*, November 3, 1928.

142 "Railroads and Highways Present Huge Problem: Tented Cities and Field Kitchens Now Cover Zone of Disaster; Bridges Being Rebuilt," *Los Angeles Times*, March 16, 1928.

143 A memo forwarded to district engineer S. V. Cortelyou on March 16 conveys the intensity of the reconstruction effort: "One of Foreman Harbey's force, a man by the name of H. H. Brown work[ed] respectively 24 hours on the 13th, 16 hours on the 14th, and 12 hours on the 15th; while practically all of the balance of the crew worked 13 hours on the 13th and 11 hours on the 14th...the force is working in two shifts in order to have the highway repaired ready for through traffic when the trestle across the Santa Clara River is completed—we hope by next Monday or Tuesday; and by this time the highway [Ridge Route] should be open from the north with the detour road at San Francisquito [Creek] temporarily surfaced and ready for use." See "Report on Flood Damage in Castaic District," memo from I. S. Voorhees to S. V. Cortelyou, Los Angeles district engineer, March 16, 1928, St. Francis Dam File, Division of Safety of Dams, Sacramento. For more on the work of Division of Highway personnel in the rebuilding effort, see S. V. Cortelyou, Los Angeles District Engineer, to C. H. Purcell, State Highway Engineer, March 17, 1928, St. Francis Dam File, Division of Safety of Dams, and "How the State Highway Forces Met

Emergency Following Dam Disaster," *California Highways and Public Works* (April 1928): 3–4, 22, 25.

144 "Thousands of Sightseers Turned Back in Attempt to Visit the St. Francis Disaster Area," *Los Angeles Times*, March 19, 1928.

145 "Board Intends to Prove Mulholland Dam Safe," *Los Angeles Times*, March 28, 1928.

146 "Rehabilitation Forces Move In," *Los Angeles Times*, March 26, 1928; also see "Jury Views Dam Site," *Los Angeles Times*, March 24, 1928. The contract with Associated General Contractors was undertaken on a "cost plus 6 percent" basis; see "Dynamite Dam Theory Backed," *Los Angeles Times*, April 7, 1928. The rapid pace of reconstruction is illustrated and described in "Reconstruction Work in Santa Clara River Valley in Full Swing," *Los Angeles Times*, April 5, 1928; "Work Army Makes Great Strides Cleaning up Wreckage in Wake of Dam Flood: Valley Rising Out of Muck," *Los Angeles Times*, April 6, 1928; and "Order Rising From Chaos," *Los Angeles Times*, April 8, 1928. Completion of the reconstruction work by the AGC contractors is reported in "Flood Area Now Restored: Contractors Close Offices in Santa Paula," *Los Angeles Times*, October 28, 1928.

CHAPTER FIVE

1 "Ventura Asks Full Redress," *Los Angeles Times*, March 16, 1928.

2 "Restoration Not Quarrels," *Los Angeles Times*, April 7, 1928.

3 "Thousands Rush to Aid in Work of Rescue and Relief," *Los Angeles Times*, March 14, 1928.

4 Starting on March 15 and running regularly for the next several weeks, the *Los Angeles Times* published a list of contributors under the headline "'Times' Flood Relief Fund." Organizations and businesses often made contributions of $100 or more; individual contributions were generally smaller, some as little as 50 cents. Over the course of two months, the *Times* acknowledged close to five hundred contributors. William Kerckhoff's $1,000 gift and the $2 "Civil War widow" donation are documented in the *Los Angeles Times*, March 15, 1928. The last recorded contribution ($5 from Mrs. John G. Cook) was acknowledged on May 15, 1928.

5 "Relief Fund Started by Examiner," *Los Angeles Examiner*, March 14, 1928; "Examiner Fund for Relief of Sufferers Growing Fast," *Los Angeles Examiner*, March 15, 1928; and "Examiner Fund for Destitute Tops $10,000," *Los Angeles Examiner*, March 17, 1928. The *Examiner* was a flagship of the Hearst newspaper chain and a rival of the Chandler family's *Los Angeles Times*.

6 "Benefit Aids Flood Fund," *Los Angeles Times*, March 22, 1928; "Theatres Give Flood Aid," *Los Angeles Times*, March 20, 1928. On March 27 Harold B. Franklin, president of the West Coast Theaters, gave a check for $9,802 to D. C. McWatters, chairman of the Los Angeles chapter of the Red Cross; see "Midnight Show Relieves Woe: Thespians' Flood Benefit Liberal," *Los Angeles Times*, March 28, 1928.

7 "San Francisco Stages Flood Show Tonight," *Chicago Tribune,* March 23, 1923; and "Bay City Fight Benefit to Net Dam Sufferers $10,000," *Los Angeles Times,* March 25, 1928.

8 "Benefit Aids Flood Fund," *Los Angeles Times,* March 22, 1928; "Thespians Play Samaritan," *Los Angeles Times,* March 21, 1928; "Mrs. Jones Shoots 74 at Flintridge," *Los Angeles Times,* April 7, 1928; "Of Interest to Women: Society," *Los Angeles Times,* March 22, 1928; and "Horemans and Cochran Meet in Cue Match," *Los Angeles Times,* March 23, 1928. Presumably most of the fund-raising efforts were legitimate, but ruses were reported. "Warning Given on Impostors," *Los Angeles Times,* March 20, 1928, reported: "a hatless man, slovenly dressed, has been canvassing the Wilshire District in an effort to collect funds [for St. Francis victims]. The impostor is described as stockily built... with a pompadour, very thin in front." Also see Harry Carr, "The Lancer: Remorseless Crooks," *Los Angeles Times,* March 20, 1928.

9 D. C. McWatters, chairman, American Red Cross, Los Angeles Chapter, to C. C. Teague, March 22, 1928, St. Francis Dam Disaster Papers, Red Cross and American Legion folder, HM 70397, Huntington Library. Also see "Files Report of Flood Fund: Red Cross Chief Tells Where Money Goes," *Los Angeles Times,* May 4, 1928. As of that date a little over $100,000 had been spent on relief work in the Santa Clara Valley; how the remaining $100,000 in donations was spent is unknown. The biggest categories of expenditures were Sanitation and Cleanup: $16,056; Household Furnishings: $14,420; Clothing: $11,886; Service Relief: $9,489; Equipment for Relief Camps: $4,475; Food: $4,194; and Maintenance of Disaster Sufferers: $3,031.

10 For example, "Dam Breaks: Los Angeles Reservoir Goes, Southland Swept by Flood," *Los Angeles Daily News,* March 13, 1928, reported: "The dynamite theory was given support by a report to deputy sheriffs by a motorist who stated that he was on the ridge road below the dam when he saw a flash that illuminated the heavens." Such a flash was almost certainly caused by sparking transmission lines felled by the flood. Also see "Did Dynamite or Quake Do It? Officials Ask," *Chicago Tribune,* March 14, 1928. See chapter 6 for discussion of the dynamite theory.

11 "As to Responsibility: Mayor and City Attorney Say Los Angeles Will Face Dam Situation in a Broad, Helpful Way," *Los Angeles Times,* March 15, 1928.

12 Ibid. Also see Peirson M. Hall to Manuel P. Servin, October 19, 1964, Peirson M. Hall Collection, Huntington Library.

13 In the week after the flood, A. J. Ford of the Department of Water and Power's Right-of-Way and Land Division was on the ground in the lower Santa Clara Valley helping in rescue and recovery work. Ford reported that on March 16 he met Harvey Van Norman and Stanley Dunham of the BWWS "at the Water Bureau Camp on the eastern outskirts of Santa Paula, where much equipment had already arrived on the scene and a full camp had been organized during the day." See "Report of A. J. Ford after Inspection of Flood Area," entry under "Santa Paula, 5:00 P.M., March 16, 1928," in St. Francis Dam Claims Records, Reports of Flood Damage Areas, A. J. Ford, March 14–September 14, 1928, LADWP Archives.

14 Charles C. Teague, *Fifty Years a Rancher* (Los Angeles: Ward Ritchie Press, 1944), 26–61, 67–96, 165–81. In this memoir, Teague describes his role in the development of Sunkist as well as his work on the Federal Farm Board during the Hoover administration. Teague, an admirer of the president, wrote that before appointment to the Farm Board he had "known Mr. Hoover quite well and we had had many conversations on the problems of agriculture and of the country generally" (168). The political power and cultural status of large landowners such as Teague is documented by Michael R. Belknap in "The Era of the Lemon: A History of Santa Paula, California," *California Historical Society Journal* 47 (June 1968): 113–40; also see Richard G. Lillard, "Agricultural Statesman: Charles C. Teague of Santa Paula," *California History* 65 (March 1986): 2–16.

15 Teague, *Fifty Years a Rancher*, 184–85.

16 "Santa Paulans Hint Law Fight," *Los Angeles Times*, March 15, 1928.

17 "Report of A. J. Ford After Inspection of Flood Area," entry under "Fillmore, 11:15 P.M., March 14, 1928"; LADWP Archives.

18 To defend their rights in the face of this attempted expropriation, in 1925 Santa Clara Valley landowners created the Santa Clara River Protective Association with Teague at the helm. Vernon M. Freeman, in *People—Land—Water: Santa Clara Valley and Oxnard Plain, Ventura County, California* (Los Angeles: Loren L. Morrison, 1968), recounts the history of the association and its evolution into the Santa Clara Water Conservation District (83–130). Freeman gives special attention to the water rights dispute with Los Angeles (90–92), but almost completely ignores the St. Francis Dam disaster in his otherwise detailed and informative book. See chapter 3 for more on this water rights dispute.

19 "Santa Paulans Hint Law Fight," *Los Angeles Times*, March 15, 1928.

20 "Ventura Asks Full Redress," *Los Angeles Times*, March 16, 1928.

21 Peirson M. Hall to Manuel P. Servin, October 19, 1964, Peirson M. Hall Collection, Huntington Library; Teague, *Fifty Years a Rancher*, 185.

22 Hall to Servín, October 19, 1964.

23 Ibid.

24 *California Historical Society Quarterly* 44 (March 1965): 63.

25 "City Ready to Pay Losses," *Los Angeles Times*, March 20, 1928.

26 *Santa Paula Chronicle*, March 19, 1928.

27 The disbursement process is described in B. S. Grant and J. E. Phillips, "Report of Restoration Work Performed by the City of Los Angeles in the Santa Clara River Valley Following the St. Francis Dam Flood of March 13, 1928," October 22, 1928, 14–16; in Aqueduct Division Reference Library Collection, Dams and Reservoirs, St. Francis Dam Collapse, June–October 1928, WP14–11:10, LADWP Archives.

28 "City Ready to Pay Losses," *Los Angeles Times*, March 20, 1928; William A. Lindauer, "City of Los Angeles General Accounting Report on St. Francis Dam Disaster" (October 31, 1930), 2, St. Francis Dam Claims Records, LADWP Archives. The city eventually appropriated $2,135,000 from the Har-

bor Department and the Bureau of Power and Light. In addition, the city sold bonds to finance reparations and by October 31, 1930, had raised $5,793,350 in six issues. The mechanics of financing claim settlements and cleanup expenses were the subject of extended political wrangling. At first it appeared that most costs would be covered by an increase in city water rates, but this prompted widespread resentment, and eventually the City Council decided that expenditures for flood damage would be covered by revenue bonds constituting a general tax obligation. In the revenue bond plan all city taxpayers, not just water users, shared in the financial burden. See "Water Body Asks Loan," *Los Angeles Times,* March 30, 1928; "City Considers Flood Finances," *Los Angeles Times,* April 17, 1928; "Rise in Water Rates Opposed," *Los Angeles Times,* April 18, 1928; "Hearing Fixed on Water Rate," *Los Angeles Times,* April 19, 1928; "Rates and Reparations," *Los Angeles Times,* April 20, 1928; "Unfair to Water Users," *Los Angeles Times,* April 21, 1928; "Protests Hold Up Action on Water Rate Advance," *Los Angeles Times,* April 22, 1928; "Water-Rate Rise Delay Announced," *Los Angeles Times,* April 24, 1928; "The Mayor's Plan," *Los Angeles Times,* May 2, 1928; "Power Bureau Cash Allotted: Another $1,000,000 Added to Restore Valley," *Los Angeles Times,* May 4, 1928; "Cryer Signs Power Loan to Aid Dam: Ordinance Giving Million to Rehabilitation Work Given Approval," *Los Angeles Times,* May 5, 1928; "Financing Plan in Disaster Due," *Los Angeles Times,* May 14, 1928; "Bonds Will Pay Flood Damages," *Los Angeles Times,* June 1, 1928; "Tax Rise near Two Million," *Los Angeles Times,* August 30, 1928; "Suit to Block Bonds Filed," *Los Angeles Times,* October 12, 1928; "Dam Relief Bonds Upheld," *Los Angeles Times,* December 1, 1929; "City Promised Tax Reduction," *Los Angeles Times,* May 25, 1929; "Bids on Bonds to be Opened," *Los Angeles Times,* July 23, 1929.

29 "Los Angeles and Santa Paula Confer on Flood Losses," *Los Angeles Times,* March 22, 1928.

30 Ibid. For an overview of the claims and reparations process, see "Restoration of the Santa Clara River Valley by the City of Los Angeles Following the Flood of March 13, 1928, Caused by the Destruction of St. Francis Dam" (July 31, 1929), 10–21, file WP19–11:11, Aqueduct Division, Reference Library Collection, Dams and Reservoirs, St. Francis Dam Collapse, 1929–1930; and Lindauer, "Accounting Report on St. Francis Dam Disaster."

31 Lindauer, "Accounting Report on St. Francis Dam Disaster," 3. The deadline of September 12 for claim submission was in accord with the Los Angeles City Charter, which stipulated a six-month limit for claims made against the city. Those seeking damages through the courts were not constrained by the deadline.

32 "Santa Paula and City Commissioners Meet Today to Form Reparations Plans," *Los Angeles Times,* March 21, 1928. Alexander Heron, chairman of the State Board of Control, played an important role in setting up the joint committees.

33 "Santa Paula and City Commissioners Meet Today to Form Reparations Plans," *Los Angeles Times,* March 21, 1928. See Outland, *Man-Made Disaster,* 232–33, for names of those who served on the committee.

34 "Santa Paula Office of Red Cross Closed," *Los Angeles Times*, June 1, 1928. J. W. Richardson announced on May 31: "so nearly complete is the American Red Cross St. Francis Dam disaster relief work here that the local office will close tonight." He also confirmed that the Newhall office had closed on April 1 and the Fillmore office on May 18.

35 "City Ready to Pay Losses," *Los Angeles Times*, March 20, 1928; Teague, *Fifty Years a Rancher*, 188–89; George B. Travis, "St. Francis Dam Disaster and Restoration Program" (August 22, 1929), 25, Charles C. Teague Papers, Bancroft Library, UC Berkeley.

36 "Santa Paula and City Commissioners Meet Today to Form Reparations Plans," *Los Angeles Times*, March 21, 1928. See also Jess Stephens to Los Angeles City Council, ca. June 30, 1929, in Jess Stephens and Lucius Green to J. E. Phillips, July 29, 1929, file WP19–11:11, LADWP Archives.

37 Teague, *Fifty Years a Rancher*, 189–90.

38 Ibid., 189–93. In a later report written by Teague's assistant George Travis, selection of the Extension Service was portrayed as a major event in the reparations process: "The Ventura County committeemen were certain a group of Los Angeles businessmen could not get a scientific picture of the damage; that it was too involved; and that further, gaining confidence of the land owners in the good judgment of Los Angeles committeemen would be well-nigh impossible. They held that the [Extension Service's] Farm Advisors were the one group in California equipped by education and experience, both to find the facts and to satisfy the farmers that the findings were both accurate and impartial. So strongly did the Ventura County committeemen hold to this opinion, that most of them were ready to turn the whole matter over to the courts if Los Angeles refused to use the Farm Advisors in this important capacity." See Travis, "St. Francis Dam Disaster and Restoration Program."

39 Teague, *Fifty Years a Rancher*, 193.

40 C. C. Teague press statement, March 24, 1928, as quoted in Travis, "St. Francis Dam Disaster and Restoration Program."

41 Teague, *Fifty Years a Rancher*, 194. The farm advisers' assessments focused on physical damage that was the immediate result of the flood. But many farmers in the valley became worried that sand and rock debris had raised the elevation of the river's natural floodplain and that, in the future, their lands would be subject to inundation during heavy storms. The city and the valley farmers discussed strategies for mitigating such possible damage for several months; by spring 1929 the city had agreed to build levees and flood control structures within the Santa Clara River basin to protect agricultural land. See "Flood Ruin Greatest in History," *Los Angeles Times*, April 19, 1929: "the agreement reached between the City of Los Angeles and the Ventura County committee has resulted from...a careful survey of the entire river area...It provides for certain jetty and [levee] work and extensive river protection work...[It] will cost $300,000 and cover an area about nineteen miles long...County Engineer Charles Petit and [Harvey] Van Norman of Los Angeles agreed on the actual work to be done." Also see "River Survey Gets Priority," *Los Angeles Times*, October 5, 1928. The "Increased River Hazard" issue is also discussed in Travis,

"St. Francis Dam Disaster and Restoration Program"; also see George L. Eastman to C. C. Teague, March 31, 1928, River Hazard Negotiations and Work Folder, HM 70405, St. Francis Dam Disaster Papers, Huntington Library.

42 Teague, *Fifty Years a Rancher,* 186.

43 Ibid., 186–87. Other Ventura County civic leaders also aggressively opposed access by lawyers to flood victims. A week after the disaster, Ventura County District Attorney James Hollingsworth took up the attack and admonished anyone planning to seek redress outside the process administered by the Joint Restoration Committee: "I cannot use language too strong to condemn the tactics of such scavengers. It is of the utmost importance that the entire damage claim be settled through the constituted parties in this county, who have been appointed by the Ventura County board of supervisors...My office will appreciate information giving the names of the attorneys who are seeking to profit by the distress of the sufferers of the flood." Quote from "Suspend All from Bar is Threat Here: Rough Traveling for Attorneys Seen by Hollingsworth," *Ventura Free Press,* March 20, 1928.

44 "Restoration Not Quarrels," *Los Angeles Times,* April 7, 1928.

45 "Report to Jess E. Stephens by E. B. Mayer of Expenditures and Detailed Cost: Restoration Work, Santa Clara Valley, 'St. Francis Dam Claim Fund,'" June 20, 1929, 1–2, St. Francis Dam Claims Records, LADWP Archives; "Restoration of the Santa Clara River Valley" (July 31, 1929), 2–4.

46 Lindauer, "Accounting Report on St. Francis Dam Disaster," 1; Los Angeles Board of Water and Power Commissioners, "Twenty-Seventh Annual Report" (1928), 47.

47 Lindauer, "Accounting Report on St. Francis Dam Disaster," 2; "Restoration of the Santa Clara River Valley" (July 31, 1929), 2–6; "Report to Jess E. Stephens by E. B. Mayer," 1–2; Stephens to Los Angeles City Council, ca. June 30, 1929; "Santa Paula and City Commissioners Meet Today to Form Reparations Plans," *Los Angeles Times,* March 21, 1928. The AGC's 6 percent cost plus contract is noted in "Dynamite Dam Theory Backed," *Los Angeles Times,* April 7, 1928. The AGC contractors based in Santa Paula were E. A. Irish, C. U. Hueser, and Wells & Bressler; the contractors in Fillmore were Thomas Haverty, Robinson-Roberts, and K. S. Littlejohn; the contractors in Bardsdale were Hall-Johnson and George Mitchell; and the contractor in Piru was P. L. Ferry. All of the AGC contractors were under the supervision of general director C. E. Bressler. See Grant and Phillips, "Report of Restoration Work," 24; LADWP Archives.

48 Stephens to Los Angeles City Council, ca. June 30, 1929, 2–3.

49 Grant and Phillips, "Report of Restoration Work," 5.

50 Ibid.

51 Lindauer, "Accounting Report on St. Francis Dam Disaster," 3A, St. Francis Dam Claim Fund, items A, B, D, F, M, TS, G, and RP. The authors offer these numbers as a reasonable approximation of the costs of the cleanup to the city. While cleanup expenses are extensively documented, inconsistencies in the data make it impossible to calculate a precise figure.

52 The settlement paid to the Camulos Ranch Corporation is noted in "Claim Approved on Damage by Flood Disaster," *Los Angeles Times*, November 5, 1928. In September 1928, the Newhall Land and Farming Company submitted a damage claim totaling $1,367,911.93; see "St. Francis Dam Claim and Demand for Damages of the Newhall Land and Farming Company filed or expected to be filed September 1, 1928 to the City of Los Angeles and Board of Water and Power Commissioners of the City of Los Angeles," folders 195, 196, Newhall Land and Farming Company Collection, Huntington Library. A few days after the original filing, a "supplement and amendment" was filed that added $7,426.48, bringing the claim total to $1,367,911.93. Although the figure was "based upon the Farm Advisors Reports made pursuant to the plans adopted by the Citizens' Restoration Committee authorized by the City of Los Angeles," this did not preclude a final settlement for a significantly smaller sum. In 1930 the Newhall Company accepted a settlement of $737,040, a payment that proved vital to the ranch's survival during the Great Depression. See Ruth Waldo Newhall, *A California Legend: The Newhall Land and Farming Company* (Valencia, Calif.: Newhall Land and Farming Company, 1992), 99–101, 116; James F. Dickason, *The Newhall Land and Farming Company: Unlocking the Productivity of the Land* (New York: Newcomen Society of the United States, 1983), 15; and Lindauer, "Accounting Report on St. Francis Dam Disaster," 16.

53 Stephens to Los Angeles City Council, ca. June 30, 1929, 3–4.

54 Joint Architectural Committee, "Report on Building Restorations" (November 1, 1928), 8), file 627.8, Museum of Ventura County.

55 Lindauer, "Accounting Report on St. Francis Dam Disaster," 3A; Outland, *Man-Made Disaster*, 182. A report prepared in November 1928 by the Joint Architectural Committee (cited above) differs in some statistical findings from the report prepared two years later by Los Angeles Chief Accountant Lindauer.

56 Lindauer, "Accounting Report on St. Francis Dam Disaster," 16. Soon after the flood, damage to Southern Pacific's facilities, bridges, trackage, and right-of-way was estimated at $200,000, not including lost revenue resulting from "loss of traffic." See "Southern Pacific Earnings Improve," *Wall Street Journal*, March 21, 1928, for the estimate of the company's flood damage.

57 Travis, "St. Francis Dam Disaster and Restoration Program," 16.

58 Teague, *Fifty Years a Rancher*, 188; *Los Angeles Times*, March 15, 1928; "Restoration of the Santa Clara River Valley" (July 31, 1929), 12.

59 *Calif. Statutes* (1913), 176. The Workmen's Compensation, Insurance, and Safety Act of 1913 was frequently referred to as the Boynton Act. On the history of workers' compensation in California, see Glenn Merrill Shor, "The Evolution of Workers' Compensation Policy in California, 1911–1990" (PhD diss., University of California, Berkeley, 1990).

60 "First Claim in Flood Filed," *Los Angeles Times*, March 21, 1928. There is no record that Pike's parents received compensation for his death, so presumably he had not been providing financial support. Pike's wife and child also died, so neither of them could file a death claim. If a deceased employee left no dependents, or if they all also died in the flood, no valid claims could be made.

61 In "Awards Made for Lives Lost in Flood," *Los Angeles Times,* May 14, 1928, it is explained that "the law provides that the death benefit under the Workmen's Compensation Act in cases of partial dependency shall be three times the amount devoted by the deceased to the support of the partial dependents."

62 "Dam Death Awards Aid Dependents," *Los Angeles Times,* May 26, 1928.

63 "Four More Get Benefits," *Los Angeles Times,* May 23, 1928.

64 "Claim No. 1801, St. Francis Dam Claim and Demand for Payment—Southern California Edison Company," September 11, 1928; St. Francis Dam Claim Records, LADWP Archives. Southern California Edison paid compensation to claimants and the company was reimbursed by the city of Los Angeles. Prior to September 1928, thirty claims for deceased Edison employees were approved. Three were denied, presumably because the claimants could not establish that they were financially dependent on the deceased employee.

65 "Awards Made for Lives Lost in Flood," *Los Angeles Times,* May 14, 1928.

66 "Restoration of the Santa Clara River Valley" (July 31, 1929), 12, 13; Stephens to Los Angeles City Council, ca. June 30, 1929, 5; "Citizens' Restoration Committee, Subcommittee on Death and Disability Claims" (July 15, 1929), 6–7, file 979.49, Museum of Ventura County.

67 See Henry F. Johnson Death Claim for Leona Johnson, file #2538, St. Francis Dam Claims Records, LADWP Archives.

68 Ibid. (Memorandum of meeting with attorney for Henry Johnson, filed October 29, 1928).

69 Ibid. (Report of Death Claim of Leona Johnson.) Statements by the eleven informants were taken on October 26, 27, and 29, 1928.

70 Ibid. (Alys Schauss, Memorandum [of meeting with Johnson's attorney Rankin], November 14, 1928).

71 Ibid. (Alys Schauss, Memorandum, December 18, 1928)

72 Ibid. (Statement of Settlement, April 23, 1928)

73 Gladys Harnischfeger Death Claim file #2285, St. Francis Dam Claims Records, LADWP Archives.

74 Ibid. (Memorandum, June 21, 1928).

75 Ibid. (Release and Receipt File #2285, Gladys Harnischfeger, October 1, 1928).

76 "Citizens' Restoration Committee, Subcommittee on Death and Disability Claims" (July 15, 1929), 6–7, 11–17, Museum of Ventura County.

77 George Travis to C. C. Teague, June 27, 1928, Death Claims Committee folder, HM 70406, St. Francis Dam Disaster Papers, Huntington Library.

78 See "List of Death Claims Settled Up to June 30th, 1928," attached to Richard Bard to George O. Eastman, June 30, 1928, St. Francis Dam Disaster Papers, Death Claims Committee folder, HM 70406, Huntington Library.

79 See A. J. Ford, "Notice of Liens Filed by Edward P. Garrett," July 11, 1928, memo attached to Jas. P. Vroman to W. B. Mathews, July 11, 1928, St. Francis Dam Claims Records, Miscellaneous Claims Settlements Correspondence, April 1928–February 1930, LADWP Archives. A. J. Ford's memo lists, along with

Lillian Curtis and Ray Rising, Jesus Torres, Juan Carrillo, H. H. Kelley, Ben V. Dornaleche, Anastacio Ramos, Pedro Samaniego, Carlos Samaniego, Andre Garcia, Aaron Coffer, Ynez Parada, and Rifino Samaniego as clients of Garrett. "Suits in Flood Deaths Filed: Total of $1,690,000 Asked in Series of Actions: Thirty-seven Complaints on List at Bakersfield," *Los Angeles Times*, August 17, 1928. Other claims filed under the auspices of "Attorney Galvin" and "Mr. Higley, attorney-at-law" are referenced in J. S. Mann to Lucius Green, September 7, 1928, Death Claims Committee folder, HM 70406, St. Francis Dam Disaster Papers, Huntington Library.

80 Claim of Ray E. Rising, Claim #2259, file L-1, D-47, file 235, L-1, D-47, St. Francis Dam Claims Records, Record of Claims Index, LADWP Archives.

81 "Citizens' Restoration Committee, Subcommittee on Death and Disability Claims," 6, 8, 11.

82 "BAR Plans Clean-Up: Unethical Acts Condemned," *Los Angeles Times*, August 3, 1928. Also see "Attorneys Here Face Accusation," *Los Angeles Times*, July 19, 1928, and "Discourage Litigation of Flood Claims," *Ventura County News*, July 13, 1928, in which a spokesman for the Subcommittee on Death and Disability Claims warned against "persons who have approached you giving promises and guarantees of large settlements if you will put your case in their hands. Their motive is the enrichment of themselves by reason of your misfortunes."

83 "Suits in Flood Deaths Filed."

84 "City Asks for Change of Courts: Los Angeles Petitions that St. Francis Damage Suits Be Removed from Kern," *Los Angeles Times*, September 25, 1928: "the cases, in so far as the motion for a change of venue is concerned, are consolidated. Lawrence Edward, as attorney for the plaintiffs, is opposing the motion and Frederick Von Schader and Kenneth K. Scott, deputy city attorneys for Los Angeles, are urging the motion."

85 After its success in forcing all claimants to abandon Bakersfield as a court venue, the city drew a hard line in negotiating claims. In April 1929 the city announced that it had denied seventeen claims (twelve for deaths, five for injuries) because "they were regarded as unreasonably high...The claimants whose demands were turned down...will have to institute court actions, unless they change their minds and accept the compromises offered." The city considered "excessive" three death claims for $100,000 each and personal injury claims for $100,000 and $50,000 each. See "City Denies Seventeen Dam Claims," *Los Angeles Times*, April 4, 1929.

86 *Ray E. Rising v. City of Los Angeles, Department of Water and Power of the City of Los Angeles et al.: Complaint in Action for Damages* (Case No. 273626), March 13, 1929, St. Francis Dam Claims Records, Claim #2259 (Rising, Ray E.), box 209174, LADWP Archives; *Ray E. Rising v. City of Los Angeles, Department of Water and Power of the City of Los Angeles et al.: Notice of Intention to Move for a New Trial* (Case Nos. 273625, 273626, 273627, 273628), June 14, 1930, LADWP Archives.

87 *Ray E. Rising v. City of Los Angeles, Department of Water and Power of the City of Los Angeles et al.: Complaint in Action for Damages*, 2.

88 "Verdict of Coroner's Jury," Coroner's Inquest, 2.

89 *Ray E. Rising v. City of Los Angeles, Department of Water and Power of the City of Los Angeles et al.: Answer* (Case No. 273625), April 30, 1929, 2–3, LADWP Archives.

90 "Dam Disaster Suit on Trial," *Los Angeles Times*, May 23, 1930,

91 "Rock Dissolved as Jury Watches Breathlessly," *Los Angeles Times*, May 27, 1930. When a conglomerate rock sample that plaintiff attorney Edwards asserted was taken from the dam's west abutment dissolved in a glass of water, Mulholland steadfastly denied that the sample was taken from the dam site. For more on the "dissolving red conglomerate," see chapter 6.

92 "Dam Break Blame Put on Faults," *Los Angeles Times*, May 29, 1930.

93 "Broken Dam's Roar Echoes: St. Francis Dam Disaster Survivor Describes Flood at Trial of $175,000 Suit Against City," *Los Angeles Times*, May 26, 1930.

94 The Rising civil trial verdict was reported in "Flood Damage being Paid Off," *Los Angeles Times*, June 6, 1930.

95 "To: Ray E. Rising and his attorneys, Lawrence Edwards and Len H. Honey, attorneys for plaintiff, Stockton, California: Notice of Intention to Move for a New Trial," June 14, 1930; Jess Stephens and Lucius Green to Erwin Werner, August 8, 1930; Erwin P. Werner to Jess Stephens and Lucius Green, June 24, 1930; Erwin P. Werner to Jess Stephens and Lucius Green, August 11, 1930; all in St. Francis Dam Claims Records, Claim #2259 (Ray E. Rising), box 209174, LADWP Archives.

96 Ray E. Rising to Lucius Green, September 11, 1930, St. Francis Dam Claims Records, Claim #2259 (Ray E. Rising), LADWP Archives; *Ray E. Rising v. City of Los Angeles, Department of Water and Power of the City of Los Angeles et al.: Satisfaction of Judgment* (Case Nos. 273625, 273627, 273628), December 20, 1930.

97 *Los Angeles Herald-Examiner*, March 10, 1963.

98 "Claim of Lillian L. Curtis, Claim #1757," St. Francis Dam Claims, Record of Claims Index, LADWP Archives; "Citizens' Restoration Committee, Subcommittee on Death and Disability Claims," 20.

99 "Citizens' Restoration Committee, Subcommittee on Death and Disability Claims," 6, 8; "Claim of Lillian L. Curtis, Claim 1757," file 68, L-1, LADWP Archives; *Los Angeles Times*, May 14, 1928.

100 "Widow Asks $260,000 for Flood Loss," *Los Angeles Times*, August 19, 1928.

101 "Claim of Lillian L. Curtis, #1757," file 68, L-1, LADWP Archives; "Suits in Flood Deaths Filed: Total of $1,690,000 Asked in Series of Actions"; "City Asks for Change of Courts: Los Angeles Petitions that St. Francis Damage Suits be Removed from Kern," *Los Angeles Times*, September 25, 1928.

102 "Claim of Lillian L. Curtis, #1757," St. Francis Dam Claims Records, Record of Claims Index; "Resolution," Claim #1757, Individual Claims File, St. Francis

Dam Claims Records, June 1928–March 1931, LADWP Archives; Lawrence Edwards to Jess Stephens and Lucius Green, June 23, 1930, St. Francis Dam Claims Records, Letters to Board of Water and Power Commissioners Authorizing Payment of Claims, November 1930–April 1933, LADWP Archives; Stephens and Green to Edwards, June 26, 1930, St. Francis Dam Claims Records; Edwards to Stephens and Green, August 16, 1930, St. Francis Dam Claims Records.

103 "Claim of Lillian L. Curtis, #1757," "Resolution"; Lillian L. Curtis to Lucius Green, November 12, 1930, Individual Claims File, St. Francis Dam Claims Records, June 1928–March 1931.

104 Marianne Tyler, "The Day the Dam Broke," *Los Angeles Herald-Examiner*, February 9, 1978.

105 A related question arises: how much money did the city pay Jess Stephens and Lucius Green for their work on the Rising case? When Stephens retired as city attorney in April 1929, the *Los Angeles Times* noted that "many of our ablest men, after giving years of faithful service in public posts, often inadequately paid, must turn, as Stephens is turning, to other fields of activity to acquire a modest surplus for their later years. And yet it may be that he will not entirely sever his relations with the big city, in keeping with the line that he has laid out for himself in the future." As predicted, Stephens did not "sever his relations" and soon, as a private attorney, he was hired to handle the Rising case. Stephens' and Green's compensation for defending the city against the Rising lawsuit is unknown, but they may have received as much for their work as private counsel as Rising won in damages; see "Jess Stephens: Lawyer–Man," *Los Angeles Times*, April 2, 1929.

106 "Report of A. J. Ford After Inspection of Flood Area," 4.

107 Travis, "St. Francis Dam Disaster and Restoration Program," 12.

108 "Citizens' Restoration Committee, Subcommittee on Death and Disability Claims," 11, 18–20. This analysis of injury claims rests on the raw data presented in this source rather than on its sometimes flawed summaries. As of July 1929, only four claims for personal injuries remained unsettled.

109 "Santa Paulans Bow in Memory," *Los Angeles Times*, March 13, 1929.

110 Describing the purpose of the memorial dinner, George Bond, president of the Southwest Improvement Club, said: "The banquet will be in the way of a thanksgiving. We are going to show that we are thankful that things are again back to normal and that Los Angeles has paid her debt." The club's membership was "composed chiefly of flood sufferers." See "Memorial Banquet Planned," *Los Angeles Times*, March 6, 1929.

111 Teague, *Fifty Years a Rancher*, 183.

112 Lindauer, "Accounting Report on St. Francis Dam Disaster," 1–3A.

113 The estimated value of the water lost in the flood and the cost incurred by the Bureau of Power and Light to buy replacement electricity are provided in "Memo to H. A. Van Norman: Summary Showing Present Status St. Francis Flood Claims and Other Expenses," June 26, 1929; this memo was prepared in response to Jason P. Spoman to A. J. Ford, June 24, 1929; both in St. Francis

Dam Claims Records, A. J. Ford Reports to Board re Claims, September 1928–June 1929; LADWP Archives. A valuation of $1.25 million for 38,000 acre-feet of water was in fact rather conservative: in 1928, domestic consumers were charged 13 cents for 100 cubic feet of water, and there are 43,560 cubic feet in an acre-foot, so they paid $56.50 per acre-foot. The retail value of a full reservoir, 38,000 acre-feet, would therefore be about $2.15 million, if all the water was sold to domestic consumers (none to industrial or agricultural customers). Domestic water rates at the time of the flood are given in "Los Angeles Water Rate Increase," *Wall Street Journal,* April 19, 1928.

114 Upon Stephens' retirement as city attorney in the spring of 1929, the *Los Angeles Times* published a glowing tribute highlighting "his work in a grave and weighty matter...the matter of adjusting the great loss of life and property that followed the breaking of the St. Francis Dam." See "Jess Stephens: Lawyer–Man," *Los Angeles Times,* April 2, 1929.

115 John M. Barry, *Rising Tide: The Great Mississippi River Flood of 1927 and How it Changed America* (New York: Simon & Schuster, 1997), 235–58.

CHAPTER SIX

1 William Mulholland testimony, Coroner's Inquest, 16.

2 Telegram from Governor C. C. Young to Representative Phil Swing, March 27, 1928, St. Francis Dam file, Division of Safety of Dams, Sacramento, California.

3 Outland, in *Man-Made Disaster,* describes how an unidentified woman in the Santa Clara Valley placed the sign in her front yard (167); the source is given as the *Santa Paula Review,* March 15, 1928. The authors of *Heavy Ground* have no reason to question the validity of this story but have not been able to locate the article or confirmation in any other sources.

4 "California Engineer to Probe Dam Collapse," *Los Angeles Herald,* March 13, 1928. According to this account, State Engineer Hyatt had stated before leaving Sacramento for the dam site, "I inspected the dam myself during construction." Hyatt did not become state engineer until September 1927, more than a year after construction was completed. There is no evidence that he ever visited the dam site prior to the disaster.

5 See notes 21–25, below.

6 "Governor Young in Flooded District," *Los Angeles Herald,* March 13, 1928. Young flew to Long Beach from San Diego and was then driven to Santa Paula.

7 See David P. Billington and Donald C. Jackson, *Big Dams of the New Deal Era: A Confluence of Engineering and Politics* (Norman: University of Oklahoma Press, 2006), for an overview of the Central Valley Project and the Shasta and Friant dams (243–82).

8 See Norris Hundley jr., *Water and the West: The Colorado River Compact and the Politics of Water in the American West* (Berkeley: University of California Press, 1975); and Billington and Jackson, *Big Dams of the New Deal Era,* 102–51. The origins of Hoover Dam can be traced to the Imperial Valley in Southern

California, when the California Development Company began drawing water from the Colorado River in 1900. Heavy floods in 1905–7 caused extensive damage in the valley (creating what is today known as the Salton Sea). The private company went bankrupt, and in 1911 the Imperial Irrigation District took control of the valley's irrigation infrastructure. The district soon began promoting construction of a large flood control and storage dam in the lower Colorado River basin. After World War I the district became allied with the U.S. Reclamation Service and, under the political leadership of Congressman Phil Swing, initiated a serious effort to win congressional support for what became the Boulder Canyon Project Act.

9 Outlines of early planning for the Colorado River Aqueduct are available in "Colorado River–Los Angeles Aqueduct Line Surveyed," *Engineering News-Record* 93 (August 14, 1924): 277; and "Los Angeles Plans 268-Mi. Aqueduct From Colorado Aqueduct," *Engineering News-Record* 94 (April 2, 1925): 560. The latter article notes that "considerable data have been collected under the direction of William Mulholland" and estimates "a net lift (exclusive of the friction head) of 1,416 feet will be necessary" to pump water through the aqueduct. A description of what became the Colorado River Aqueduct was published in *Engineering News-Record* in May 1928, a mere two months after the St. Francis Dam collapse. The article's introduction explains: "The water situation in Los Angeles for approximately 50 years has been studied closely and continually by William Mulholland, chief engineer and general manager, Bureau of Water Works and Supply, and when he pointed out the need for a new source, a bond issue of $2,000,000 was promptly voted in 1925 for investigations and surveys. With these funds...extensive surveys and preliminary plans for the Colorado Aqueduct have been carried out." See H. A. Van Norman and E. A. Bayley, "Colorado River–Los Angeles Aqueduct Project," *Engineering News-Record* 100 (May 31, 1928): 850–54. The final design and right-of-way is described in F. E. Weymouth, "A Great Job Looms in the Colorado Aqueduct Project," *Engineering News-Record* 108 (June 16, 1932): 847–50.

10 Between the initiation of the Boulder Canyon Project as a legislative proposal in the early 1920s and authorization (by Congress and President Coolidge) in December 1928, the city of Los Angeles and other Southern California communities assumed an essential role in promoting the project. As early as July 1921, Los Angeles had expressed interest in helping build Boulder Canyon Dam in return for control over the hydroelectric power plant; see Beverly Moeller, *Phil Swing and Boulder Dam* (Berkeley: University of California Press, 1973), 24). By 1924 this interest had expanded into a formal water claim for 1,500 cubic feet per second of Colorado River flow. With this claim, the City of Los Angeles served as the catalyst for the Colorado River Aqueduct and for what evolved into the Metropolitan Water District of Southern California (MWD). In terms of Congressional approval for the project, the MWD proved vitally important because it would constitute a major customer for hydroelectric power generated at the dam—and thus help ensure that the federal treasury would be reimbursed for construction costs; see Bissell, *The Metropolitan Water District of Southern California: History and First Annual Report* (Los Angeles: Metropolitan Water District of Southern California,

1939), 38–39. When fifty-year leases governing use of Boulder Dam power were authorized in 1930, 64 percent of the dam's power was reserved for use in Southern California: 36 percent went to the Metropolitan Water District, about 9 percent to the Southern California Edison Company and other private power companies, and about 18 percent to the city of Los Angeles and other municipally owned utilities. Arizona and Nevada were each allotted 18 percent of the dam's power; see Kleinsorge, *Boulder Canyon Project* (Stanford, Calif.: Stanford University Press, 1941), 138–66. Also see "How Boulder Dam Will Refinance Colorado River Project: Statement to Colorado River Commission Regarding Present and Probable Future Power Demands, Power Supply and Cost per K.W.H. for Southern California by Los Angeles Bureau of Power and Light," October 16, 1928; box 407, Southern California Edison Collection, Huntington Library.

11 Catherine Mulholland, *William Mulholland*, 171, 179.

12 Ibid., 280, 283–84.

13 Bissell, *Metropolitan Water District of Southern California*, 36.

14 Catherine Mulholland, *William Mulholland*, 301–5.

15 Ibid., 316–17.

16 At his inaugural Young intoned: "The prospects are very bright that the Congress at its present session will furnish the needed relief for the south by passing the bill for the dam at Boulder Canyon. California will certainly do all she can toward this end by making clear her attitude through representatives of this administration in Washington...I feel assured that this Legislature will also meet the acute need of the south for an adequate domestic water supply by authorizing the formation of a metropolitan water district such as may permanently solve her difficulties along this line." See "The Governors' Gallery," California State Library website, http://governors.library.ca.gov/addresses/26-young.html.

17 "Southland Needs Dam, Says Young," *Los Angeles Times*, March 13, 1928.

18 "Boulder Dam Favorably Reported to Congress," *Sacramento Bee*, March 15, 1928.

19 J. D. Galloway to Chester H. Rowell, June 4, 1928; and Galloway to Rowell, June 6, 1928; both letters in J. D. Galloway Papers, folder 85, WRCA, UC Riverside.

20 The day the dam collapsed, a preliminary hearing was underway in Independence for "six men charged with criminal conspiracy and with the dynamiting" of the aqueduct at Cottonwood Creek; see "Dynamiting Described, Witness Tells of His Deed," *Los Angeles Times*, March 13, 1928.

21 "100 Believed Dead in Aqueduct Dam Break," *Los Angeles Examiner*, March 13, 1928.

22 "Dam Breaks: L.A. Reservoir Goes, Southland Swept by Flood," *Los Angeles Daily News*, March 13, 1928; "Big Dam Flooded Kills 170, 592 Missing, 80 Injured," *Los Angeles Herald*, March 14, 1928. Farther afield, the *Chicago Tribune* trumpeted dynamite or an earthquake as the cause; see "Did Dynamite or Quake Do It Officials Ask," *Chicago Tribune*, March 14, 1928.

23 "Mulholland Lays Break to Ground Shift," *Los Angeles Examiner*, March 14, 1928.

24 "Inspection of St. Francis Dam, March 19, 1928," memo by Arthur P. Davis, attached to Arthur P. Davis to John R. Freeman, March 28, 1928, box 54, John R. Freeman Papers, MIT.

25 The rope-map-dynamite theory is described in "Dynamiting Theory as Map of Dam, Rope Found," *Los Angeles Herald*, March 20, 1928; "St. Francis Dam Blown Up, Charge as Quiz Opened," *Los Angeles Express*, March 20, 1928; "Evidence Pointing to Blast is Found, is Claim of Probers," *Los Angeles Express*, March 20, 1928; "Dam Blasting Theory Strengthened," *Los Angeles Examiner*, March 21, 1928; and "Dynamiting Inquiry is Demanded," *Los Angeles Herald*, March 28, 1928. Also see "Mulholland on Stand at Coroner's Inquiry Defends Work and Hints at Dynamiting," *Los Angeles Times*, March 22, 1928.

26 "Denies Dam Dynamited," *Los Angeles Record*, March 24, 1928. Ranchers in the Owens Valley also vehemently denied that their protests against the aqueduct played any role in the St. Francis tragedy. "Owens Valley Resents Slurs," *Los Angeles Times*, March 24, 1928, reports on the Bishop Chamber of Commerce's spirited rejection of allegations that representatives of Owens Valley interests had attacked the dam.

27 "Red Herring!" *Los Angeles Record*, March 22, 1928. The discovery of a box of dynamite in the flood zone near Fillmore incited speculation, but the box was traced back to the Southern California Edison camp at Kemp, which maintained a supply of dynamite to aid in foundation blasting for power line construction. See "Dynamite in Valley Traced, Body Found," *Los Angeles Examiner*, March 24, 1928, where it was reported that Edison's stash of explosives "was scattered when the torrent swept away the encampment." It should also be noted that in the week before the collapse (around March 8), a small amount of dynamite had been used in construction of a new road leading up to the top of the west side abutment. At the Coroner's Inquest BWWS construction superintendent J. H. Bouey testified that a small amount of blasting was done about 300 feet downstream from the dam. Small holes were drilled to a depth of about four feet into the red conglomerate formation and each filled with a stick and a half of dynamite; asked why he relied on blasting to loosen the rock, Bouey replied: "Did you ever try to move any of that stuff with a grader? It is pretty near impossible to move it without shooting it." No meaningful relationship between this minor blasting and the failure of the dam was ever argued. See J. E. Bouey testimony, Coroner's Inquest, 441–43.

28 For example, former City Council member, and later judge, Peirson Hall stated in the 1960s that he had long believed the dam had collapsed because of a dynamite attack. See Outland, *Man-Made Disaster*, 216, and also Peirson's review of the first edition of *Man-Made Disaster*, published in *California Historical Society Journal* 45 (March 1965): 61–64, available in Peirson Hall Papers, box 10, Huntington Library. Hall expressed a belief that the dam was dynamited, and that the perpetrators "perished in the flood." In her book, *Rivers in the Desert: William Mulholland and the Inventing of Los Angeles* (New York: HarperCollins, 1993), 198–205, Margaret Leslie Davis appears to embrace the theory that a dynamite attack brought down the dam.

29 "Cal. Engineer to Probe Dam Collapse," *Los Angeles Herald,* March 13, 1928.

30 "Anchoring Weak in Big Dam, Claim," *Los Angeles Herald,* March 15, 1928. In "State Orders Dam Quiz," *Los Angeles Examiner,* March 16, 1928, Hyatt is quoted: "The right abutment of the dam was anchored in red conglomerate rock which should not have been used as a foundation for a dam of this size without a deep cut-off wall, which was not included in the St. Francis project." Also see "State Engineer Blames City for Disaster," *Los Angeles Examiner,* March 16, 1928.

31 Edward Hyatt to R. F. del Valle, March 17, 1928, St. Francis Dam file, Division of Safety of Dams, Sacramento. Copies of this letter were also sent to Mulholland, C. C. Teague, the *Los Angeles Times,* the *Los Angeles Examiner,* and the *Los Angeles Express.*

32 Edward Hyatt to H. A. Van Norman, March 16, 1928, St. Francis Dam file, Division of Safety of Dams, Sacramento. Hyatt concludes this "personal" letter with an expression of concern for Mulholland: "I did not feel that I know Mr. Mulholland well enough to send him personal sympathy at the outset. However, I want you to know, both for him and yourself, that my deepest sympathy goes to the engineers connected with the work, who must share in the grief occasioned by this terrible calamity."

33 "State Orders Dam Quiz."

34 See telegram from C. C. Young to A. J. Wiley, March 15, 1928; and telegram from A. J. Wiley to C. C. Young, March 15, 1928; St. Francis Dam file, Division of Safety of Dams, Sacramento, California. No payment would be made for professional services, but commission members were reimbursed for travel, food, and lodging.

35 For a good review of his career dating back to his graduation as an engineer from Delaware College in 1882 and his move west to Idaho in 1883, see "Memoir of Andrew Jackson Wiley," *Transactions of the American Society of Civil Engineers* 96 (1932): 1577–85. He died unexpectedly on October 8, 1931, at age sixty-nine.

36 Governor Young's instructions to the commission were formally conveyed at its first meeting on March 19 in Los Angeles; they were subsequently printed in the Governor's Commission *Report,* 5.

37 Ibid., 16.

38 Edward Hyatt, Memorandum St. Francis Dam, undated, ca. March 26, 1928, St. Francis Dam file, Division of Safety of Dams, Sacramento, California.

39 "Commission of Governor Opens Inquiry," *Los Angeles Examiner,* March 20, 1928. Wiley's desire to maintain a low public profile stands in contrast to the investigative committee sponsored by the Los Angeles City Council that was chaired by Bureau of Reclamation Commissioner Elwood Mead. When Mead arrived in Los Angeles on March 20 he was greeted by a photographer and his picture (with council member Pierson Hall) appeared with the caption "Elwood Mead Arrives in L.A. to Head Dam Disaster Inquiry," *Los Angeles Express,* March 20, 1928. In the end, Mead's commission essentially parroted

what the Governor's Commission report had previously concluded and the City Council–sponsored investigation became little more than a footnote.

40 "Keyes Sends Experts to Probe Dam Collapse," *Los Angeles Herald*, March 16, 1928.

41 "Grand Jury Dam Inquiry to Depend on Coroner," *Los Angeles Times*, March 25, 1928. This article also reported that "Keyes will not ask quiz unless Nance's verdict urges criminal investigation."

42 "Inquest Set Wednesday for Flood Deaths," *Los Angeles Examiner*, March 18, 1928.

43 The citizens on the jury were Oliver Bowen, W. H. Eaton, I. C. Harris, Henry Holabird, Sterling Lines, Z. N. Nelson, Blaine Noice, C. D. Walz, and Ralph Ware. In the inquest transcript they are identified only as "a juror" when any one of them asks a question. They all signed the final verdict, the only time they are identified by name in the lengthy transcript. They are also identified in "Water Board to Demand Grand Jury Quiz on Dam: Mulholland on Stand at Coroner's Inquiry Defends Work and Hints at Dynamiting," *Los Angeles Times*, March 22, 1928.

44 Ray Rising testimony, Coroner's Inquest, p. 2.

45 Julia Rising and her three children had been buried in Oakwood Cemetery in the San Fernando Valley the previous weekend. See "Homage to Dead in Valley," *Los Angeles Times*, March 20, 1928. Davis, in *Rivers in the Desert*, claims that Julia Rising's "sheet-shrouded body" was brought to the inquest so that Ray Rising could "behold his wife for the first time since she had slipped away from his grasp in the deadly current" (187–88). Davis' book provides a fanciful account of the inquest that is difficult, and at times impossible, to reconcile with the court reporter's transcript.

46 Ray Rising testimony, Coroner's Inquest, 1–3; Nance's reading of victims' names, 3–5; Dr. Frank Webb testimony, 5–6.

47 When Mulholland was asked, "How was the site of the St. Francis Dam selected?" he responded: "We observed that there was a large water space at the camp; we had a camp in there when we were building the aqueduct. Contained there, I think, for six years. I was very familiar with the topography there and I have had a habitation there that long; I virtually made my habitation there because I lived more or less on the aqueduct while it was under construction. I am speaking now of the Los Angeles Aqueduct. I was the Chief Engineer on that and designer and constructor of it"; Coroner's Inquest, 7.

48 Ibid., 16.

49 Ibid., 23.

50 Ibid., 25.

51 Ibid., 32.

52 "Water Board to Demand Grand Jury Quiz on Dam: Mulholland on Stand at Coroner's Inquiry Defends Work and Hints at Dynamiting," *Los Angeles Times*, March 22, 1928.

53 "Praise Accorded to Mulholland," *Los Angeles Times*, March 22, 1928.

54 "Mulholland Sobs Story at Dam Disaster Inquest," *Los Angeles Herald*, March 21, 1928. This article also claimed that "after Mulholland suffered a near collapse on the witness stand he was dismissed for the day."

55 Nat Fisher testimony, Coroner's Inquest, 37.

56 For example, this exchange: "Q. [District Attorney Keyes]: Do you know anything about any soil being brought from an adjacent field there and mixed in this thing, in the concrete? A. [Dunham]: Absolutely not. Q. Was there any such thing that took place? A. No." Or this: "Q. [Coroner]: If it was shown there was clay, lumps of clay, large lumps several inches in diameter in that concrete now, how would you account for that? A. Might slip by the man watching. Q. And dirt in large quantities, if found in the concrete, how would it get there? A. It would have to just slip by, but I haven't seen any of it myself"; Stanley Dunham testimony, Coroner's Inquest, 50, 53. While it is likely that some small amount of dirt or clay was mixed into the St. Francis concrete, this line of questioning was largely pointless, because concrete gravity dams do not require particularly strong concrete in order to be safe; their stability depends on weight (or mass) and not on the compressive strength of the concrete.

57 H. A. Van Norman testimony, Coroner's Inquest, 87–127.

58 Construction of the city's north outfall sewer is described in "Making the Concrete Pipe for Los Angeles Outfall Sewer," *Engineering News-Record* 93 (August 21, 1924): 296–98, where it is noted that "the work is being done under the general supervision of H. A. Van Norman, consulting engineer for the city of Los Angeles, in charge of sewer construction." This article appeared just as foundation excavation for the St. Francis Dam was underway, with the initial casting of concrete only a few weeks in the future.

59 Near the end of the inquest, Mulholland's chauffeur, George Vejar, was questioned about the March 12 visit. He testified about where he parked the car (a "Marmon sedan"), and where Mulholland and Van Norman walked. Any expectation that he would contradict his boss's testimony about "muddy" leakage on the west side abutment was not fulfilled; Coroner's Inquest, 722–26.

60 Several witnesses offered testimony about conditions at the dam and in the canyon below prior to the flood, including R. E. Atmore (Coroner's Inquest, 165–72), G. O. Dorsett (172–77), Chester Smith (177–83, 185–88), Hugh Nichols (183–85), and Anna Scott (215–19).

61 Outland, *Man-Made Disaster*, 70–76.

62 Clyde C. Ruble (Coroner's Inquest, 412–14) explained why the adit was closed on the afternoon of March 12. Ruble was the superintendent in charge of power plants for the BLP; he lived in Los Angeles and was not in the canyon on the night of March 12.

63 "Statement, Etc. of Archie J. Eley. Oct. 26, 1928" in "Report of Death Claim Leona Johnson #2538," St. Francis Dam Claims Records, LADWP Archives.

64 Clyde C. Ruble, the Bureau of Power and Light superintendent in charge of power plant operations, and Earl Martindale, electrical engineer in charge of Power House No. 2, both testified about Mathews' post-disaster demeanor. Mathews was distraught over the death of his brother, Carl, and Carl's family,

and this likely affected his behavior and his testimony. See Coroner's Inquest, 411–18, 418–21. The deaths of Carl Mathews and his extended family (a total of seven) are noted in *The Intake,* April 1928, 5.

65 "Van Norman is Grilled," *Los Angeles Record,* March 22, 1928. Harvey Van Norman apparently offered to accompany the jurors on their site visit, but Coroner Nance quashed the proposal, seeking to avoid any politicking or the appearance thereof by the city or the water department. The trip is also noted in "Jury to Visit Dam Site," *Los Angeles Express,* March 22, 1928.

66 "Jury Views Dam Site: Inspectors Test Canyon Rock," *Los Angeles Times;* "Jury Analyzes Samples of Dam Rock," *Los Angeles Herald*; "Take Red Samples," *Los Angeles Examiner*; "Experts May Blame Site for the Disaster," *Los Angeles Examiner*; all of these articles published March 24, 1928.

67 D. C. Mathews testimony, Coroner's Inquest, 202–3.

68 Ibid, 205.

69 "Tells of Tony's Fear of Dam," *Los Angeles Record,* March 26, 1928.

70 D. C. Mathews testimony, Coroner's Inquest, 208–9.

71 Edward Hyatt, "Memorandum St. Francis Dam," ca. late March 1928; St. Francis Dam Disaster file, Division of Safety of Dams, Sacramento, California.

72 "Summon Dam Designer for Disaster Quiz," *Los Angeles Express,* March 24, 1928. Wiley's desire to complete the report in a week was signaled in "Call Forty in Inquest," *Los Angeles Times,* March 20, 1928, which reported that the committee's "work will consume about a week and will be completed here [in Los Angeles]." Wiley met this self-imposed deadline.

73 Hyatt, "Memorandum St. Francis Dam."

74 Governor's Commission *Report.* The printed edition of the report (p. 19) includes a letter dated April 20, 1928, from Young to the commission members thanking them for their service. Presumably the publication became publicly available shortly thereafter. A copy of the report also appeared in the April 1928 issue of *California Highways and Public Works* (a magazine termed the "Official Journal of the Department of Public Works, State of California") but only three illustrations were included; see "Text of the Report to Governor Young on Causes of the St. Francis Dam Failure," *California Highways and Public Works* (April 1928): 6–7; 25–31.

75 Governor's Commission *Report,* 1–18. The report is organized in a series of sections with the headings "Description of St. Francis Dam," "Failure of the Dam," "Methods Followed in Constructing the Dam," "Geological Conditions at the Dam Site," "Conditions at the Dam After Failure," "Causes of the Failure," and "Conclusions."

76 Ibid., 16. The Governor's Commission also described how, similar to the demonstration given during Mathews' inquest testimony, small samples of the conglomerate could disintegrate when immersed in water, while noting that this characteristic "is probably local and confined to a belt" near the schist/conglomerate contact plane (12).

77 Ibid.

78 Ibid.

79 Ibid.

80 Edward Hyatt, "Memorandum St. Francis Dam."

81 J. David Rogers estimates that the east side landslide encompassed 500,000 cubic yards of disintegrated schist; "A Man, a Dam," 53.

82 Governor's Commission *Report*, 16.

83 Ibid.

84 Ibid., 18.

85 Walter L. Huber to John R. Freeman, March 21, 1928, box 54, Freeman Papers, MIT.

86 Edward Hyatt to Phil D. Swing, March 26, 1928; St. Francis Dam Disaster file, Division of Safety of Dams, Sacramento.

87 Telegram from C. C. Young to Phil D. Swing, March 27, 1928; St. Francis Dam Disaster file, Division of Safety of Dams, Sacramento.

88 Ibid.

89 Telegram from Phil D. Swing to Governor C. C. Young, March 27, 1928; St. Francis Dam Disaster file, Division of Safety of Dams, Sacramento.

90 The report of the City Council committee chaired by Elwood Mead was published in "The St. Francis Dam Failure," *New Reclamation Era* (May 1928): 61–71. Those signing the committee's report were General Lansing Beach, consulting engineer L. C. Hill, consulting engineer D. C. Henny, and the Bureau of Reclamation's chief engineer, R. F. Walter. Public notice of the report, and its concurrence with the Governor's Commission, appears in "Mead Committee is in Agreement with Others on St. Francis Failure," *Engineering News-Record* 100 (April 23, 1928): 675.

91 "All St. Francis Dam Reports Agree as to Reason for Failure," *Engineering News-Record* 100 (April 19, 1928): 639.

92 Mulholland testimony, Coroner's Inquest, 373–74.

93 Ibid., 376–77.

94 Ibid., 378.

95 Ibid., 379–80.

96 Ibid., 622–23. At the conclusion of this testimony Mulholland sought to remind the jury that he had been recognized as a leader in the world of seismology, boasting that: "I might say I was honored by having conferred upon me the LL.D degree of the University of California, so don't call me colonel anymore, I am a doctor"; 624.

97 "Wilbur Fisk McClure," *Transactions of the American Society of Civil Engineers* 91 (1927): 1106–9.

98 Jackson describes the lengthy approval process required for the Littlerock Dam in *Building the Ultimate Dam*, 198–205.

99 "Memorandum Inspection Trip of Little Rock-Palmdale Irrigation District by A. F. McConnell, August 7, 1922." A subheading on the memo indicates: "Order for Inspection Trip Given by Mr. McClure and transmitted by Mr. Bailey," McClure's deputy engineer. Memo in Littlerock Dam file, Division of Safety of Dams, California Department of Water Resources.

100 A letter from McClure to the Littlerock Creek Irrigation District, dated June 5, 1924, indicates formal acceptance of the completed Littlerock Dam; Littlerock Creek Irrigation District files, Littlerock Calif. Sources for material in this paragraph include the following memos in Littlerock Dam file, Division of Safety of Dams: "Little Rock Creek Dam. Test on Cement Specimen No. 9238. Test by Raymond G. Osborne, Los Angeles, September 22, 1922"; "Little Rock Creek Dam, E. C. Eaton, November 27, 1922"; "Little Rock Creek Dam, Memorandums of Trips on December 14th and 21st, 1922, E. C. Eaton, 12/24/22"; "Little Rock Dam, E. C. Eaton, September 11, 1923"; and "Little Rock Dam, E. C. Eaton, January 22nd, 1924." Also see Jackson, *Building the Ultimate Dam,* 306–7, n. 75.

101 "Blame Me if Anyone in Dam Crash—Mulholland," *Los Angeles Examiner,* March 28, 1928.

102 Mulholland testimony, Coroner's Inquest, 383.

103 Ibid., 384. In the third week of the inquest, Mulholland again took the stand to stress that the water he observed leaking down the west side abutment on March 12 was "clear." He also returned to the stand because in his initial testimony he had characterized the abutment as "saturated" with water (p. 14); for the record he wish to clarify that it was not saturated but simply covered with some flow on the surface; Coroner's Inquest, 582–89.

104 "Ready to Face Music—Mulholland," *Los Angeles Herald,* March 27, 1928; "Mulholland Unflinching," *Los Angeles Times,* March 28, 1928.

105 "Blame Me if Anyone in Dam Crash."

106 Mulholland testimony, Coroner's Inquest, 376.

107 Ezra Scattergood testimony, Coroner's Inquest, 409–10.

108 The tension between Scattergood and the Bureau of Water Works and Supply did not dissipate after Mulholland left the BWWS in November 1928 and Van Norman became chief engineer. Throughout the 1930s the rivalry continued as Van Norman and his allies in city government sought to place the Bureau of Power and Light under his supervision. But Scattergood fought for his independence, benefiting from the support of John R. Haynes. Peace did not come until Samuel B. Morris took charge as chief engineer and general manager of the Department of Water and Power in 1944. See Ostrom, *Water and Politics,* 100–102.

109 James P. Brome testimony, Coroner's Inquest, 591.

110 Ibid., 595.

111 Ibid., 600–601.

112 Ibid.

113 R. F. del Valle testimony, Coroner's Inquest, 615.

114 Ibid., 616–17.

115 Ibid., 619–20.

116 See "Coroner Verdict Delayed," *Los Angeles Times,* March 30, 1928, where it was reported that "engineers will give report on dam Monday." Also see "Report on Dam Not Ready," *Los Angeles Times,* April 3, 1928. On April 2, Nance announced the second delay ("Gentlemen, there is nothing for us to do but take an adjournment until Wednesday morning at 9:30 A.M."), and the inquest was suspended until April 4; Coroner's Inquest, 549.

117 Edward Mayberry, Coroner's Inquest, 555–58, offered evidence on difficulties with the assumption that the water gauge showed increasing leakage in the hours before the collapse. R. R. Proctor, a BWWS employee who helped operate the water gauge, offered testimony on how it functioned; see Coroner's Inquest, 638–40.

118 Dean Keagy testimony, Coroner's Inquest, 424–32; Katherine Spann testimony, 432–35; Ace Hopewell testimony, 641–46; Otto Steen testimony, 646–54; Henry Silvey testimony, 667–70. See Outland, *Man-Made Disaster,* 207–10, for more on the problems posed by the west-side leakage hypothesis.

119 Charles Lee testimony, Coroner's Inquest, 654–67.

120 Allan Sedgwick testimony, Coroner's Inquest, 495–524, 602–14, 768–79; Edward Mayberry testimony, 545–54, 697–722; W. G. Clark testimony, 558–72, 779–84; Charles Leeds testimony, 572–90; L. Z. Johnson testimony, 524–46, 554–58, 542.

121 Charles Leeds testimony, Coroner's Inquest, 578. Outland also notes Leeds' memorable quote in *Man-Made Disaster* (1963), 210. The Mayberry Committee's final report is attached as part of the Los Angeles County Coroner's Inquest; see "Report to Mr. Asa Keyes, District Attorney, Los Angeles County, Calif. On The Failure of the St. Francis Dam by Edward L. Mayberry, Walter C. Clark, Charles T. Leeds, Allan F. Sedgwick, Louis E. Johnson, April 4th, 1928," box 13, Courtney Collection, Huntington Library.

122 "Seven Attack Report on Dam," *Los Angeles Times,* April 6, 1928.

123 Charles Lee testimony, Coroner's Inquest, 659.

124 Charles H. Lee, "Theories of the Cause and Sequence of Failure of the St. Francis Dam," *Western Construction News* 3 (June 25, 1928): 405–8.

125 "Dynamited Dam Theory Backed," *Los Angeles Times,* April 7, 1928. This article also featured a subheadline: "Earth Movement View Point Obtains Support."

126 "Mulholland Lays Break to Ground Shift," *Los Angeles Examiner,* March 14, 1928.

127 Hemborg testimony, Coroner's Inquest, 624–41.

128 Sedgwick expressed skepticism about Hemborg's survey; Coroner's Inquest, 637.

129 "Seven Attack Report on Dam" reported that "testimony indicating a considerable earth movement at the dam site was introduced into the hearing through Water Bureau Surveyor Hemborg."

130 Frank Reiber testimony, Coroner's Inquest, 685–86. For discussion of Reiber's testimony, also see Outland, *Man-Made Disaster*, 191–92.

131 Edward Mayberry and Allan Sedgwick of the Mayberry Committee were not impressed by Reiber's dynamite theory, and both took the witness stand to raise questions and objections. See Edward Mayberry testimony, Coroner's Inquest, 697–720; and Allan Sedgwick testimony, 768–79.

132 "Dynamite Dam Theory Backed," *Los Angeles Times*, April 7, 1928.

133 Halbert P. Gillette, "Three Unreliable Reports on the St. Francis Dam Failure"; "Some Lessons Taught by the St. Francis Dam Failure"; and "The Cause of the St. Francis Dam Failure," all in *Water Works* 67 (May 1928), 177–78, 178–79, 181–86. Soon after the inquest's conclusion Gillette publicly aired his critique of the three seemingly official investigating committees—the Governor's Commission, the Los Angeles City Council Committee, and the Los Angeles district attorney's Mayberry Committee. Gillette lambasted the investigating teams for hasty and faulty research; based upon his own fieldwork and research (which convinced him that the red conglomerate was not as weak as publicly portrayed), he described how only the mechanics of an east-abutment-first collapse sequence made sense of post-failure conditions at the site.

134 Halbert P. Gillette testimony, Coroner's Inquest, 750–53.

135 Ibid., 761.

136 Edward C. Starks testimony, Coroner's Inquest, 726–36. Also see Outland, *Man-Made Disaster*, 192.

137 Starks testimony, Coroner's Inquest, 727.

138 Ibid., 729.

139 Ibid., 732.

140 Zattu Cushing testimony, Coroner's Inquest, 799–800.

141 S. B. Robinson interjection, Coroner's Inquest, 800.

142 Cushing testimony, Coroner's Inquest, 813.

143 William Bright testimony, Coroner's Inquest, 816.

144 Nance statement to jury, Coroner's Inquest, 817–18. Nance also directed that the jury consider only evidence presented at the inquest and that "all collateral information, hearsay and preconceived impressions shall be disregarded by you in arriving at your verdict."

145 E. J. Dennison statement to jury, Coroner's Inquest, 818–19.

146 "Verdict of Coroner's Jury," Coroner's Inquest, 1–11. The "Verdict of Coroner's Jury" is a separately paginated section.

147 Ibid., 2–5. The jurors said that the dam likely collapsed first on the red conglomerate of the west side because a "preponderance of expert opinion favors the conclusion." Nonetheless, they expressed ambivalence about this judgment—they had heard testimony that the schist forming the east abutment was "a weak material, badly shattered, very susceptible to seepage of water, and to slippage along the planes of cleavage"—and they hesitated to conclude that they fully understood "the exact sequence" of the collapse (5, 10–11).

148 Ibid., 4.

149 Ibid., 2–3, 10–11.

150 Ibid., 2–3.

151 "Coroner Puts Dam Break Blame on Water Board—Collapse Found Due to Foundation; Verdict Says Tests Inadequate; State Check Urged in Future," *Los Angeles Times*, April 13, 1928; and "The Coroner Verdict," *Los Angeles Times*, April 13, 1928.

152 "Honest Error Made in Site, Verdict Says," *Los Angeles Examiner*, April 13, 1928. The *Examiner* also quoted attorney K. K. Scott "of the Water Bureau" as stating that the verdict was satisfactory but "perhaps a little more sweeping than expected."

153 "Resign Now!" *Los Angeles Record*, April 13, 1928.

154 Rogers, in "A Man, a Dam," offers an analysis of the disaster aligned with the work of the Grunskys, Willis, and Lee but also offers new insight into how the east side landslide formed a temporary embankment dam that held back the reservoir for a brief but significant period after the failure began.

155 "Carl Ewald Grunsky," *Transactions of the American Society of Civil Engineers* 100 (1935): 1591–95; C. E. and E. L. Grunsky, "St. Francis Dam Failure at Midnight, March 12–13, 1928," *Western Construction News* 3 (May 25, 1928): 314–20; Bailey Willis, "Report on the Geology of St. Francis Damsite, Los Angeles County, California," *Western Construction News* 3 (June 25, 1928): 409–13, 409; Outland, *Man-Made Disaster*, 26.

156 The Grunskys' impending investigation of the disaster for the "Ventura County Supervisors" is briefly noted in "State Flood Inquiry to Work on Report," *Los Angeles Times*, March 23, 1928. Their work apparently received no other notice in the Los Angeles press.

157 Grunsky and Grunsky, "St. Francis Dam Failure," 314–20, 314; Willis, "Report on the Geology of St. Francis Damsite," 409–13.

158 Grunsky and Grunsky, "St. Francis Dam Failure," 319, 320; Willis, "Report on the Geology of St. Francis Damsite," 411, 412, 413.

159 Grunsky and Grunsky, "St. Francis Dam Failure," 316, 319.

160 Frank Reiber drew attention to the crushed ladder at the inquest, and it is possible that the Grunskys first became aware of it then. At any rate, in their illustrated article the Grunskys were the first to give the crushed ladder wide publicity; see Frank Reiber testimony, Coroner's Inquest, 681–82.

161 Lee presented a paper on the St. Francis disaster at the May 8, 1928, meeting of the Society of Engineers in San Francisco. Soon thereafter he was asked to present a paper for the June 19, 1928, meeting of the San Francisco section of the American Society of Civil Engineers. The basic text of this presentation was published as "Theories of the Cause and Sequence of Failure of the St. Francis Dam," *Western Construction News* 3 (June 25, 1928): 405–8, 407; 408; Charles H. Lee Papers, file MS/76/1 98L14C, WRCA, UC Riverside.

162 Lee, "Theories of the Cause and Sequence of Failure," 407.

163 Lee offered Mulholland's assistant W. W. Hurlbut an opportunity to review what he planned to present at the San Francisco ASCE meeting: "I would appreciate it if you would look it over and make any suggestions which you think...desirable." See Charles Lee to W. W. Hurlbut, June 15, 1928, Lee Collection, WRCA, UC Riverside. There is no evidence that Hurlbut responded. However, in his published paper, Lee did respect the interest of the BWWS in promulgating the west abutment "earth movement" theory that bureau surveyor Hemborg had testified about at the Coroner's Inquest. In his *Western Construction News* article, Lee concludes his lengthy discussion of how the collapse started at the east abutment with the brief caveat: "As a contributing cause the general earth movement at the west abutment cannot be overlooked" (408).

164 Ibid., 407, 408.

165 Ibid., 408.

166 Arizona's longstanding antipathy to the Colorado Compact and the Boulder Canyon Project is a major theme in Hundley, *Water and the West.* Outland also emphasizes Arizona's opposition to Boulder Dam in relation to political problems engendered by the St. Francis disaster. Moeller, in *Phil Swing and Boulder Dam,* describes the tenuous status of the Boulder Canyon Project's authorization in early 1928 (107–12). Also see "Threat Made of Filibuster: Ashurst Says Boulder Bill Shall Not Pass," *Los Angeles Times,* March 31, 1928.

167 "A City Above the Law," *Argonaut,* March 24, 1928. Copy available in box 406, Southern California Edison Collection, Huntington Library.

168 "Boulder Board Fight Renewed," *Los Angeles Times,* March 20, 1928.

169 For example, see "Action Likely in Congress on Boulder Dam," *Chicago Tribune,* April 4, 1928, which reports that a filibuster in the Senate is "promised." Also see "Plan Filibuster on Boulder Dam Bill," *Wall Street Journal,* April 30, 1928, where it was reported that "a bitter filibuster is being organized in the Senate against the Boulder Canyon dam bill. All-night sessions are in prospect."

170 "Quiz Demanded on Los Angeles: Bureau of Light and Power Assailed by Douglas," *Los Angeles Times,* May 10, 1928. Douglas had been outspoken in his opposition to Boulder Dam even before the St. Francis disaster; see L. W. Douglas, "The Boulder Dam Project," *Washington Post,* March 11, 1928.

171 "Smoot Warns of Boulder Dam, a Death Trap," *Chicago Tribune,* May 1, 1928.

172 Moeller, *Phil Swing and Boulder Dam,* 113–15. Efforts to block a vote in the House of Representatives (ultimately unsuccessful) are described in "Anti-Boulder Forces to Act: Group Plans Call Today on House Leaders," *Los Angeles Times,* May 11, 1928; "'Buck' Passers Toy with Dam: Rules Committee May Grant Bill Right of Way," *Los Angeles Times,* May 12, 1928; "Boulder Dam Vote Likely: Indications Point to House Passing Bill this Week," *Los Angeles Times,* May 20, 1928; "Boulder Dam Weathers Day's Battle in House," *Los Angeles Times,* May 25, 1928. Passage of the House bill is noted in "Final Drive for Senate Dam Vote Opens Today," *Los Angeles Times,* May 28, 1928.

173 "Boulder Canyon Dam Project Speech by Hon. Hiram W. Johnson of California of the Senate of the United States, Tuesday May 22, 1928" (Washington, D.C.:

Government Printing Office, 1928), 19–20. Johnson fancied himself as a great Progressive, seeking to portray "the menace of the Power Lobby" as it sought to control the nation's natural resources: "The situation is really the age-old struggle of those who hold that what God gives belongs to the people and those that hold that it belongs to the few." See Hiram W. Johnson, "The Boulder Canyon Project," *Community Builder* 1 (March 1928): 41–51; available in box 407, Southern California Edison Collection, Huntington Library.

174 Johnson, "Boulder Canyon Dam Project Speech," May 22, 1928, 20–21.

175 "Congress Ends Session with Senate in Uproar over Boulder Dam Bill," *New York Times,* May 30, 1928. Moeller describes the raucous scene in the Senate: "At 5:30 P.M. on 29 May the Senate adjourned in wild disorder. There were threats of fist fights on the Senate floor, while onlookers shouted and hooted in the gallery"; *Phil Swing and Boulder Dam,* 115–16.

176 The House-approved bill included an amendment calling for an independent engineering investigation of the entire proposed project; this amendment paved the way for a joint resolution with the Senate. See Moeller, *Phil Swing and Boulder Dam,* 114–15; and "Boulder Dam Bill Amended by House in Five-Hour Fight," *Washington Post,* May 25, 1928, which reported that Swing "initiated one of the most important changes effected in the bill, an amendment to provide for a commission of engineers to work in conjunction with the Secretary of the Interior in perfecting final plans for the huge dam project." The Sibert commission first met in Washington, D.C., in late July and carried out its work through the fall. See "Boulder Dam Commission Holds First Meeting," *Engineering News-Record* 100 (August 2, 1928): 182.

177 Over the summer, attacks against Boulder Dam continued, with the Los Angeles City Council dismissing them as "propaganda." See "Council Rips Anti-Boulder Propaganda," *Los Angeles Record,* July 2, 1928; and "Boulder Dam Propaganda Denounced," *Los Angeles Examiner,* July 3, 1928. In the latter article the "Power Trust" was blamed for spreading misinformation and the *San Francisco Bulletin* was blasted as a "traitor."

178 "Board Reports on Boulder Dam," *Engineering News-Record* 101 (December 6, 1928): 857; and "Board of Engineers Favors Black Canyon for Dam in Colorado River," *Southwest Builder and Contractor* 72 (December 7, 1928): 32–33.

179 The bureau's actions proved legally viable because language requiring the limit of thirty tons per square foot did not appear in the final Boulder Canyon Project legislation signed by President Coolidge. The way the design could be considered to meet the thirty-ton criterion—using the "trial-load method" of analysis—is described in Elwood Mead, "Memorandum to the Secretary re the Meeting of the Consulting Engineers to Approve Detail Plans of Boulder Dam," December 28, 1929; Bureau of the Reclamation, RG 115; General Administration and Project Records, Project Files, 1919–1945; entry 7; box 490; folder 301.1, Colorado River Project, Board & Engineering Reports on Construction Features, 1929, National Archives, Denver. The Colorado River Board acknowledged acceptance of the bureau's rationale for not revising the design in U.S. Department of the Interior, *Boulder Canyon Project Final Reports, Part IV–Design and Construction* (Denver: Bureau of Reclamation, 1941), 25.

180 For example, the board recommended an overall budget for the project of at least $165 million; $125 million had been stipulated in the bills considered in spring 1928. See "Boulder Canyon Board Picks Black Canyon," *New York Times,* December 4, 1928.

181 For more on the Power Trust and its effect on passage of the Boulder Canyon Project Act, see Donald J. Pisani, *Water and American Government: The Reclamation Bureau, National Water Policy, and the West, 1902–1935* (Berkeley: University of California Press, 2002): 237–41, 261–63; and Moeller, *Phil Swing and the Boulder Dam,* 111–12.

182 A *Wall Street Journal* article published shortly before release of the Sibert Commission report reflects the changes that had taken place in the political landscape between May and November 1928. The article reported Wyoming Governor Frank C. Emerson as announcing: "differences among seven states affected by the possible erection of a dam at Boulder [or] Black Canyon on the Colorado River [have] been practically settled, with indications that the states were ready to support the bill...I believe most people in this district [the Southwestern United States] are looking at the project with open minds and will be influenced by the report of the investigating committee of engineers appointed by Secretary [of the Interior Hubert] Work. This report will be completed shortly and will disclose the feasibility of a dam at either Black or Boulder Canyons." Utah Senator Reed Smoot and Arizona's legislators never embraced the Boulder Canyon Project Act, but by the end of 1928 other political leaders in the Colorado River basin were willing to align with California in supporting it. See "Says States in Accord on Colorado River Dam: Gov. Emerson of Wyoming Reports States Concerned Ready to Withdraw Opposition to Swing-Johnson Bill," *Wall Street Journal,* November 22, 1928.

183 Pisani, *Water and American Government,* 261.

184 The Senate approved the legislation on December 14. Johnson's success in the Senate is celebrated in "Dam Victory Hailed by Leaders in L.A.," *Los Angeles Herald,* December 15, 1928. Circumstances had changed dramatically since May and Senator Smoot now perceived the futility of trying to block the bill. He refrained from organizing a filibuster and cast no ballot when the final vote was taken.

185 See "Boulder Bill Backed by House Committee," *New York Times,* December 18, 1928, which reported: "The last congressional obstruction to the Swing-Johnson Boulder Canyon dam bill appeared to be crumbling today with the unanimous decision of the House Irrigation Committee to ask the House to agree to changes made in the measure by the Senate. Shortly after the decision by the committee Representative Douglas, Democrat, of Arizona, one of the most vigorous opponents the measure has had in the House, announced that he would not seek to have the House disagree with the Senate revisions...Representative Swing, Republican, of California, co-author of the bill, said most of the changes were designed to make the proposal conform to the report of the board of engineers which studied it last summer." Also see Moeller, *Phil Swing and Boulder Dam,* 120–22. The House passed the bill 166 to 122.

186 "L.A. Rejoices as Coolidge Signs Dam Bill," *Los Angeles Express*, December 21, 1928. Also see "President Approves Bill Authorizing Black Canyon Dam: A Sum of $165,000,000 Approved for a Dam and Power House and All-American Canal," *Engineering News-Record* 101 (December 27, 1928). Also see Moeller, *Phil Swing and Boulder Dam*, 119–22.

187 Two articles published in the *New York Times*, "Debate Renewed on Boulder Dam: Old Dispute Between States is Revived When Johnson Calls Up the Measure," December 6, 1928; and "Senators Hopeful on Boulder Dam," December 8, 1928, describe objections to the proposed Boulder Canyon Project Act, but neither mentions the St. Francis disaster.

188 "Sixth Report on St. Francis Dam Offers New Theories," *Engineering News-Record* 100 (June 7, 1928): 895.

189 "Sliding Versus Scouring," *Engineering News-Record* 100 (June 7, 1928): 879.

190 C. E. Grunsky and E. L. Grunsky, "The Grunsky Report on the Failure of the St. Francis Dam," *Engineering News-Record* 101 (July 26, 1928): 144.

191 "Essential Facts Concerning the Failure of the St. Francis Dam: Report of the Committee of the Board of Direction," *Proceedings of the American Society of Civil Engineers* 55 (October 1929): 2147–63.

192 As part of this discussion E. L. Grunsky reiterated the conclusions of the report written with his father in the spring of 1928 and published in *Western Construction News*. Once again, he laid out the problems posed by a west-side-first failure hypothesis and explained how all physical evidence pointed toward initial east side failure. But America's civil engineering community exhibited little interest in the Grunsky-Willis-Lee analysis; see E. L. Grunsky, discussion of "Essential Facts Concerning the Failure of the St. Francis Dam: Report of the Committee of the Board of Direction," *Proceedings of the American Society of Civil Engineers* 56 (May 1930): 1026–31.

193 "Boy Killed in Plunge from Remnant of St. Francis Dam," *Los Angeles Times*, May 28, 1928. Perhaps there were other such incidents, but the authors of *Heavy Ground* are not aware of them; Parker's death appears to mark the sole fatality resulting from a visit to the site of the fractured dam. Earlier that month city officials had requested that sightseers not visit the dam by driving up San Francisquito Canyon, as reported in "Motorists Asked to Shun Dam," *Los Angeles Times*, May 13, 1928: "The Automobile Club of Southern California has been requested by the city authorities to urge its members and all motorists not to go upon this road. Recently it has been found necessary to prohibit sightseers from this section altogether, due to the highway being crowded with work trucks [going to Power House No. 2] and unsafe in spots for ordinary drivers."

194 "St. Francis Dam Ruins Blown Up," *Los Angeles Times*, May 11, 1929.

195 Ibid. The demolition is also described in "Standing Section of St. Francis Dam Razed with Dynamite," *Engineering News-Record* 103 (July 11, 1929): 51.

196 While both the city of Los Angeles and the civic leadership of Ventura County were ready to put the disaster behind them, the people of the Santa Clara Valley who had suffered and lost loved ones did not so easily forget. Memory of

the flood is what later spurred Charles Outland to write *Man-Made Disaster,* and such memories became a part of family lore for many others in the valley. The authors of *Heavy Ground* have sought to give voice to at least some of the victims, but there remain many stories that can best be told by descendants of those who directly experienced the tragedy.

CHAPTER SEVEN

1 M. M. O'Shaughnessy, "Discussion," published with A. H. Markwart and M. C. Hinderlider, "Public Supervision of Dams: A Symposium," *Transactions of the American Society of Civil Engineers* 98 (1933): 853.

2 Fred Noetzli, ibid., p. 865.

3 The article "Anchoring Weak in Big Dam, Claim," *Los Angeles Herald,* March 15, 1928, quotes State Engineer Edward Hyatt: "This improper construction would not have been possible under state supervision...At the next session of the legislature I will ask that laws be passed placing all such [municipal] projects under direct state supervision." While it is doubtful that Hyatt said this verbatim, the story does show how quickly the press turned to the implementation of new dam safety legislation. A new state law is also discussed in "State Supervision Needed to Safeguard Great Public Works," *Los Angeles Examiner,* March 20, 1928; the *Fresno Bee* of March 28, 1928, in "Dam Report is Indictment of Shocking Negligence," made the following plea: "Inspection of all dams should be made by the state immediately...By all means let California protect its people."

4 Governor's Commission *Report,* 18.

5 Edward Hyatt to M. R. McKall, April 7, 1928, St. Francis Dam file, Division of Safety of Dams, Sacramento.

6 Ibid.

7 Norwood Silsbee (secretary-treasurer of the ASCE Sacramento Section) to C. B. Sadler (secretary-treasurer of the ASCE San Diego Section), June 19, 1928, folder 25, Hiram N. Savage Papers, WRCA, UC Riverside. Members of the committee included Hiram Savage (San Diego), Walter Huber (San Francisco), George Pollack (Sacramento), and Frederick Howell (Los Angeles).

8 For example, see F. D. Howell to George Pollack, Hiram Savage, and W. L. Huber, September 21, 1928; folder 25, Savage Papers, WRCA, UC Riverside.

9 W. L. Huber to George Pollack, December 3, 1928; for more on the possible use of consulting boards by the state engineer, see D. L. Bissell to H. N. Savage; both in folder 25, Savage Papers, WRCA, UC Riverside.

10 Samuel B. Morris to Edward Hyatt and Leon Whitsett, November 27, 1928, folder 389 ("Dam Safety Regulation"), box 89, Morris Papers, Huntington Library. Morris was not a member of the ASCE committee but, as chief engineer of the Pasadena Water Department, he paid close attention to legislative action relating to dam safety.

11 The composite act was developed from Senate Bill 685 and Assembly Bill 723, both submitted on January 18, 1928. See Edward Hyatt to Samuel B. Morris, February 6, 1929, folder 389, Morris Papers, Huntington Library.

12 The complete text of the new law is presented in *California Statutes*, chap. 766 (1929): 1505–14.

13 See Jackson, *Building the Ultimate Dam*, 241–42, for discussion of the 1931 state-sponsored advisory committee, convened by Hyatt and chaired by Huber, to study multiple-arch dam technology.

14 H. L. Jacques, "Bouquet Canyon Dam Built for Los Angeles Aqueduct," *Engineering News-Record* 112 (June 21, 1934): 810–13; H. A. Van Norman, "Bouquet Canyon Reservoir and Dam," *Civil Engineering* 4 (August 1934): 393–97. With its proposal for a storage dam in the upper Santa Clara River watershed, the city of Los Angeles also acknowledged the interests of C. C. Teague and the Santa Clara Water Conservation District. The city had made a claim on San Francisquito Creek flood flow with respect to the St. Francis Dam (see chapter 3), but no comparable claim was made on the flood flow of Bouquet Creek. See H. A. Van Norman to the Honorable Board of Water and Power Commissioners, June 30, 1930, file WP01–97:1, LADWP Archives, where Chief Engineer Van Norman advises his superiors: "it is important that the conference with the Santa Clara Valley people be the first step in the proceeding so that people there will feel that we are going about it in an open and frank manner and seeking their cooperation...It is suggested that we propose to the people of the Santa Clara Valley that we jointly ask the State Division of Water Rights to determine the yield of said watershed [above Bouquet Canyon Dam] and work out a schedule of quantities of water that should be released [into the creek below the dam]."

15 O'Shaughnessy, comments on "Public Supervision of Dams, A Symposium," *Transactions of the American Society of Civil Engineers* 98 (1933): 853. In a letter to Hyatt of September 1928, George Dillman made a similar point: "I am not personal, but some of your predecessors were not able men and some of your successors may not be." In the same letter, Dillman wryly observed that "had the position of dam inspector [for all California dams] been created prior to the failure of the St. Francis Dam, I hazard a guess that Mulholland would have been a candidate for the job, with a good chance of his getting it." See George Dillman to Edward Hyatt, September 13, 1928, Supervision of Dams file, 1928, Public Utility Commission Records, California State Archives, Sacramento.

16 See "Memoirs of Edward Hyatt," *Transactions of the American Society of Civil Engineers* 120 (1956): 1568–69. After Wilbur F. McClure's death in 1926, Paul Bailey served briefly as the state engineer; upon his resignation at the end of August 1927, Edward Hyatt succeeded him. Immediately before this promotion Hyatt served as chief of the Division of Water Rights within the California Department of Public Works. See *Biennial Report of the Division of Water Rights* (Sacramento: California State Printing Office, November 1, 1926).

17 A. H. Markwart, “Recommendation for Legislation and Application of Law,” in “Public Supervision of Dams, A Symposium,” *Transactions of the American Society of Civil Engineers* 98 (1933): 828–35.

18 For more on the political economy of New Deal dam building, see Billington and Jackson, *Big Dams of the New Deal Era.*

19 Fred Noetzli, “An Improved Type of Multiple Arch Dam,” *Transactions of the American Society of Civil Engineers* 87 (1924): 410; Noetzli, “Discussion,” published with Markwart and Hinderlider, “Public Supervision of Dams: A Symposium,” 865. Noetzli died of heart disease in May 1933 at the age of forty-six; see “Fred A. Noetzli, Dam Authority, Taken by Death,” *Los Angeles Times,* May 26, 1933. His professional accomplishments are memorialized in “Fred Adolph Noetzli,” *Transactions of the American Society of Civil Engineers* 99 (1934): 1496–97.

20 R. B. Jansen, “Review of the Baldwin Hills Reservoir Failure,” *Engineering Geology* 24 (December 1987): 7–81; Kenneth Reich, “’71 Valley Quake a Brush with Catastrophe,” *Los Angeles Times,* February 4, 1996.

21 N. J. Schnitter, “The Evolution of the Arch Dam: Part Two,” *Water Power and Dam Construction* 28 (November 1976): 19.

22 Jackson discusses Huber’s involvement with the Littlerock Dam in *Building the Ultimate Dam,* 201–2.

23 Walter L. Huber to John R. Freeman, March 21, 1928, box 54, Freeman Papers, MIT.

24 H. W. Dennis, G. A. Elliot, and Walter L. Huber (Chair) to Edward Hyatt, September 15, 1932, folder 630 (“Multiple Arch Dam Advisory Committee”), Walter L. Huber Papers, WRCA, UC Riverside. The committee’s work began in spring 1931 and concluded with a report to the state engineer in late summer 1932. Noetzli was very familiar with multiple arch dam technology and, in January 1924, had visited Eastwood’s Littlerock Dam in the company of J. B. Lippincott, commenting that he was “greatly impressed by this excellent piece of work...and the first class quality of concrete.” See Fred Noetzli to Paul Bailey, February 4, 1924, in Littlerock Dam file, Division of Safety of Dams, California Department of Water Resources. It is not known why Walter Huber, and not Noetzli, was selected to be a member of State Engineer Hyatt’s Multiple Arch Dam Advisory Committee.

25 F. W. Hanna, “Designing a High Storage Dam for the Mokelumne Project,” *Engineering News-Record* 100 (March 15, 1928): 444–47.

26 Research into the way uplift acts on dams continued for decades after the St. Francis collapse. Whether uplift existed or posed a threat was not the issue, but rather, the theoretical means of calculating uplift pressure in relation to the horizontal cross-section of the dam, or foundation (that is, the “area factor”), and how various technologies, such as cutoff trenches, grouting, drainage wells, etc., reduced uplift pressure (that is, the “intensity factor”). See, for example, Julian Hinds, “Upward Pressures under Dams: Experiments by the United States Bureau of Reclamation,” *Transactions of the American Society of*

Civil Engineers 93 (1929): 1527–82; D. C. Henny, "Stability of Straight Concrete Gravity Dams," *Transactions of the American Society of Civil Engineers* 99 (1934): 1041–1123; Arthur Casagrande, "Seepage Through Dams," *Journal of the New England Water Works Association* 51 (June 1937): 131–70; H. de B. Parsons, "Hydrostatic Uplift in Pervious Soils," *Transactions of the American Society of Civil Engineers* 93 (1929): 1317–66; E. W. Lane, "Security from Under-Seepage on Earth Foundations," *Transactions of the American Society of Civil Engineers* 100 (1935): 1235–1351; L. F. Harza, "Uplift and Seepage under Dams on Sand," *Transactions of the American Society of Civil Engineers* 100 (1935): 1352–1406; Ivan E. Houk and Kenneth B. Kenner, "Masonry Dams, A Symposium: Basic Design Assumptions," *Transactions of the American Society of Civil Engineers* 106 (1941): 1115–30; Serge Leliavsky, "Experiments of Effective Uplift Area in Gravity Dams," *Transactions of the American Society of Civil Engineers* 112 (1947): 444–87; L. F. Harza, "The Significance of Pore Pressure in Hydraulic Structures," *Transactions of the American Society of Civil Engineers* 114 (1949): 193–289. Harza states: "by deductive reasoning it is shown that under and within any concrete structure,...hydrostatic uplift acts over the entire horizontal area, instead of over one-half or two-thirds as commonly assumed" (193). He thus makes clear that the theoretical magnitude of the area that uplift acts on—not just the existence of uplift—is the issue.

27 See Jackson, *Building the Ultimate Dam*, 198–205, 241–42, on the way regulation affected innovation in multiple-arch dam design.

28 O'Shaughnessy, comments on "Public Supervision of Dams: A Symposium," 853.

CHAPTER EIGHT

1 "Mulholland Resigns Post," *Los Angeles Times*, November 14, 1928.

2 J. B. Lippincott, memo proposing Mulholland Memorial Fountain, July 30, 1937, folder HM 4846, R. F. del Valle Papers, Huntington Library.

3 Outland, *Man-Made Disaster*, 167; Catherine Mulholland, *William Mulholland*, 326–27.

4 "William Mulholland Still a Big Man," *Western Construction News* 3 (April 10, 1928): 223.

5 John R. Freeman to Joseph B. Lippincott, April 4, 1928, box 54, Freeman Papers, MIT.

6 John R. Freeman to Caleb Saville, March 29, 1928, box 54, Freeman Papers, MIT.

7 Arthur P. Davis to John Freeman, March 28, 1928; and Arthur P. Davis, "Inspection Report of St. Francis Dam, March 19, 1928," both in box 54, John R. Freeman Papers, MIT.

8 On March 19, 1928, Mulholland appeared before the Board of Water and Power Commissioners to request a "leave of absence" while the investigation was underway. The board resolved to "decline to grant such request and [urge]

the chief to remain on the job he has faithfully filled for half a century." See LADWP Archives, "St. Francis Dam Collapse, March 1928," file WP01–48:1. Also see "Leave of Absence Plea by Mulholland Refused," *Los Angeles Examiner*, March 20, 1928.

9 The *Los Angeles Record*, May 17, 1928, reported: "Members of the City Council and other officials left today by automobile for the Boulder Dam site and the Black Canyon dam site...where they will make an inspection tour." The group's return trip via the "aqueduct route" is noted in the *Los Angeles Record*, May 21, 1928. Mulholland was not a participant. Also see Peirson Hall Collection, Scrapbook 1928, Huntington Library.

10 "Mulholland Resigns Post," *Los Angeles Times*, November 14, 1928. His resignation was also noted in "Mulholland Retires After 50-Year Service at Los Angeles," *Engineering News-Record* 101 (November 22, 1928): 785; this article provided a brief review of his career but did not mention the St. Francis disaster. An editorial published in the *Los Angeles Times*, November 15, 1928, lauded his career, also without mentioning St. Francis. In "Council Praises Mulholland as He Resigns His Post," *Hollywood Citizen*, November 14, 1928, it was reported: "Friends of Mr. Mulholland conceded today that the tragedy which grew from the St. Francis Dam break hastened his desire to retire." It was also reported than Harvey Van Norman would be his "probable successor," but that "other names prominently mentioned are Arthur Davis and A. J. Wiley."

11 Van Norman's appointment as chief engineer and general manager of the Bureau of Water Works and Supply was announced on November 27. See "Van Norman to Succeed His Chief," *Los Angeles Times*, November 28, 1928.

12 "Address by Senator del Valle on Behalf of the Board of Water and Power Commissioners on the Occasion of the Retirement November 13, 1928 of William Mulholland, Chief Engineer and General Manager Bureau of Water Works and Supply," published in *The Intake* 5 (December 1928): 1–2.

13 At the dinner, a poem entitled "The Builder" was read by C. A. Dykstra, the BWWS director of personnel and efficiency who also served as the evening's toastmaster. Short speeches were offered by Ezra Scattergood, Harvey Van Norman, board president John Richards, and other departmental employees. As a retirement gift, Mulholland was given a rocking chair.

14 When the MWD convened a board of consultants to formally review the proposed right-of-way for the Colorado River Aqueduct, Mulholland was included among the city officials that greeted the board. But the *Los Angeles Times* photo documenting the event featured only the three consultants (including A. J. Wiley), MWD chairman William Whitsett, and MWD chief engineer F. E. Weymouth; see "River Aqueduct Routes Studied," *Los Angeles Times*, November 29, 1929.

15 "Rock Dissolved as Jury Watches Breathlessly," *Los Angeles Times*, May 27, 1930. When a conglomerate rock sample that the plaintiff claimed came from the dam's west abutment dissolved in a glass of water, Mulholland held that the sample could not have come from the dam site. Mulholland's testimony that a possible earth movement caused the dam's collapse was rebutted by

plaintiff witness F. L. Ransome, who testified that the disaster was caused by "faulty foundations," and that that no earth movement occurred that could have caused the failure. Ransome had served as a geologist on the Governor's Commission that investigated the collapse, and his testimony adhered to its report. See "Dam Break Blame Put on Faults," *Los Angeles Times,* May 29, 1930. On the Ray Rising civil trial, see chapter 5.

16 The Ray Rising verdict is reported in "Flood Damage being Paid Off," *Los Angeles Times,* June 6, 1930.

17 "Aqueduct Need Told by Expert: William Mulholland Warns Citizens Committee," *Los Angeles Times,* July 14, 1931. When the MWD board met later in the month to set the date for the aqueduct bond election, Mulholland made an appearance in support of the bond, proclaiming that "the average citizen, the man and the woman who own their own homes in Los Angeles I find without exception in favor of starting the project at once. They realize the vital necessity of this work. We must not be alarmed by the cries of the chronic kicker who is always opposed to every development regardless of its necessity." See *Los Angeles Times,* July 25, 1931.

18 Catherine Mulholland, *William Mulholland,* photograph dated February 17, 1932, following p. 136.

19 "Blast Seen by Throng: Work Starts on Aqueduct Bore," *Los Angeles Times,* May 14, 1933.

20 "Mulholland...Man of Broad Vision," *Southern California Business* 8 (September 1929): 15.

21 Catherine Mulholland, *William Mulholland,* 330.

22 "William Mulholland: Maker of Los Angeles," *Western Construction News* 8 (August 1933): 330.

23 See Catherine Mulholland's biography for a description of Mulholland's dream (330). For more on Father John J. Crowley, see Hoffman, *Vision or Villainy,* 261–62.

24 H. L. Jacques, "Bouquet Canyon Dam Built for Los Angeles Aqueduct," *Engineering News-Record* (June 21, 1934): 810–13.

25 "Water in New Reservoir: Mayor and High Officials See Torrent Turned into Bouquet Canyon Basin," *Los Angeles Times,* March 29, 1934. In contrast to St. Francis, where Mulholland and the city made claims to 5,000 acre-feet per year of flood runoff from San Francisquito Creek (see chapter 3), at Bouquet Canyon the city made no claim to any floodwater that might be captured by the dam. The city worked with the Santa Clara Water Conservation District to ensure that the new dam and reservoir did not transgress upon the water rights of farmers in the Santa Clara River Valley. See H. A. Van Norman to Honorable Board of Water and Power Commissioners, June 30, 1930, WP01–97:1. LADWP Archives.

26 Catherine Mulholland, *William Mulholland,* 330; "Mulholland Near Death: Former Chief Engineer of City Water Department Failing Rapidly," *Los Angeles Times,* December 21, 1934.

27 Catherine Mulholland, *William Mulholland,* 350.

28 "Mulholland of Aqueduct Fame Dies: City's Flags at Half-Staff," *Los Angeles Times,* July 23, 1935. His body lay in state in City Hall prior to burial; a lengthy list of honorary pallbearers appears in "City Honors Mulholland," *Los Angeles Times,* July 24, 1935. The active pall bearers included H. A. Van Norman, W. W. Hurlbut, J. E. Phillips, and Ezra Scattergood,

29 "Thousands in Last Tribute to William Mulholland," *Colorado River Aqueduct News* 2 (August 5, 1935): 2. This weekly publication produced by the MWD also reported that "it is believed that the thousands of aqueduct engineers and construction men who knew Mr. Mulholland will gratefully accept this opportunity to show their respect for the great man who envisioned and set under way the plans for the Colorado River Aqueduct."

30 "Leaders Join in Tribute to Aqueduct's Builder," *Los Angeles Times,* July 23, 1935.

31 For tributes following his death that did not mention St. Francis Dam, see *Transactions of the American Society of Civil Engineers* 101 (1936): 1604–8, and *Civil Engineering* 9 (March 1939): 199. Also see "Statement on the Death of William Mulholland by Chairman W. P. Whitsett" and "Statement on the Death of William Mulholland by General Manager F. E. Weymouth," both in *Colorado River Aqueduct News* 2 (August 5, 1935): 2. Weymouth praised the "genius and tireless public service of William Mulholland," while Whitsett was more florid: "William Mulholland has gone to join the immortal company of giants among men—super-men whose vision and practical genius have made it possible for the remainder of us to live in this land and enjoy its beauty and its golden fruits."

32 "The Colorado River Aqueduct," *Engineering News-Record* 121 (November 24, 1938): 637. When the aqueduct was completed, MWD Chairman W. P. Whitsett made special mention of Mulholland at the October 14, 1939, dedication ceremony: "It was William Mulholland who saw the vital need of a still greater supply of water to meet the future requirements not only of Los Angeles but all of this metropolitan area. And it was he who pointed to the Colorado as the one available source for the additional supply we must have. Under his direction the first engineering studies and surveys for the Colorado River Aqueduct were set under way in 1923. When the Metropolitan Water District was established in 1928 it took up the work that had been carried forward…by William Mulholland and his able engineering associates and assistants." See "A Tribute to Aqueduct Builders," *Colorado River Aqueduct News* 6 (October 25, 1939): 2.

33 See chapter 2 for more on the Weid Canyon/Mulholland Dam.

34 "Mulholland Dam Backed by Earthfill Against Downstream Face," *Engineering News-Record* 112 (May 3, 1934): 558–60. The original spillway was at 746 feet above sea level; in 1932 it was permanently lowered to 715. This article provides an excellent review of various engineering studies of Mulholland Dam undertaken after the St. Francis disaster; it also makes clear that many Hollywood residents sought to eliminate the dam as a component of the city's storage system.

35 A photograph of Mulholland Dam being covered by a thick earthen blanket appears in "Mulholland Dam Above Los Angeles, Reinforced," *Engineering*

News-Record 112 (March 8, 1934): 335. In another issue (May 3, 1934), an editorial called "Psychologic Safety Factor" stated: "Situations in which public opinion may prove a factor in engineering design are exceptional, yet unquestionably there are times, particularly in the case of water storage projects, when such a situation must be considered a factor to be applied after the technical requirements have been satisfied" (576).

36 See "Memorial for Mulholland to be Objective of Drive," *Los Angeles Times,* July 22, 1937; "Mulholland Fund Sought," *Los Angeles Times,* September 3, 1937; "Mulholland Plaque Shown," *Los Angeles Times,* July 21, 1938; "Committee Chooses Fountain for Mulholland Memorial," *Los Angeles Times,* July 20, 1939; "Mulholland Memorial," *Los Angeles Times,* July 21, 1939.

37 Memorandum Proposing Mulholland Memorial Fountain, by J. B. Lippincott, July 30, 1937, folder HM 4846, R. F. del Valle Papers, the Huntington Library.

38 "Mulholland Memorial Week to Start Today in Schools," *Los Angeles Times,* October 16, 1939.

39 "Mulholland Fund Aided by Schools," *Los Angeles Times,* December 6, 1939.

40 The celebration is previewed in "Mulholland Fete Tonight: Fifty-foot Illuminated Jets to Commemorate Aqueduct Builder," *Los Angeles Times,* August 1, 1940. This article provides the text of the memorial inscription and a list of the distinguished attendees, including J. B. Lippincott, Harvey Van Norman, W. P. Whitsett, newspaper publisher Harry Chandler, and Mayor Fletcher Bowron; granddaughter Catherine Mulholland also attended. The fountain was designed to "circulate 2,250 gallons of water a minute, and then return [the flow] to either of the big mains [connected to the fountain] and then continue as part of the city's water supply." Photographs and a description of the dedication, including the text of the plaque, are provided at waterandpower.org/museum/Mulholland_Monuments.html.

41 Carey McWilliams, *Southern California Country: An Island on the Land* (New York: Duell, Sloan & Pearce, 1946); and Remi A. Nadeau, *The Water Seekers* (Garden City, N.Y.: Doubleday, 1950).

42 Robert Towne, *Two Screenplays: Chinatown and The Last Detail* (New York: Grove Press, 1994), provides a good record of dialogue and screen directions used in *Chinatown,* although it includes a few scenes and dialogue that do not appear in the film as released, some of which relate to Mexican victims of the flood caused by the collapse of the fictitious VanderLipp Dam.

43 "Silver Torrent Crowns the City's Mighty Achievement," *Los Angeles Times,* November 6, 1913. This was Mulholland's concluding comment at the ceremonial opening of the Los Angeles Aqueduct on November 5, 1913; see Catherine Mulholland, *William Mulholland,* 245–46. As his granddaughter acknowledged in her biography, this became "the best remembered moment of the occasion."

APPENDIX A

1 See Governor's Commission *Report,* 7, 33, for data on the reservoir level from the spring of 1926 until the dam's collapse in March 1928.

2 William Mulholland testimony, Coroner's Inquest, 35–36.

3 Harvey Van Norman testimony, Coroner's Inquest, 93.

4 Neither the Governor's Commission *Report* nor the official reports published by the Board of Water and Power Commissioners refer to the surface area of the St. Francis reservoir when filled to elevation 1,835 feet. Outland, in *Man-Made Disaster,* indicates the surface area of the full reservoir to be 600 acres (249), and the post-disaster report submitted to District Attorney Asa Keyes indicates that "the area of water at spillway [level] was over 600 acres." See Robert Mayberry et al., "Report to Mr. Asa Keyes, District Attorney, Los Angeles County, Calif. on the Failure of the St. Francis Dam, April 4, 1928," 8; box 13, Richard Courtney Collection, Huntington Library. When the dam's spillway crest was raised ten feet from elevation 1,825 to 1,835 feet prior to June 1925, it was reported (see chapter 3) that the reservoir storage capacity increased by 6,000 acre-feet (from 32,000 to 38,000 af); this equates to a surface area of 600 feet, with 600 acres × ten feet = 6,000 acre-feet. To be conservative, this analysis of reservoir outflow estimates the surface area of a full reservoir to be 615 acres.

5 The Governor's Commission *Report* states: "The five outlet pipes each 30 inches in diameter were controlled by sliding gates fastened to the upstream face of the dam" (7).

6 For more on how to calculate outflow from pipes, see E. A. Hoskins, *A Textbook of Hydraulics* (New York: Henry Holt, 1906), 30–38; and H. E. Babbitt and J. J. Doland, *Water Supply Engineering,* 3rd ed. (New York: McGraw Hill, 1939), 32–33.

7 The depths of the outlet pipes are derived from information in the Governor's Commission *Report,* 7, and from a detailed photo of the upstream face; Huntington Library, HM 70402, plate xxv.

8 The assistance of Arthur Kney, Head of the Department of Civil and Environmental Engineering at Lafayette College, who checked the calculations presented in this appendix, is gratefully acknowledged.

APPENDIX B

1 Governor's Commission *Report,* 16.

2 Rogers, "A Man, a Dam," 42–43, 58.

3 See Governor's Commission *Report,* 7, 33, for a description and graphic record of reservoir levels from March 1926 through March 1928.

4 For more on the construction and location of the east side tunnel, see chapter 3.

5 This discussion has focused exclusively on use of a concrete gravity design for the St. Francis site. If we consider dam technology more broadly, use of a reinforced concrete buttress design with an inclined upstream face, comparable to the 150-foot-high Mt. Dell Dam built by Salt Lake City (1917–25; see chapter 2), would have provided a means of obviating the destabilizing effect of uplift. It is also plausible that a massively dimensioned earth/rockfill

embankment dam with a concrete cut-off wall extending deep into the foundation could have been safely built at the St. Francis site.

6 There is a voluminous literature on technological disasters and on the reasons why seemingly well-designed systems and structures can fail. Perhaps the best-known book in this genre is the sociologist Charles Perrow's *Normal Accidents: Living with High Risk Technologies* (New York: Basic Books, 1984). Historian and economist Mark Aldrich focuses specifically on railroad-related disasters in *Death Rode the Rails: American Railroad Accidents and Safety, 1828–1965* (Baltimore: Johns Hopkins University, 2006), but he has a broad perspective and offers a significant counterpoint to what he terms the "unhistorical argument linking complex technologies to disasters" presented by Perrow (431). In telling the story of the St. Francis Dam disaster, the authors of *Heavy Ground* have not attempted to tie the disaster to broad-based sociological theories of organization, managerial hierarchy, and technological complexity. Others may wish to take on this task, but in so doing, they should pay close attention to the history of the disaster and to the specific historical contexts in which Mulholland and others acted. The St. Francis Dam failed not because it represented some "normal accident" that arose from the implications of technological complexity in a modernizing world. In the authors' view, the dam failed for very specific reasons involving decisions made by Chief Engineer Mulholland and his staff in the process of its design, construction, and operation.

Index

Page numbers in **bold** refer to figures or illustrations. Dam names and place-names refer to California unless specified otherwise.